DEDICATED

To

Our late father, mother, elder brothers & their wives, our beloved wives & sons who are the inspiration and backbone of our past, present, and future.

A GUIDE TO PERFORMANCE AND EFFICIENCY ASSESSMENT OF INDUSTRIAL EQUIPMENT

For Professionals, Academicians and Students

A. K. DAS & P. K. DAS

INDIA • SINGAPORE • MALAYSIA

ISBN 979-8-89233-382-5

CONTENTS

ABOUT THE BOOK

This book is written as a guide to industrial professionals, young engineers, entrepreneurs, industrialists, and other stakeholders who are attached to process industries in different forms through industrial/process equipment for several human needs. But, the performance and efficiency of the equipment are not really taken care of during the operations and processes, which may be due to the dearth of proper knowledge, procedure or ignorance. Because of that, a large quantity of energy remains unutilized or wasted causing excess energy costs and subsequently generation of a huge quantity of carbon footprint directly and indirectly which could be saved by proper performance and efficient management, and hence, our Nature Earth could be sustainable.

In this book, we have highlighted the performance and loss of efficiency of such industrial equipment during running. Attempt has been made to disseminate our sound, in-depth knowledge, and long experience achieved from several industries while working in different fields. The book explains the actual energy needed for performance, the reason for energy loss, and the scope of energy savings which can be possible by proper energy management.

This book will also be apprehensible for all students of diploma, undergraduate, post graduate, researchers, and academicians in the stream of electrical, mechanical, chemical, power, and all other engineering courses as a textbook as well as a reference book.

Dr. A. K. Das

P. K. Das

ABOUT THE AUTHORS

Dr. A.K. Das: Presently is the Professor & Head of the Department of Renewable Energy Engineering, at the Maulana Abul Kalam Azad University of Technology, West Bengal (Formerly the West Bengal University of Technology), India.

In addition, he is looking after the International Student Centre (ISC), MAKAUT, WB as Director.

Earlier, he was also the Director and Coordinator of Partnership Cell (now CCPTR), MAKAUT, WB for about 2 years.

Prof. Das did his BSc (Honors) in Chemistry, BTech & MTech in Chemical Engineering, and obtained his Ph.D. (Tech) from Calcutta University.

He has industrial experience of more than 24 years of his career in different MNCs and Govt. undertaking companies in a senior management cadre.

Before joining MAKAUT, WB, he held the post of Registrar at MCKV Institute of Engineering (an Autonomous Institute), Liluah, Howrah, WB and prior to joining MCKVIE, he was the Professor & Head of the Chemical Engineering Department and Dean (Academic & Planning) of Durgapur Institute of Advanced Technology and Management (an Affiliated College of MAKAUT), Durgapur, WB for 9 years.

He has the credit for publishing many papers in different National and International journals and presenting many papers at National, International conferences & seminars, delivering lectures, and in attending a nos. of seminars and workshops in different Institutes and Organizations.

Many of his present and past students are engaged in academics and research in Indian Premier Institutes and Abroad.

He guided one research scholar for the Ph.D. degree from NIT Durgapur and is guiding three more research scholars in NIT Durgapur & MAKAUT, WB at present.

He is the co-author of four books on Power Plants for industrial professionals and engineers.

Prof. Das is the Fellow of IE(I), and the Life-Member of many Professional bodies like IIChE, CAII (Kolkata), and SESI, New Delhi.

He is a Consulting Engineer of MSME, Durgapur, Certified Internal auditor & Water auditor and also the Chartered Engineer of IEI(India).

P.K. Das: A highly knowledgeable and long experienced engineer cum manager with a strong technical background both in Processes and Power plants. He acquired Engineering Qualification from The Institution of Engineers (I), Kolkata both in Chemical Engineering and Mechanical Engineering in the years 1981 &1988 respectively. He also Possesses a Boiler Operation Engineering (BOE) competency certificate from the Directorate of Boilers, Kolkata. He completed PG Diploma in Materials Management in the year 1989. He became a Fellow of The Institution of Engineers (I) in the year 2003.

He acquired vast experience and profound knowledge in Chemical Process Plants and Power Plants for over 14 years in the Government sector and 33 years in the private sector. During such a long career, he has gained experience in Project Execution, Commissioning, Operation &Maintenance of Process Plants and Power plants.

He was associated with renowned companies like Hindustan Fertilizer Corporation of India Ltd, Indian Charge Chrome Ltd, Dishergarh Power Plant, General Electric, and Thermax Limited. He commissioned over twenty Power Plants in a captive Range from 4MW to 150MW comprise of all types of boilers having fuel of all kinds. Mentioned worthy achievement is that he was involved in the commissioning and O&M of the biggest solar thermal Plant in India of 100MW at Pokhran, Rajasthan. He was in Ayoki Fabricon Pvt. Ltd. as Head(O&M) post-retirement and presently he is the Advisor of Klein Constro Development Pvt. Ltd, Pune.

PREFACE

The Twenty-first century is the era for hot discussions on global warming, greenhouse gas emission, extraordinary climate change, natural calamities, etc. The United Nations Framework Convention on Climate Change (UNFCCC) was formed in 1994 to stabilize greenhouse gas emissions, global warming and to protect the earth from the threat of climate change. Conference Of Parties (COP) members have been meeting every year since the year 1995. United Nations Development Programme (UNDP) plays a critical role in helping countries to achieve the Sustainable Development Goals (SDGs). Countries have united to combat the well-established cause of global warming and are implementing several action plans to mitigate and reverse global warming. The action plans like the ban on the usage of fossil fuels in industries, power generation plants, and transportation sectors transitioning to renewable energy and alternative energy sources that do not emit greenhouse gases such as solar, wind, biofuels, hydrogen, etc.

All the stated action plans are now at the macro level which will have a drastic impact on the reduction of GHG emissions resulting in considerable mitigation of global warming. It is estimated that by enhancing Efficiency, GHG emmision can be reduced by 20%.

The other important actions, which also play an important role are conserving energy and an efficient way of operating all industrial, commercial, transportation, and domestic activities.

This book, ***A Guide to Performance and Efficiency Assessment of Industrial Equipment*** is written to explain - how the efficiency of equipment is derived, what are the causes of deviated efficiency and the remedial measures to increase efficiency. The content is written targeting industrial activities where performance and efficiency have also taken center stage in the present stage of global warming.

It is well known that an efficient and safe way of operation in the industry brings manifold benefits to industrialists and entrepreneurs in terms of product cost, growth & profit. The efficient way of operation also supports the sustainable environment and economic growth of any country.

Performance and Efficiency of any equipment, system, and plant as a whole can be measured in several ways such as absolute efficiency of any equipment and its deviation from design efficiency, specific energy consumption by the equipment or sub-section i.e., energy consumption per unit output (kW/unit), and specific energy consumption by the plant as a whole per unit output and by the comparison of the same with that of best efficient running plant in the world.

Once the efficiency is derived and any deviation is observed, the next step is the implementation of corrective measures to reduce the deviations.

The cause of deviation may be due to wrong or changed O&M practices or the wrong selection of equipment. It is our endeavour to point out the root cause of deviations in the efficiency of equipment, sub-sections, and the plant as a whole.

The equations and relations, that have been used in this book, are common industry practices and these will be helpful for plant personnel to derive the efficiency of equipment, sub-sections, and the plant as a whole. While explaining any subject commonly used units have been used for ease of understanding.

As every country is moving for a paradigm shift from fossil fuel energy sources to renewable energy sources, we have also tried to highlight the renewable energy sources and the working principle of equipment involved along with their efficiency.

The descriptive part of the book has been written in a lucid manner so that personnel engaged in all different energy sectors including industries, can understand the concepts, and feel comfortable to grasp the knowledge and implement the same for the noble cause of overall sustainability.

If our small attempt through this book helps campaigning global warming awareness and its mitigation, it will be our tiny contribution toward the sustainability of the environment.

Finally, criticisms and suggestions for improvement of the book are always welcome.

Dr. A.K. Das & P. K. Das

ACKNOWLEDGEMENT

During our life long tenure both in industries and academics, we felt a gap of understanding of importance of energy for economic development of country and the sustainable growth among the engineering professionals. Most of the industrial organizations, commercial business centres, institutes, offices are very much reluctant to look into the future and concern mainly on present benefits. It is one of the causes for the steady increase in environmental risk and environmental impacts. There is also a dearth of knowledge and awareness that the energy is source of any kind of activities to sustain on this Earth and hence, there is no energy, no activity.

Taking this scope, we took the challenges to write this book where we shared our experiences and knowledges to spread among those engineers who have the appetite to learn this subject.

Meanwhile, our colleagues encouraged us to write such a book and also gave lot of inputs during compilation of the book contents. During the writing of this book all our colleagues also enquired often the updated status which further boosted our energy to complete the book as early as possible.

We are very much grateful to all my colleagues, senior colleagues, and my entire friend community who have supported me directly or indirectly to complete the document.

Lastly, we thank to Notion press who has agreed to publish our book like previous books.

Dr. A. K. Das

P.K. Das

SECTION 01

FUNDAMENTAL LAWS, UNITS, MEASUREMENTS & BASIC RELATIONS IN ENGINEERING

1.0 Matter & Classical physics

1.1 Introduction:

Fundamental laws & equations have been discussed in this section as refreshing. Conventional notations are used to expalain the equations.

1.2 Statics and dynamics related to the matter:

i. Newton's law of motion

   a) Property of inertia, Inertia = M x r (Kg-m), Moment of inertia = M x r^2 (Kg-m^2)

      [M-mass, r-radius (average distance from center to furthest part)]

   a) Body gains momentum on velocity; Momentum= M x v (Kg m/s)

   b) Rate of change of momentum is force; F= M x $\Delta v/t$ (Newton- N) [Δv- change in velocity]

   c) Action & reaction forces are equal; $F_1=F_2$ [F_1-force of action, F_2–force of reaction]

ii. Rate of change of distance is velocity; v = s/t (m/s) [s- distance travelled]

iii. Rate of change of velocity is acceleration; a =$\Delta v/t$= s/t^2 (m/s^2)

iv. Causes friction force while moving over a surface; Frictional force = k x applied force [k- coefficient of friction]

v. Body movement in:

   a) Linear motion: s = u t, v = u+2at, $v^2=u^2+2as$, s=ut+1/2 at^2 [u &v -initial & final velocity, a-acceleration, s-distance, t-time].

   b) Rotational motion: $\theta=\omega$ t, $\omega_2=\omega_1+2\alpha$ t, $\omega_2^2=\omega_1^2+2\alpha\theta$, $\theta=\omega_1$ t +1/2 α t^2 [θ-angular distance in radian, ω_1 & ω_2 -initial & final angular velocity (rad/s), α-angular acceleration (radian/ s^2). t-time(s)].

c) Simple Harmonic motion: $S= S_m \sin \theta = S_m \sin (\omega \times t)$ [S-amplitude at any time, S_m-max amplitude, θ- angle at any amplitude & t-time].

d) Gyratory motion: motion in two directions simultaneously. Earth's motion is gyratory motion.

vi. Force of gravity: The force that pulls all bodies on earth towards its Centre: $F= M \times g$; $g = 9.81 m/s^2$

vii. Gravitational force: All bodies pull each other. Gravitational force depends on the mass of the objects and the square of the distance between them. $F= G \times (M_1 M_2)/d^2$ [G-Constant- $(6.6743 \pm 0.00015) \times 10^{-11}\ m^3 / (kg\ s^2)$. M_1 &M_2 = Body masses, d= distance between the bodies]

viii. Energy can neither be created nor be destroyed; can take other forms. $E_1=E_2$.When, [E_1&E_2 – Energy before and after any activity]

ix. Archimedes principle: Buoyancy Force applied by water on a floating body; Upward force (F) =Quantity of displaced water (W) x g (N)

2.0 Liquid

2.1 Physical property

i. Low density, solvent in nature, takes the shape of the container, incompressible, low in comparison to solid.

ii. Expansion of volume on heat: $\Delta V= V \times \gamma \times \Delta T$ [When, γ = coefficient of volumetric expansion]

iii. Water has high specific heat (Specific heat= 1 Kcal/ (Kg °C).

iv. Density varies with temperature: $\rho_{t=} \rho_0(1- \gamma \Delta T)$; ρ varies as $1/\sqrt{T}$; water is having the highest density at 4 °C at 1000 kg/ m^3.

v. Liquid phase to gas can be changed on heating; Heat required for phase change is the latent heat of evaporation; for water its 540 Kcal/Kg.

2.2 Statics and dynamics related to liquid

i. Gains potential energy due to altitude= ρ g h

ii. Flows through any conduit: its characteristics are described as:

a) Reynold's number: Inertia force/ viscous force= $\rho D v/\mu$

b) Laminar flow: N_{RE}= <2000

c) Transition flow: N_{RE}= 2000-4000

d) Turbulent flow: N_{RE}= >4000

iii. Energy remains constant during flow: Bernoulli's theorem;

   a) Energy Equation: $P/\rho + \frac{1}{2} v^2 + g \times h$

   b) Pressure equation: $P + \frac{1}{2} \rho v^2 + \rho g h$

iv. Exerts pressure in all directions in a closed vessel while force is applied on it; called Pascal's law; $P = F/A$ (N/ m^2 or Pa).

v. Viscosity: Force applied by liquid while flows; Viscous force, $F = \mu . dv/dy$

vi. Osmosis:

   a) Osmotic pressure: Osmotic pressure is the pressure required to stop water from diffusing through a membrane by osmosis. It is determined by the concentration of the solute in the solution. Water diffuses into the area of higher concentration from the area of lower concentration until the concentration is uniform throughout.

   b) Osmotic pressure can be calculated using the equation: $\pi = CRT$; When, π denotes the osmotic pressure, C is the molar concentration of the solute (in the solution), R is the gas constant, T is the temperature.

vii. Surface tension: $T = F/L$ (N/m); when, L is the length of the surface.

   a) Intermolecular forces between dissimilar surfaces, called cohesive forces, draw the surface particle to the bulk liquid mass to bring them together.

   b) Surface tension is defined as the force that is required (against the cohesive force) to increase the area of the liquid along the surface.

viii. Liquid Imparts frictional force while flowing, Darcy's equation; $F = f\, L/D \times (v^2/2g)$.

   a) $f = 64/Re$ for laminar flow

ix. Total pressure in flow = static pressure + dynamic pressure; $P = P_0 + 1/2\, \rho\, v^2$.

x. Follows continuity equation: $A_1V_1 = A_2V_2$

xi. Eddy formation: during turbulent flow eddy is formed

xii. Heat transfer property- Convection.

2.3 Reactivity

i. Remains in ionic form

ii. Liquid reacts with Solid, Liquid &Gas

iii. Ability to break salts into ions in solution

3.0 Gaseous Substances

3.1 Physical properties

i. Low density

ii. Mixes with other gas and sparsely with liquid

iii. Takes shape of the container, Low boiling point

iv. Compressible & Expandable

   a) Adiabatic/ Isentropic: PV^{γ}= Constant

   b) Isothermal: PV= constant

   c) Polytropic: PV^{n}= constant

v. Expandable: J-T effect- due to change in internal energy (cooling) $\mu_{JT} = (\partial T/\partial P)_H$

vi. Density varies with temperature: P= ρ RT; ρ =P/RT

vii. Diffusivity

   a) Diffusion is a natural and physical process, which happens on its own, without stirring or shaking the solutions. Liquid and gases undergo diffusion as the molecules are able to move randomly. The molecules collide with each other and change their direction.

   b) Graham's law states that a gas's diffusion or effusion rate is inversely proportional to the square root of its molar mass. This law is in equation form: $r \propto 1/\sqrt{(M)}$ or. $r(M)^{1/2}$ = constant [r- rate of diffusion]

3.2 Statics and dynamics related to gas

i. Follows gas laws

   a) Bayle's law: $V \propto 1/P$

   b) Charles' law: $V \propto T$

   c) Gay Loussac's law: $P \propto T$

   d) Henry's law: $P \propto C$ (or) $P = K_H C$

   e) Dalton's law: $P_t = P_1+P_2+P_3 +......$ [P_t – total pressure, P_1-Partial pressure element-1, P_2- Partial pressure element-2 ... etc.]

   f) Avogadro's law: 1 mole of gas occupies 22.4 Liter at NTP, constant R=0.28

ii. Bernoulli's theorem is applicable

3.3 Reactivity

i. Mostly oxidant

a) Fluorine: The most reactive

b) Inert gases: Gases that do not undergo chemical reactions under specific conditions such as oxidization. These include argon, carbon dioxide, helium, and nitrogen.

4.0 Light

4.1 Properties of light

i. Reflection

a) Angle of incidence & angle of reflection are the same

b) Remains in the same plane

ii. Refraction

a) Snell's law: Sin i / Sin r = n [i & r are angle of incident and angle of refraction respectively]

b) Remains in the same plane

iii. Dispersion

a) light breaks into 7 colors in a prism

b) Velocity of light varies from medium to medium as per its refractive index; (n)=c/v

iv. Diffraction: light bend while passing through tiny hole/slit

v. Scattering of light

a) Scattering is the process of absorption and then re-emission of light energy. In scattering, light spreads in all directions. The air molecules, a size smaller than the wavelength of the incident light, absorb the energy of incident light and re-emit it without change in its wavelength. The intensity of scattered light depends on the particles' size and is a function of the wavelength of the light ray. Light scatters by particles suspended in a gas or liquid- the combination of absorption, reflection & refraction.

b) Tyndall effect: The Tyndall effect (Tyndall scattering) is the scattering of light by particles in a colloid or else particles in a very fine suspension. Example: flour suspended in water appears to be blue because only scattered light reaches the viewer and blue light is scattered by the flour particles more than red.

c) The Raman effect: The effect of appearance of extra spectral lines near the wavelength of the ***incident light***. The Raman lines in the scattered light are weaker than the light at the original wavelength. The Raman-shifted lines occur both at longer and shorter wavelengths than the original light; the lines at the shorter wavelengths are usually very weak.

d) Rayleigh scattering: It results from the electric polarizability of the particles. The oscillating electric field of a light wave acts on the charges within a particle, causing them to move at the same frequency. The particle, therefore, becomes a small radiating dipole whose radiation is seen as scattered light. Blue sky is due to Rayleigh scattering.

e) Mie Scattering: In this type of scattering the size of the particle is more than the wavelength of the light. Hence, there is a non-uniform scattering. It is also an elastic type of scattering. Example-Cloud white due to this, the fog and water droplet color is also the result of this type of scattering.

vi. Interference - It is the phenomenon in which two waves superpose to form the resultant wave of the lower, higher, or same amplitude.

X-ray diffraction, Bragg's law: When the X-ray is incident onto a crystal surface, its angle of incidence, θ, will reflect with the same angle of scattering, θ. And, when the path difference, d is equal to a whole number, n, of wavelength, λ, constructive interference will occur.

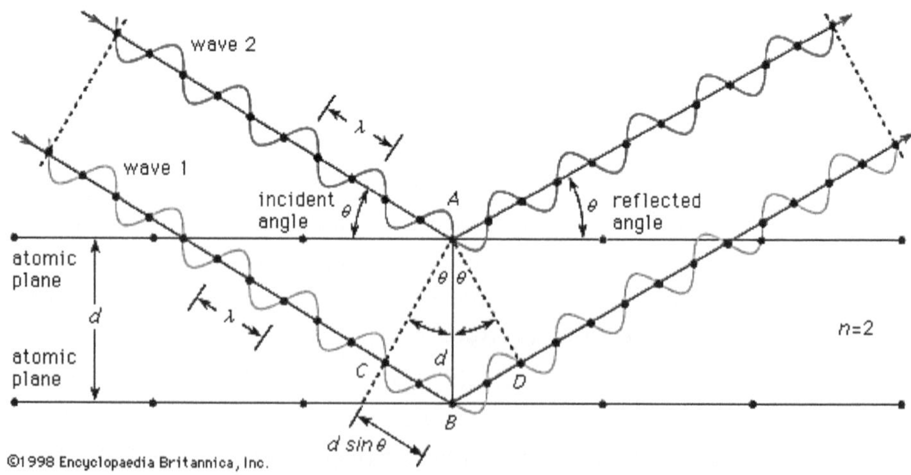

▲ **Fig-1: X-ray diffraction**

Thus, $n\lambda = 2d \sin \theta$, which is the Bragg law. As may be seen from the diagram, when $n = 2$, there is only one wavelength along path CB; also, the reflected angle will be smaller than that say, $n = 3$. Waves reflected through an angle corresponding to $n = 1$ are said to be in the first order of reflection, the angle corresponding to $n = 2$ is the second order, and so on.

5.0 Metal –its property, cause & effect

5.1 Physical & Mechanical Properties

i. Solid or liquid
   a) Solid Metal
   b) Metalloid
   c) Liquid Metal
ii. High density: more than water
iii. Produce sound on hammering
iv. Shines on polishing
v. Expandability on applied force: strain = $\Delta L/L$
vi. Expandability on heat:
   a) Linear Expansion: $\Delta L = L \times \propto_L \times \Delta t$
   b) Area expansion: $\Delta A = A \times \beta_A \times \Delta t$
   c) Volume expansion: $\Delta V = L \times \gamma_v \times \Delta t$
vii. Elasticity- Hooks law: Stress $\propto$ Strain
viii. Internal stress is developed when an external force is applied; force developed/ area is called stress; types- tensile, compressive, and shear stress.
ix. Can be joined by welding, fusion, forging
x. Alloy can be formed with other metal

5.2 Heat Transfer Properties

Conduction: $q = K A \Delta t/L$ [K=coefficient of conduction, A-Area, Δt- temperature difference and L-thickness]

5.3 Electro-Magnetic Properties

i. Electron conduction, called electricity

   a) Conductor allows an electron to flow

   b) Non-conductor does not allow electrons to flow

   c) Semiconductors sparsely allow electrons to flow

ii. Imparts resistance to current flow: R =V/I [R-Resistance, V- volt, I-current]

iii. Eddy current formation- while magnetic flux is imposed or current flows through a solid body a current is generated inside the metal body.

iv. Magnetic property can be induced on some metal, called magnetic metal

v. Electrons from metal can be drifted by:

   a) Magnetic force- Faraday's law of induction & Lenz's law: $e=-N\, d\varphi/dt$ [N- number of coils, $d\varphi/dt$- the rate of change of magnetic flux]

   b) Heat-Seebeck effect:

      1. Generated voltage (V) is the Seebeck voltage.

      2. It is related to the difference in temperature (ΔT) between the heated junction of dissimilar metal and the open junction by a proportionality factor (α) called the Seebeck coefficient, $V = \alpha\Delta T$.

   c) Sunray - Photoelectric effect: Photoelectric effect is the phenomenon of the ejection of electrons from a material upon exposure to electromagnetic radiation of a frequency greater than the threshold frequency of the metal.

      1. $E_{photon} = h\nu = KE_{electron} + \Phi$ [Φ= work function, ν- frequency of light, $h\nu = E_{photon}$; h- Planck's constant]]

      2. The Sun emits electromagnetic radiation with frequencies from 10^{18} Hertz down to 10^{4} Hertz.

   d) Pressure: Piezoelectric effect

      The nature of the piezoelectric effect is closely related to the occurrence of electric dipole moments in solids.

   e) Applied Voltage: On application of voltage, electrons flow in a conductor.

5.4 Chemical Properties

i. Reacts with Oxidant, an Oxidant is:
   a) Supplier of ion
   b) Supplier electron
   c) Oxygen

ii. Acts as reductant (reducing agent)

iii. Reacts with Acid & Alkali; Compound formation after reaction.

iv. Reactivity property
   a) The elements of lower valence are more reactive than those which have higher valence.
   b) Metals with a big atomic size and non-metals with a small atomic size are more active than others.
   c) In a group, the reactivity of metals ***increases from top to bottom and that of non-metals decreases from top to bottom***.
   d) Metalloids form compound ions (SO_4^{-2}, CO_3^{-2})
   e) Metal reactivity relates to the ability to lose electrons (oxidize), form basic hydroxides, and form ionic compounds with non-metals.
   f) For any two metals in the series, the metal placed higher in the series can displace lower metals from their salt solution.
   g) Reactivity series is a list of metals arranged in decreasing order of their reactivity. Most reactive metals are at the top while the least reactive metals at the bottom.

1) Potassium is a highly reactive metal
   a) more vigorously it reacts with other substances
   b) more easily it loses electrons to form positive ions (cations)
   c) displaces other metals in displacement reactions.

2) Copper is a less reactive metal

6.0 Thermodynamics

Thermodynamics is the study of the relations between heat, work, temperature, and energy. The laws of thermodynamics describe how the energy in a system changes and whether the system can perform useful work on its surroundings.

i. Thermodynamic property:
   a) Enthalpy
   b) Entropy
ii. Thermodynamic Systems comprises of:
   a) System boundary
   b) Surroundings

1. Isolated System – An isolated system ***cannot exchange energy and mass*** with its surroundings. The universe is considered an isolated system.
2. Closed System – Across the boundary of the closed system, ***the transfer of energy takes place but the transfer of mass doesn't take place***. Refrigerators, and compression of gas in the piston-cylinder assembly are examples of closed systems.
3. Open System – In an open system, the mass and energy both may be transferred between the system and surroundings. A steam turbine is an example of an open system.

6.1 Thermodynamic processes

There are four types of thermodynamic processes that have their unique properties, and they are:

i. **Adiabatic Process** – A process where no heat transfers into or out of the system occurs.
ii. **Isochoric Process** – A process where no change in volume occurs and the system does no work.
iii. **Isobaric Process** – A process in which no change in pressure occurs.
iv. **Isothermal Process** – A process in which no change in temperature occurs.

6.2 Thermodynamic Equilibrium

At a given state, all properties of a system have fixed values. Thus, if the value of even one property changes, the system's state changes to a different one. In a system that is in equilibrium, no changes in the value of properties occur when it is isolated from its surroundings.

i. Thermal equilibrium: When the temperature is the same throughout the entire system, we consider the system to be in thermal equilibrium.

ii. Mechanical equilibrium: When there is no change in pressure at any point of the system, we consider the system to be in mechanical equilibrium.

iii. Chemical equilibrium: When the chemical composition of a system does not vary with time, we consider the system to be in chemical equilibrium.

iv. Phase equilibrium: Phase equilibrium in a two-phase system is when the mass of each phase reaches an equilibrium level.

A thermodynamic system is said to be in thermodynamic equilibrium if it is in chemical equilibrium, mechanical equilibrium, and thermal equilibrium and the relevant parameters cease to vary with time.

6.3 Thermodynamic Potentials

Thermodynamic potentials are quantitative measures of the stored energy in a system. Potentials measure the energy changes in a system as they evolve from the initial state to the final state. Different potentials are used based on the system constraints, such as temperature and pressure.

Different forms of thermodynamic potentials along with their formula are tabulated below:

a) Internal Energy: $U = TdS - PdV + \ldots\ldots$

b) Helmholtz free energy: $F = U - TS$

c) Enthalpy: $H = U + PV$

d) Gibbs Free Energy: $G = U + PV - TS$

6.4 Laws of Thermodynamics

Thermodynamics laws define the fundamental physical quantities like energy, temperature and entropy that characterize thermodynamic systems at thermal equilibrium. These thermodynamics laws represent how these quantities behave under various circumstances.

There are four laws of thermodynamics and are given below:

a) Zeroth law of thermodynamics

b) First law of thermodynamics

c) Second law of thermodynamics

d) Third law of thermodynamics

i. **Zeroth Law of Thermodynamics**

The Zeroth law of thermodynamics states that if two bodies are individually in equilibrium with a separate third body, then the first two bodies are also in thermal equilibrium with each other.

ii. **First Law of Thermodynamics**

The first law of thermodynamics, also known as the ***law of conservation of energy***, states that energy can neither be created nor destroyed, but it can be changed from one form to another.

iii. **Second Law of Thermodynamics**

The second law of thermodynamics states that ***the entropy in an isolated system always increases***. Any isolated system spontaneously evolves towards thermal equilibrium—the state of maximum entropy of the system.

The entropy of the universe only increases and never decreases. Many individuals take this statement lightly and for granted, but it has an extensive impact and consequence.

No system or machine is 100% efficient while converting energy from one form to another.

iv. **Third Law of Thermodynamics**

The third law of thermodynamics states that ***the entropy of a system approaches a constant value*** as the temperature approaches absolute zero. The entropy of a pure crystalline substance (perfect order) at absolute zero temperature is zero. This statement holds true if the perfect crystal has only one state with minimum energy.

Third Law of Thermodynamics Examples:

Let us consider steam as an example to understand the third law of thermodynamics step by step:

1. The molecules within it move freely and have high entropy.
2. If one decreases the temperature below 100 °C, the steam gets converted to water, when the movement of molecules is restricted, decreasing the entropy of water.
3. When water is further cooled below 0 °C, it gets converted to solid ice. In this state, the movement of molecules is further restricted and the entropy of the system reduces more.

4. As the temperature of the ice further reduces, the movement of the molecules in them is restricted further and the entropy of the substance goes on decreasing.
5. When the ice is cooled to absolute zero, ideally, the entropy should be zero. But in reality, it is impossible to cool any substance to zero.

7.0 Electricity

7.1 Static electricity

i. Static Charge
   a) Positive charge
   b) Negative charge
ii. Opposite charges attract each other
iii. Same charges repel with each other
iv. Charge can be generated by friction with two similar bodies.

7.2 Current Electricity

Characteristics of current electricity:

i. The flow of electrons vis-à-vis charge.
ii. Charge-flow per second is current
iii. Driving force/energy, called voltage, is required for flowing electrons
iv. Seeks easy path to flow: gets earthed through earth connection
v. Can flow through conductor only
vi. Follows Ohm's law, Kirchhoff's Current Law and Kirchhoff's Voltage Law.

It is of two kinds- DC & AC.

7.3 Direct Current

Characteristics of Direct current (DC)

i. Flows through the core of the conductor
ii. Generates electro-magnetic flux (field) in a fixed direction; follows Right-hand thumb rule
iii. Induces magnetism on magnetic material.

7.4 Alternating Current

Characteristics of AC:

i. Current flows in SHM with a high frequency of 50 -60 HZ

ii. Energy flows through the outer surface of the conductor

iii. Generates self-induced voltage in the reverse direction to counter flow

iv. Induces alternating magnetism on magnetic material

v. Generates alternating electrical flux around the conductor

8.0 Magnetism

Magnetism is the force exerted by magnets when they attract or repel each other. Magnetism is caused by the motion of electric charges.

i. **Properties of magnet**

They are:

Attractive Property – Magnet attracts ferromagnetic materials like iron, cobalt, and nickel.

Repulsive Properties – Like magnetic poles repel each other and unlike magnetic poles attract each other.

Directive Property – A freely suspended magnet always points in a north-south direction.

Magnetic field- It is the region around a magnet, in which the force of attraction or repulsion produced by the magnet can be detected, is magnetic field.

Magnetic flux density-Magnetic flux density is the amount of magnetic flux in an area taken perpendicular to the magnetic flux's direction. It is denoted by the symbol B and it is measured in the units of Tesla.

ii. **Magnetic field intensity/ Field strength/magnetic intensity**

The Magnetic Field Intensity or Magnetic Field Strength is represented as vector H and is defined as the ratio of the MMF needed to create a certain Flux Density (B) within a particular material per unit length of that material. Magnetic field intensity is measured in units of amperes/metre.

It is given by the formula: $H= B\ \mu/M$

When,

a) B is the magnetic flux density

b) M is the magnetization

c) μ is the magnetic permeability

The SI unit of magnetic field intensity is Tesla. One tesla (1 T) is defined as the field intensity generating one newton of force per ampere of current per metre of conductor.

9.0 Electro- Magnetic phenomena

i. It is caused due to the combination effect of electrical field and magnetic field

ii. Magnetic field pushes electron at right angle to generate electrical current vis a-vis voltage: Faraday's law of Induction

iii. Rotating magnetic flux (by rotor) generates alternating current/ voltage on magnetic material (stator) due to induction- Alternator

iv. Rotating electrical flux (stator) generates rotating magnetic flux (due to induction) on magnetic Material (rotor)-Motor

v. Static electrical alternating flux induces static alternating electrical flux on magnetic material- transformer

vi. Leakage electric flux also induces air to generate voltage.

3-phase AC current

i. Phases are apart with each other by 120 deg.

ii. Line voltage (V_L): voltage between phases

iii. Phase voltage (V_P): voltage between phase and neutral

connection	Phase Voltage	Line voltage
Star connection	$V_{P=} V_L \div \sqrt{3}$	$V_L = \sqrt{3} \times V_P$
Delta Connection	$V_P = V_L$	$V_L = V_P$
Line voltage 11KV	Star: 11/√3=6.35 KV Delta: 11 KV	Star: 6.35x√3=11 KV Delta: 11 KV
Power	Star: $\sqrt{3} \times V_P$ x I x pf Delta: V_P x √3 x I x pf	

iv. Electrical field is capsuled in the armor but the magnetic field spreads around the cable

v. Transmission by connections: a) star/star b) star /delta c) delta/delta.

10.0 Detail discussion on laws

Gas Laws

The gas laws deal with how gases behave with respect to pressure, volume, temperature, and quantity. An important characteristic of gases is that, despite wide differences in chemical properties, all the gases more or less obey the gas laws.

Pressure: Gases are the only state of matter that can be compressed very tightly or expanded to fill a very large space. Pressure is force per unit area, calculated by dividing the force by the area on which the force acts. The earth's gravity acts on air molecules to create a force, that of the air pushing on the earth. This is called atmospheric pressure.

The units of pressure that are used are Pascal (Pa), standard atmosphere (atm), and torr. 1 atm is the average pressure at sea level. It is normally used as a standard unit of pressure. The SI unit though, is the Pascal. 101,325 Pascal equals 1 atm.

For laboratory work the atmosphere is very large. A more convenient unit is the torr. 760 torr equals 1 atm. A torr is the same unit as the mmHg (millimetre of mercury). It is the pressure that is needed to raise a tube of mercury 1 millimetre.

The Gas Laws: Pressure-Volume-Temperature Relationships.

i. **Boyle's Law: The Pressure-Volume Law**

Boyle's law or the pressure-volume law states that the volume of a given amount of gas held at constant temperature varies inversely with the applied pressure when the temperature and mass are constant;

$V \alpha 1/P$.

Another way to describe it is by saying that their products are constant. $PV = C$.

When the pressure goes up, the volume goes down. When the volume goes up, the pressure goes down.

From the equation above, this can be derived: $P_1V_1 = P_2V_2 = P_3V_3$, etc.

This equation states that the product of the initial volume and pressure is equal to the product of the volume and pressure after a change in one of them under constant temperature.

For example, if the initial volume was 900 mL at a pressure of 760 mmHg, at what pressure the volume is 450 mL?

$P_1V_1 = P_2V_2$; $P_2 = (V_1/V_2)\ P_1 = 760 x 900 / 450 = 1520$ mmHg.

ii. **Charles' Law: The Temperature-Volume Law**

This law states that the volume of a given amount of gas held at constant pressure is directly proportional to the Kelvin temperature. $V \alpha T$; $V / T = C$.

As the volume goes up, the temperature also goes up, and vice-versa. $V_1 / T_1 = V_2 / T_2 = V_3 / T_3$ etc.

iii. **Gay-Lussac's Law: The Pressure Temperature Law**

This law states that the pressure of a given amount of gas held at constant volume is directly proportional to the Kelvin temperature; $P \alpha T$; $P / T = C$.

With an increase in temperature, the pressure goes up.

Initial and final pressures and temperatures under constant volume can be calculated.

$P_1 / T_1 = P_2 / T_2 = P_3 / T_3$ etc.

iv. **Avogadro's Law: The Volume Amount Law**

This law states the relationship between volume and amount of gas in moles when pressure and temperature are held constant.

If the amount of gas in a container is increased, the volume increases and vice-versa, $V \alpha n$

Using a constant it can be expressed as: $V / n = C$

This means that the volume-amount fraction will always be the same value if the pressure and temperature remain constant. $V_1 / n_1 = V_2 / n_2 = V_3 / n_3$ etc.

v. **The Combined Gas Law**

Combining laws of Boyle, Charles' and Gay-Lussac: $V \alpha T/P$

The volume of a given amount of gas is proportional to the ratio of its Kelvin temperature and its pressure. Using a constant, it can be expressed as: $PV / T = K$ [K = Constant].

As the pressure goes up, the temperature also goes up, and vice-versa. Also, initial and final volumes and temperatures under constant pressure can be calculated.

$P_1V_1 / T_1 = P_2V_2 / T_2 = P_3V_3 / T_3$ etc.

10.1 Derivations from The Ideal Gas Laws

The previously stated laws all assume that the gas being measured is an ideal gas, a gas that obeys the gas laws exactly. But over a wide range of temperature, pressure, and volume real gases deviate slightly from ideal. Since, according to Avogadro, the same volumes of gas contain the same number of moles, it is now possible to determine the formulas of Volume of gas, gaseous elements, and their Molecular masses.

i. **Volume of gas**

The ideal gas law is: PV = n RT

When n is the number of moles and R is a constant called the universal gas constant= 0.0821 L-atm / Mole-K. It means, the Volume of 1-mole gas at 1 atm pressure and at 1 K temperature is 0.0821 Litre or, 82.1 cc or, 22.4 Litre at 1 atm pressure & at 0 deg C. [22.4/273=0.82]

Example:

A balloon is filled with about 1000 mole of H_2. If the outside temperature is 21^0 C and the atmospheric pressure is 750 mm Hg, what is the volume of the balloon?

Volume = n RT/P =1000 x 0.0821 x (21+273)/(750/760) = 24459 L.

ii. **Molecular weight of a gas and its chemical formula**

If the definition of the mole is included in the equation, the result is:

PV = n RT; PV = [gm(m) / molecular weight (M)] RT [gm = gram of gas; gm/ molecular weight= mole (n)].

Or, Molecular weight (M) = gm x RT / PV.

This equation provides a method of determining the Molecular Weight of a gas if mass, temperature, volume and pressure of the gas are known (or can be determined).

Example:

A 100 mg sample of a compound with the empirical formula CHF_2 is vaporized into a 256 mL flask at a temperature of 22.3 ° C. The pressure in the flask is measured to be 70.5 mmHg. What is the molecular formula of the compound?

i. P= 70.5 mmHg = 70.5/760 atm=0.093 atm

ii. V= 256 ml= 0.256 L

iii. m= 100mg= 0.1 gm.

iv. T= 22.3 +273= 295.3 K

Molecular weight = m x RT / PV

= 0.1 x 0.0821 x 295.3/(0.093x0.256) = 102 gm/ mole.

Molecular weight of CHF_2 = 12+1+19x2= 51.0 g / mole; Facto r = 102 / 51.0 = 2;

Hence, chemical formula: $C_2H_2F_4$.

iii. **Density of gas**

PV = n RT= (m/molecular wt.) RT

P = m/ (V x molecular wt.) RT= (m/V) x RT/molecular weight

P = d x RT/ molecular wt. [When, density (d) = (m / V)]

Density (d)= P x Molecular wt. / RT.

Example:

Let us compare the density of He and air (average molecular wt.= 28 g/ mole) at 25.0 °C and 1.0 atm.

d_{He} = (4.003 g / mole)(1.00 atm) / (0.0821 L-atm / mole-K)(298 K) = 0.164 g / L
d_{air} = (28.0 g / mole)(1.00 atm) / (0.0821 L-atm / mole-K)(298 K) = 1.14 g / L.

iv. **Comparison of density of a gas at different conditions**

Example

Compare the density of air at 25 ^{0} C and air at 1807 °C and 1.0 atm.

d_{air} at 25 ^{0} C = (28.0 g / mole) (1.00 atm) / (0.0821 L-atm / mole-K)(298 K) = 1.14 g / L
d_{air} at 1807 ^{0} C = (28.0 g / mole)(1.00 atm) / (0.0821 L-atm / mole-K) (2080 K) = 0.164 g / L.

v. **Dalton's law of Partial Pressure**

Dalton's Law of Partial Pressures states that the total pressure of a mixture of non-reacting gases is the sum of their individual partial pressures.

$P_{total} = P_a + P_b + P_c + \ldots$ [When, P_a, P_b, P_c are partial pressure of gases a, b &c respectively]

Or, $P_{total} = n_a RT / V + n_b RT / V + n_c RT / V + \ldots$ [When, total volume= V]

Or, $P_{total} = (n_a + n_b + n_c + \ldots) RT / V$

The pressure in a flask containing a mixture of 1 mole of 0.20 mole O_2 and 0.80 mole N_2 would be the same as the same flask holding 1 mole of O_2.

Partial pressures are useful when gases are collected by bubbling through water (displacement). The gas collected is saturated in water vapor which contributes to the total number of moles of gas in the container.

Example

A sample of H_2 was prepared in the laboratory by the reaction:

Mg (s) + 2 HCl (aq.) = $MgCl_2$(aq.) + H_2(g)

456 mL of gas was collected at 22.0 ^{0} C. The total pressure in the flask was 742 torr. How many moles of H_2 were collected? The vapor pressure of H_2O at 22.0 °C is 19.8 torr. [1 torr = 1 mm Hg].

i. V= 456 mL= 0.456 L

ii. P_{total}=742 torr.

iii. P_{H20}= 19.8 torr

iv. PH_2=742-19.8=722.2torr=722.2/760atm.=0.9503 atm.
$n_{H2} = P_{H2}V / RT$; $n_{H2} = (0.9503 \text{ x}0.456) / (0.0821\text{x } 295) = 0.0179$ mole H_2.

11.0 Non-Ideal Gases

The ideal gas equation (PV= n RT) provides a valuable model of the relations between volume, pressure, temperature and number of particles in a gas.

Real molecules do have volume and do attract each other. All gases depart from ideal behaviour under conditions of low temperature (when liquefaction begins) and high pressure (molecules are more crowed so the volume of the molecule becomes important).

J. D. van der Waals proposed his equation, known as the van der Waals equation. As there are attractive forces between molecules, the pressure is lower than the ideal value. To account for this the pressure term is augmented by an attractive force term a/V^2. Likewise, real molecules have a volume. The volume of the molecules is represented by the term b. The term b is a function of a spherical diameter d known as the van der Waals diameter. The van der Waals equation for n moles of gas is:

$[P + n (a /V)^2] (V - n b) = n RT$.

The values for a & b, below, are determined empirically:

Molecule	a (liters²-atm / mole²)	b (liters / mole)
H_2	0.2444	0.02661
O_2	1.360	0.03183
N_2	1.390	0.03913
CO_2	3.592	0.04267

Molecule	a (liters^2-atm / mole^2)	b (liters / mole)
Cl_2	6.493	0.05622
Ar	1.345	0.03219
Ne	0.2107	0.01709
He	0.03412	0.02370

12.0 Detail Discussions on Fluid Characteristics

1. Surface Tension

Intermolecular forces, called cohesive forces, draw the surface particle to the bulk liquid mass to bring them together. It is defined as the force that is required (against the cohesive force) to increase the area of the liquid along the surface.

Surface tension (T) is mathematically expressed as the ratio of the surface force F to the length L as:

T=F/L, when F= force in N and L= length (m). Its unit= N/m.

2. Viscosity

Viscosity is a measure of a fluid's resistance (shear or tensile stress) to deformation or flow. It is described as the internal molecular friction of a moving fluid.

The shear stress (τ) is proportional to velocity gradient (v/L).

$\tau \alpha v/L$; $\tau = \mu v/L$; $\mu = \tau \times L/v = N/m^2 \times m \times s \times 1/m = N\text{-}s/m^2 = Pa\text{-}s$.

i. Types of Viscosity

There are two ways to measure a fluid's viscosity as follows:

a) **Dynamic Viscosity or Absolute Viscosity or simply viscosity**

The absolute unit is named after Poiseuille as Poise (P). 1P= dyne-s/cm^2; 1 Pa-s = 10 P [1dyne= 1gm x 981 cm/s^2].

b) **Kinematic Viscosity**

Kinematic viscosity is more useful than absolute or dynamic viscosity.

Kinematic viscosity= Absolute viscosity/ density = $N\text{-}s/m^2 \div kg/m^3$

$= kg \times 1m/s^2 \times 1/m^2 \times s \times m^3/kg = m^2/s$.

It being a bigger unit, commonly used unit = cm^2/s, also called Stokes (St).

ii. **Effect of viscosity of fluid on different conditions**

a) **Effect of temperature on Viscosity of liquid**

Viscosity is first and foremost a function of fluid. The viscosity of water at 20 °C is 1.0020 milli Pa-s (which is conveniently close to 1 by coincidence alone). Most ordinary liquids have viscosities on the order of 1 to 1,000 mPa- s, while gases have viscosities on the order of 1 to 10 μ Pa- s.

Viscosity varies with temperature and it (for simple liquid) decreases with increasing temperature. As temperature increases, the average speed of the molecules in a liquid increases and the amount of time they spend "in contact" with their nearest neighbors decreases. Thus, as temperature increases, the average intermolecular forces decrease.

b) **Effect of temperature on viscosity of the gas**

The viscosity of gases increases as temperature increases and is approximately ***proportional to the square root of temperature***. This is due to the increase in the frequency of intermolecular collisions at higher temperatures. Since, most of the time the molecules in a gas are moving freely through the void, anything that increases the number of times one molecule is in contact with another will decrease the ability of the molecules as a whole for coordinated movement.

c) **Effect on Pressure**

Viscosity is normally independent of pressure, but liquids under extreme pressure often experience an increase in viscosity.

iii. **Importance of Viscosity**

Viscosity is used to find Reynolds' number (Re) for fluid flow, the number reveals whether flow is laminar or turbulent flow. Based on the type of flow head loss can be found out.

Reynolds' number (Re) = ρ d v/μ = d v x (ρ/μ) = dv/ν (nu) [When, ρ= density of fluid, d = diameter of flow path v= Fluid velocity, μ = Absolute viscosity, ν (nu)= kinematic viscosity]. If Re<2000, it is laminar flow, > 4000, it is turbulent flow. In between, it is called transient flow.

To be noted during head loss calculation, equation for turbulent flow is also used for transient flow.

Example: The kinematic viscosity of a hydraulic fluid is 0.0001 m^2/s. If it is flowing in a 30-mm diameter pipe at a velocity of 6 m/s, what is the Reynolds number? Is the flow laminar or turbulent?

Re=0.03x 6/0.0001= 1800; hence, the flow is laminar.

13.0 Head loss equation (Darcy-Weisbach formula)

$H_f = flv^2/(2gd)$ (m) [f= friction factor, l= length of pipe (m), d= diameter of pipe(m), Head loss is in meter.

Friction factor (f) can be calculated for laminar flow after finding Reynolds' number as:

f= Re/ 64. Hence, The formula can be modified as: $H_f = [(Re/64) \times l \times v^2/(2gd)]$

For Turbulent flow, the Darcy- Weisbach formula can be used, taking from a standard chart.

14.0 Friction

Frictional force refers to the force generated by two surfaces that contacts and slide against each other.

The maximum amount of friction force that a surface can apply upon an object can be easily calculated with the use of the given formula: $F_{Fric.} = \mu \times F_{Norm}$. When, μ= Friction coefficient (certain% of applied force).

The normal force is the support force exerted upon an object that is in contact with another stable object. The normal force can be simply described in most cases by the following formula: $F_{Norm} = mg$.

15.0 Thermodynamic Processes

Basic equation for relation among heat, internal energy & work done:

$Q=U+W$; $dQ = dU + dW = dU + P\,dV = dU + V\,dP$

When, Q = Energy supplied, U= internal energy, W= work done, P= pressure and dQ, dU, dP, dV & dW represent change in the stated quantity.

i. **Constant Volumetric Process Also Called Isochoric Process**

   No work done, only change in internal energy; $dQ = dU$

ii. **Constant Temperature Process or Isothermal Process**

   No change in temperature during process. PV= Constant; when, P=Pressure, V= Volume.

   Work Done: $\int PdV = n\,RT \times \int_{Vi}^{Vf} dV/V = n\,RT \ln (V_f/V_i)$.

iii. **Constant Pressure Process**

   No change in pressure during the process; also called isobaric process.

   $dQ = dU + P\,dV = dH$, H= Enthalpy.

   Work done: P dV Joule [P= Pressure (Pa), dV= change in volume m^3].

iv. **Adiabatic Process**

No heat transfer during process; PV^{γ} = constant, when $\gamma = C_p / C_v$

Work done (W): $K \int P\, dV = K \int_{vi}^{vf} dv/v^{\gamma} = K [(V_f^{1-\gamma} - V_i^{1-\gamma})/(1-\gamma)]$ [K= n RT].

v. **Polytropic Process**

An analogy to adiabatic process when, $\gamma = n$; PV^{n} = constant; $n > \gamma$.

Work done (W): $K \int P\, dV = K \int_{vi}^{vf} dv/v^{n} = K [(V_f^{1-n} - V_i^{1-n})/(1-n)]$.

vi. **Isentropic Process or Reversible Adiabatic process**

Process in which entropy remains constant is isentropic process.

vii. **Throttling Process**

Enthalpy remains constant before and after Throttling. $H_1 = H_2$; H_1 & H_2 are enthalpies before & after the process respectively.

16.0 Heat of solution

When preparing dilutions of concentrated sulfuric acid, a great amount of heat is released in the dissolving process. This is due to the reason that heat is generated during solution.

Enthalpy changes when a solute undergoes the physical process of dissolving into a solvent. Calcium chloride releases heat when it dissolves according to the equation below.

$CaCl_2(s) \rightarrow Ca^{2+}(aq) + 2Cl^{-}(aq) + 82.8kJ$

Ammonium nitrate absorbs heat from the surroundings when it dissolves.

$NH_4NO_3(s) + 25.7kJ \rightarrow NH^{+4}(aq) + NO^{-3}(aq)$

The molar heat of solution (ΔH_{soln}) of a substance is the heat absorbed or released when one mole of the substance is dissolved in water. For calcium chloride, $\Delta H_{soln} = -82.8$kJ/mol(heat release) & for ammonium nitrate, $\Delta H_{soln} =$ 25.7kJ/mol (heat absorbs).

17.0 Detail discussion on Force- Work-Power- Energy

17.1 Notations

Notations used for deriving the relations:

i. Body Mass (M) =Kg

ii. Distance / Length (d)= meter (m)

iii. Height (H) = meter (m)

iv. Time (t)= Sec (s)

v. Diameter (d) = meter (m)

vi. Radius (r) = meter (m)

vii. Rotational speed (n)= per second

viii. RPM (N) = Per Minute

ix. Velocity (v)= m/s

x. Acceleration (a)= m/s^2

xi. Force = Newton = N

xii. Torque (T)= N-m

xiii. Work = Joule (J)

xiv. Power = Watt (W)

xv. Energy= Joule, Watt-hour (J, W-Hr.)

xvi. Enthalpy (energy)= Kcal, KJ

17.2 Basic Force- Work- Power – Energy

i. Force = Mass x Acceleration= mx a = Kg x m/ s^2 = Newton (N)

ii. Work= Force x Distance= N x m= N-m= Joule

iii. Power = Force x Velocity = Work/ Time= N-m/s= Watt

iv. Energy= Power x Time = N-m/s x s= Joule or Watt- second or Watt –hour.

17.3 Work & Power due to Linear Movement

i. Work= Force x Linear Movement = N x m Joule

ii. Power= Work/ s = Force x Movement/ Time= N x m/s= Joules/s = Watt

17.4 Work & Power for Movement against Gravity /Potential Power

i. Work=Mass x g x Height= M x g x H = kg x 9.81 x m/ s^2 x m= 9.81 N-m = 9.81 Joule.

ii. Power = Mass flow rate x g x Height= M/s x g x H = kg/s x 9.81 x m/ s^2 x m = 9.81 N-m/s = 9.81 Joule/s = 9.81Watt.

17.5 Force-Work- Power due to Rotational Movement

i. Torque (T)= Force x Radius of rotation = F x r = N-m[r = radius of rotation]

ii. Power = Torque x Angular velocity= T x ω = T x 2 x π x n= N-m/s= Watt. [n= rotation per second]

iii. Or, Power= (F x r) 2x π x n = N x m/s= Watt

17.6 Velocity Power

i. **For Linear Velocity(v)**

a. Velocity work / Kinetic energy = ½ x M x v^2 = kg x [m^2 /(s x s)]= [kg x m /(s x s)] x m =N-m = Joule

b. Velocity Power = ½ x M/s x v^2 = kg/s x m x m/(s x s)=N-m/s = Joule/s= Watt

c) **For angular velocity** (ω)

i. Velocity work=½ x M x (r x ω) ^{2}, [v = r x ω)] = ½ x M x r^2 x ω^2 = ½ x M x r^2 x (2π x n)^2 = kg x m/s^2 x m = N-m= Joule.

ii. Velocity power = kg/s x m x m/s^2= N-m/s= Joule/s= Watt [kg/s= mass flow rate]

17.7 Pressure –Volume work and Power

a) **Constant Pressure Process, Volume changes**

i. Work = P x ΔV = N/m^2 x m^3 = N-m= Joule

ii. Power = P x ΔV /s = N/m^2 x m^3/s = N-m/s= Joule/S= Watt [ΔV /s = Flow rate]

b) **Constant Volume Process, Pressure changes**

i. Work = V x ΔP = m^3 x N/m^2 = N-m= Joule

ii. Power = V x ΔP /s = m^3 x N/m^2 x 1 /s = N-m/s= Joule/s= Watt; ΔP /s = Pressure change rate.

17.8 Impulse Force & Power

a) Impulse = mass x rate of change of velocity= M x ΔV/t = Kg x m/s x 1/s= N

b) For Linear movement due to impulse:

i. Work= mass x rate of change of velocity x distance= N-m= Joule

ii. Power = mass/s x rate of change of velocity x distance= N-m/s= Joule/s= Watt

c) For rotational movement due to impulse:

i. Impulse Work = Impulse Force x Distance= N x 2π x r = N-m

ii. Power = Impulse x Distance/ s =N x 2π x r x n = N-m/s= Joule/s= Watt. [n = rps]

17.9 Reaction Force & Power

a) Pressure changes, Velocity remains constant during Reaction force

   i. $V = \sqrt{(2 \times g \times \Delta P)}$, m/s

   ii. Impulse force due to reaction = m/s x kg/s = kg x m/s^2 =N [kg/s= mass flow rate]

   iii. Work= Force x Distance travelled (d)= N-m = Joule

b) For linear Movement, W= N x m= Joule

   i. Power = Force x traveling velocity = N x v = N x m/ s = Watt

c) For Rotational Movement

   i. Work =Force x distance travelled= N x 2π x r Joule

   ii. Power = N x 2π x r x n Joules/ s= Watt. [n=rotation/s]

17.10 Chemical & Heat Energy

i. Energy = Power x time= P x t = Watt-s, Joule

ii. Chemical & Heat Energy= Kcal or K-Joule

iii. For Electrical, Energy = Watt-s, Watt-hr., kWh, MWH

17.11 Relation between Kcal & kWh

i. 1 Kcal = 4.18 KJ; If divided both sides of the equation by second.

ii. 1Kcal/s. = 4.186 KJ/s=4.186kW.

iii. 1 Kcal x hr/s=4.186 kWh [Multiplied both side by hour]

iv. 1Kcalx3600x s/s =4.186 kWh;1kWh=3600/4.186 Kcal=860 Kcal

17.12 Power / Energy Conversion & inter-relation

Bernoulli's equation-For steady state non-compressible fluid:

i. Energy Equation: $P_1 / \rho + v_1{}^2/ 2 + gZ_1 = P_2 / \rho + v_2{}^2/ 2 + gZ_2$ (Joule)

ii. Pressure equation: $P1 + \rho v_1{}^2/ 2 + \rho gZ_1 = P_2 + \rho v_2{}^2/ 2 + \rho gZ_2$ (N/m^2= Pa)

iii. Potential power/ energy to Velocity energy/ power

   $Mgh=1/2 Mv^2$; $v^2=2gh$; $v = \sqrt{(2gh)}$; g= m/s^2, h= m, v =m/s

iv. Pressure power/Energy to Velocity power/Energy

   $P / \rho = v^2/ 2$; $v^2 = 2 P/ \rho$; $v = \sqrt{(2 P/ \rho)}$ (m/s)

v. **Chemical Energy to Electrical energy: Battery**

A battery is a device that converts chemical energy contained within its active materials directly into electric energy by means of an electrochemical oxidation-reduction (redox) reaction.

Redox reactions can be balanced using the half-reaction method. The standard cell potential is a measure of the driving force for the reaction. ($E_{cell} = E_{cathode} - E_{anode}$].

The flow of electrons in an electrochemical cell depends on the identity of the reacting substances, the difference in the potential energy of their valence electrons, and their concentrations. The potential of the cell under standard conditions (1 M for solutions, 1 atm for gases, pure solids or liquids for other substances) and at a fixed temperature (25°C) is called the ***standard cell potential*** (E_{cell}). Only the difference between the potentials of two electrodes can be measured. The potential of the Standard Hydrogen Electrode (SHE) is defined as 0 V under standard conditions. The potential of a half-reaction measured against the SHE under standard conditions is called its ***standard electrode potential***. By convention, all tabulated values of standard electrode potentials are listed as standard reduction potentials.

The overall cell potential is the ***reduction potential of the reductive half-reaction minus the reduction potential of the oxidative half-reaction*** ($E_{cell} = E_{cathode} - E_{anode}$). The standard cell potential is a measure of the driving force for a given redox reaction. If E_{cell} is positive, the reaction will occur spontaneously under standard conditions. If E_{cell} is negative, then the reaction is not spontaneous under standard conditions, although it will proceed spontaneously in the opposite direction.

18.0 Mass and Energy Equations

i. **Conservation of Mass**

The conservation of mass relation for a closed system undergoing a change is expressed as M_{sys} = constant or $d(M_{sys}/dt) = 0$, which is a statement of the obvious that the mass of the system remains constant during a process.

ii. **Conservation of Momentum**

The product of the mass and the velocity of a body is called the linear momentum or just the momentum of the body, and the momentum of a rigid body of mass m moving with a velocity V is mV.

Newton's second law states that the acceleration of a body is proportional to the net force acting on it and is inversely proportional to its mass, and that the rate of change of the momentum of a body is equal to the net force acting on the body. Therefore, the momentum of a system remains constant when the net force acting on it is zero, and thus the momentum of such systems is conserved. This is known as the conservation of momentum principle. In fluid mechanics, Newton's second law is usually referred to as the linear momentum equation.

iii. **Continuity Equation: $\rho_1 V_1 A_1 = \rho_2 V_2 A_2$**

Mechanical energy of a flowing incompressible fluid per unit mass = $P/\rho + V^2/2 + gz$

For steady incompressible fluid; $P_1/\rho + V_1^2/2 + gz_1 = P_2/\rho + V_2^2/2 + gz_2$.

iv. **Static, Dynamic, and Stagnation Pressures**

The Bernoulli equation states that the sum of the flow, kinetic, and potential energies of a fluid particle along a streamline is constant. Therefore, the kinetic and potential energies of the fluid can be converted to flow energy (and vice versa) during flow, causing the pressure to change. This phenomenon can be made more visible by multiplying the Bernoulli equation by the density ρ, Each term in this equation has pressure units, and thus each term represents some kind of pressure: $P + \rho v^2/2 + \rho gz$

a) P is the static pressure (it does not incorporate any dynamic effects); it represents the actual thermodynamic pressure of the fluid. This is the same as the pressure used in thermodynamics and property tables.

b) $\rho V^2/2$ is the dynamic pressure; it represents the pressure rise when the fluid in motion is brought to a stop isentropically.

c) ρgz is the hydrostatic pressure, which is not pressure in a real sense since its value depends on the reference level selected; it accounts for the elevation effects, i.e., of fluid weight on pressure. The sum of the static, dynamic, and hydrostatic pressures is called the total pressure. Therefore, the Bernoulli equation states that the total pressure along a streamline is constant.

The sum of the static and dynamic pressures is called the stagnation pressure, and it is expressed as $P + \rho V^2/2$. The stagnation pressure represents the pressure at a point when the fluid is brought to a complete stop isentropically.

SECTION

02 ROTATING EQUIPMENT

1.0 Centrifugal Pump Operation, Performance and Efficiency

1.1 Introduction

A centrifugal pump operates through the transfer of rotational energy from one or more driven rotors, called impellers. The action of the impeller increases the fluid's velocity and pressure directs it towards the pump outlet via diffuser. The diffuser converts the velocity head to the pressure head.

1.2 Types of centrifugal pump

The major centrifugal pump variations include the following:

a) Radial

b) Axial

1.3 Differences between Radial & Axial Centrifugal Pump

The major distinction between a radial and an axial centrifugal pump is in their orientation. By design, a radial centrifugal pump permits an outward motion of the liquid channeled through it. The pumped liquid is pressurized and exited through downstream piping.

By comparison, axial pumps generate fluid motion via a lifting effect of their impeller vanes.

Different parts of the centrifugal pump are shown in **Fig- 2.1**

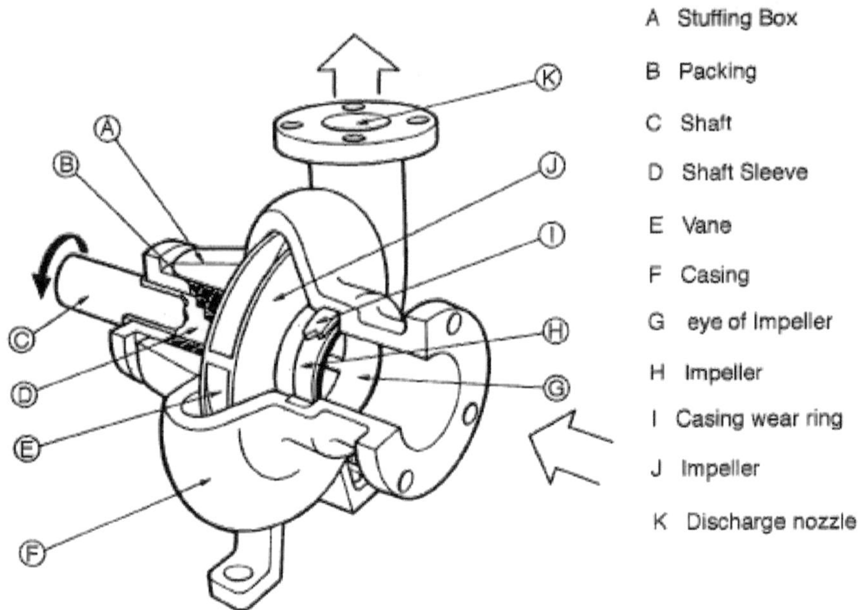

▲ **Fig-2.1: Cross-sectional view of centrifugal pump**

1.4 Common Industrial Centrifugal Pump Applications

Centrifugal pumps are used in wide range of application like domestic, commercial, and industrial application. Examples of centrifugal pump applications include but are not limited to:

a) Water supply for power plant for different purposes

b) Water supply for residential areas

c) Fire protection systems

d) Sewage/slurry disposal

e) Food and beverage manufacturing

f) Chemical manufacturing

g) Oil and gas industrial operations

1.5 Hydraulic Pump Power

The ideal hydraulic power to drive a pump depends on, Fig-2.2:

a) The mass flow rate

b) Liquid density

c) The differential height

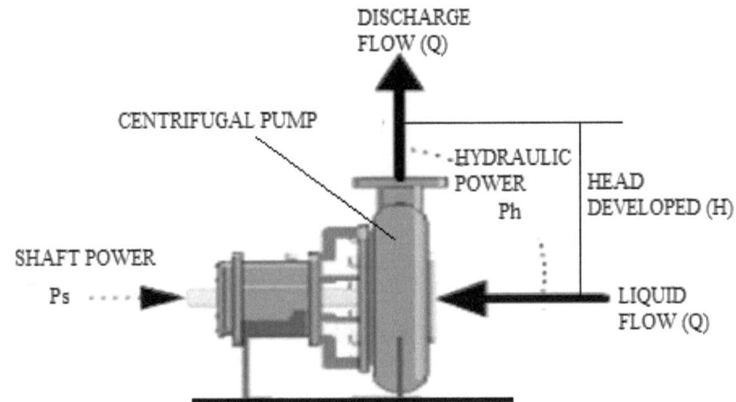

▲ **Fig-2.2: Schematic Diagram showing hydraulic & shaft power of Centrifugal Pump**

Hydraulic power may be:

a) A static lift from one height to another

b) The total head loss component of the system

It can be calculated in the following ways:

i. $\mathbf{P_h = Q \times \rho \times g \times h / (3.6 \times 10^6) = Q \times p / (3.6 \times 10^6)}$ **kW**

When,

a) P_h = Hydraulic power (kW)

b) Q = flow (m^3/h)

c) ρ = density of fluid (kg/m^3)

d) g = acceleration of gravity (9.81 m/s^2)

e) h = differential head (m)

f) p = differential pressure (N/m^2, Pa) = $\rho \times g \times h$

ii. **$P_h = Q \times P/1000$ kW**

When,

a) Q= Flow (m^3/s)

b) P= Pressure (N/m^2 or Pa)

Example: A Pump develops a differential head 15 Kg/ cm^2 with a flow of water 100 m^3/ hr. Calculate Hydraulic Power. Density of water= 1000kg/m^3.

a) Q = 100 m^3/hr. = 100/3600 m^3/s = 1/36m^3/s= 100/3600 x1000= 27.77 kg/s.

b) Differential Pressure= 15 Kgf / cm^2= 15x10 m Head = 15 x 0.98 x 10^5 N/m^2.

Hydraulic Power,

i. $P_h = Q \rho g h / (3.6 \times 10^6) = 100 \times 1000 \times 9.81 \times 150/(3.6 \times 10^6) = 40.87$ kW

ii. Alternative-1, P_h = [m^3/s x N/ m^2]/1000 = 1/36 x 15x 0.98 x 10^5 /1000= 40.82 kW

iii. Alternative-2, P_h = Kg/s x g x h / 1000 = 27.77x 9.81 x 150/1000 = 40.87 kW.

iv. Alternative-3, P_h = kg/s x h/102=27.77 x 150/102=40.83 kW.

1.6 Hydraulic Horse Power

The hydraulic Horse Power can be calculated as: P_h (HP) = P_h (kW) / 0.746

When,

a) P_h (HP) = hydraulic horsepower in (HP)

b) P_h (kW) = hydraulic horsepower in (kW)

Example – Horse Power for Pumping Water

100 m^3/h of water is pumped a head of 20 m. Find Out Hydraulic power in kW & HP.

Solution: P_h = (100 m^3/h) x (1000 kg/m^3) (9.81 m/s^2) (20 m) / (3.6x 10^6) kW= 5.45 kW= 5.45/0.746 HP= 7.3HP.

1.7 Pump Shaft Power

To generate hydraulic power shaft is required to have more power than hydraulic power. Excess power is required to overcome the losses inside pump. This power is called Pump Shaft power which depends on pump efficiency. Pump efficiency = hydraulic power (delivered by pump)/shaft power (delivered by motor).

Hence, shaft power= hydraulic power/ Pump efficiency.

$P_s = P_h / \eta_p$

When,

a) P_s = Pump Shaft power (kW)

b) P_h = Pump hydraulic Power (kW)

c) η_p = Efficiency of Pump

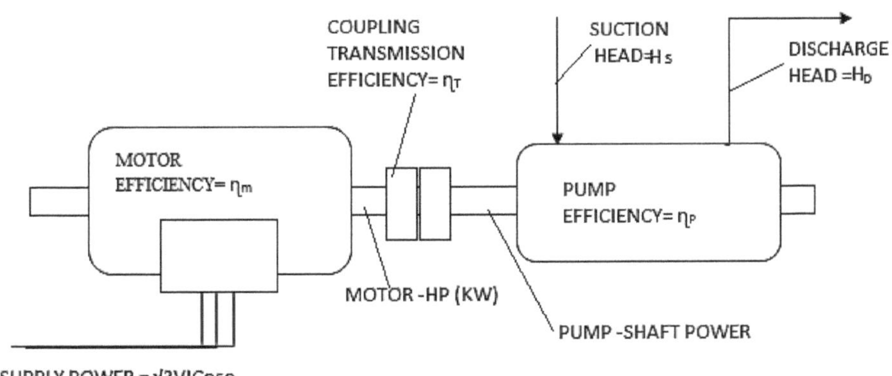

▲ **Fig-2.3: Pumping system with motor**

Example: A pump supplies water with a flow of 500 m^3/ hr. at a discharge pressure 4.5 kg/cm^2 having suction pressure of 2.5 kg/ cm^2. Find supply power considering pump efficiency= 75%, Transmission efficiency= 98%, Motor efficiency=95%.

i. Developed pressure head = 4.5-2.5= 2 kg/ cm^2 = 2 x0.98 x 10^5 N/m^2 (Pascal)=20 m

ii. Flow= 500 m^3/ hr. = 500/3600 m^3/s

iii. Pump Hydraulic Power (kW) = N/m^2 x m^3/s ÷ 1000 =2x 0.98x 10^5 x 5/36 x1/1000 =27.21 kW

iv. Supply power (kW) = 27.21/ (0.75 x 0.98 x 0.95) = 38.97.

v. A loss of approximately= (38.97-27.21/38.97x100= 30%.

Alternative approach,

Pump Hydraulic Power (kW) = 500/3.6 (kg/s) x 9.81(g) x 2x10/1000(m) = 27.25 kW

Supply power = 27.25/(0.75x0.98x0.95) = 38.97 kW.

1.8 Losses in centrifugal pumps

Sl. no	Loss due to	Internal effect	Change in parameter
1	Low RPM (N)	i. Low Flow; as F prop to N ii. Low head; as H prop to N^2 iii. Low Power; as P prop to N^3	Decrease in discharge pressure Reduction in ampere
2	More clearance	i. H-Q curve will be below the design curve ii. Power consumption increases iii. NPSH (suction pressure) increase iv. Efficiency decrease	Increase in ampere at fixed flow Suction pressure increases
3	Friction loss	i. Increase in discharge temperature ii. Loss in head iii. Loss in efficiency iv. Increase in power consumption	Discharge temperature is more than design value Increase in power consumption at constant flow
4	Density	$P = Q \times H \times \rho \times g / (3600 \times \eta)$ If other parameters remain constant, P proportional to ρ	Motor load increase Discharge pressure increase
5	Viscosity	i. Small reduction in flow, ii. Significant reduction in head or pressure, iii. Substantial increase in power draw.	

i. **Losses in a Centrifugal Pump**

a) Mechanical friction power loss due to friction between the fixed and rotating parts in the bearing and stuffing boxes.

b) Disc friction power loss due to friction between the rotating faces of the impeller (or disc) and the liquid.

c) Leakage and recirculation power loss. This is due to loss of liquid from the pump and recirculation of the liquid in the impeller. The pressure difference between impeller tip and eye can cause recirculation of a small volume of liquid, thus reducing the flow rate.

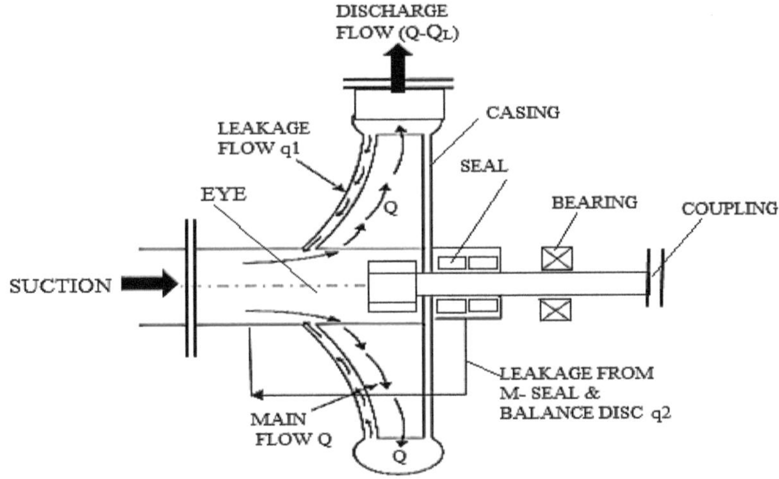

▲ **Fig-2.4: Single stage Centrifugal Pump**

ii. **Hydraulic Loss (head loss)**

Following are the head losses in a centrifugal pump.

a) Loss at the entry of eye (shock) due to imperfect angle of entry. This happens due to the reasons: Pump is running at low load &Pump is running at higher load

b) Friction loss in the blade passage

c) Circulation loss due to whirl

d) Frictional loss in casing.

iii. **Hydraulic power Loss Calculation**

Let,

Head supplied by the drive (Motor)= $H_{i,}$ Hydraulic Loss =$H_{f.}$

Then, hydraulic Loss% = H_f / H_i

And, Hydraulic Efficiency $\eta_{hyd} = (H_i - H_f)/ H_i$.

Normally for centrifugal pump head loss due to:

a) Shock and swirl loss is 7%

b) Friction in Hydraulic surfaces such as impellers, casing and balance disc friction loss is 5%

Hydraulic efficiency (η_h)is generally = 80-90%.

iv. **Leakage Loss**

This loss is due to refer fig-2.4:

i. The clearance between Casing & Impeller (q1)

ii. Balance disc Loss (q2)

Let, total Flow discharge at Impeller Tip (No Volume loss) = Q, Leakage loss between Casing & Impeller=(Q_L)= q1 + q2

Then, Loss% = Q_L/ Qx100

i. Volumetric efficiency η_{vol}=(Q-Q_L) /Qx100%

Normally, for centrifugal pump leakage losses are:

ii. Clearance leakage loss is 2%

iii. Balance mechanism leakage is 2%

Then, Volumetric efficiency= 96%.

v. **Mechanical losses**

Following are the mechanical losses in a centrifugal pump:

i. Frictional loss in bearing

ii. Gland Packing or Mechanical seal

Normally, for centrifugal pump Mechanical loss=1-4%

Hence, Mechanical efficiency (η_{mech}) is generally= 96-99%.

1.9 Overall Pump Efficiency

Overall Pump Efficiency=Power gained by the fluid/ Power supplied at pump shaft= P_{fluid} /P_{shaft}

a) $\eta_{pump} = \eta_h \times \eta_{vol} \times \eta_{mech}$

b) η_{pump} = 100- (hydraulic loss% + volumetric Loss%+ Mechanical Loss%).

1.10 Calculation of deviation

Example:

A multistage centrifugal pump has the running parameters as suction pressure 3.4 kg/ cm^2, discharge pressure 104 kg/cm^2 with a flow of 90 m^3/hr. It has a balancing leak off flow as 10 LPM and minimum circulation flow of 20 m^3/ hr. The motor is having supply of 6.6 KV and drawing a current of 35 amp with a power factor of 0.78. Design pump efficiency is 89%, motor efficiency is 98% and transmission efficiency is 99%. Consider water density as 999 kg/ m^3.

Calculate:

i. Pump hydraulic power, deviation in Pump Efficiency

ii. Motor Power and deviation in motor efficiency.

Given:

i. Suction Pressure =3.4 kg/cm^2, discharge pressure= 104 kg/cm^2

ii. Developed head = 104-3.4= 100.6 kg/ cm^2 = 1006 m = 100.6 x 10^5 Pa

iii. Flow = 90 x 999/3600 kg/s=24.97 $\approx$ 25 kg/s

iv. Hydraulic power = 25 x 9.8 x 1006/1000 kW=246.47 kW.

v. Power supplied = √3 x 6.6 x 35 x0.78= 312 kW, corrected power (at 0.8 Pf) = 312x 0.78/0.8=304.2kW

vi. Motor shaft power = Power supplied x motor efficiency=304.2 x 0.98= 298.11kW

vii. Considering transmission efficiency (99%), power supplied to pump shaft= 298.11 x 0.99= 295.13kW.

viii. Hence, pump efficiency at running condition = Pump Hydraulic Power/ Pump shaft power x 100 =246.47/295.13x100 = 83.5%.

ix. Deviation in Pump efficiency = (87-83.5)/87 x100= 4.02%.

1.11 Calculation of additional loss

Hydraulic power loss = (87-83.5)% x 246.47 kW =8.6 kW

If x kg/s is additional loss through ARV, then total Volumetric loss = (x +10/60) kg/s

Hence, (x+10/6) = 8.6 x1000/ (9.81 x1006); x= 0.7 kg/s = 0.7x 3.6 = 0.27 m^3/ hr.=1% of pump capacity.

2.0 Reciprocating Pump

Introduction: Reciprocating pump is a positive displacement pump where certain volume of liquid is sucked into enclosed volume and is discharged using pressure. Reciprocating pumps are more suitable for low volumes of flow at high pressures.

2.1 Components of Reciprocating Pump

The main components of reciprocating pump, Fig- are as follows:

a) **Suction Pipe-**Suction pipe connects the source of liquid to the cylinder of the reciprocating pump. The liquid is suck by this pipe from the source to the cylinder.

b) **Suction Valve-**Suction valve is non-return valve which means only one directional flow is possible in this type of valve. This is placed between suction pipe inlet and cylinder. During suction of liquid it is opened and during discharge it is closed.

c) **Delivery Pipe-**Delivery pipe connects cylinder of pump to the outlet source. The liquid is delivered to desired outlet location through this pipe.

d) **Delivery Valve-**Delivery valve also non-return valve placed between cylinder and delivery pipe outlet. It is in closed position during suction and in opened position during discharging of liquid.

e) **Cylinder-**A hollow cylinder made of steel alloy or cast iron. Arrangement of piston and piston rod is inside this cylinder. Suction and release of liquid is takes place in this so, both suction and delivery pipes along with valves are connected to this cylinder.

f) **Piston and Piston Rod-**Piston is a solid type cylinder part which moves backward and forward inside the hollow cylinder to perform suction and deliverance of liquid. Piston rod helps the piston to its linear motion.

g) **Crank and Connecting Rod-**Crank is a solid circular disc which is connected to power source like motor, engine etc. for its rotation. Connecting rod connects the crank to the piston as a result the rotational motion of crank gets converted into linear motion of the piston.

h) **Strainer-**Strainer is provided at the end of suction pipe to prevent the entrance of solids from water source into the cylinder.

i) **Air Vessel-**Air vessels are connected to both suction and delivery pipes to eliminate the frictional head and to give uniform discharge rate.

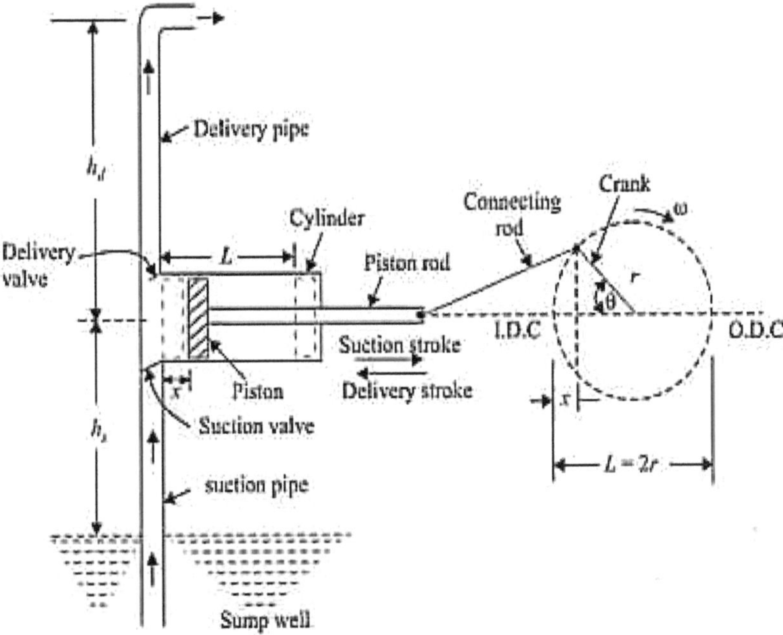

▲ **Fig-2.5: Components of a Reciprocating pump**

2.2 Working of Reciprocating Pump

The working of reciprocating pump is as follows:

a) The piston connected to the connecting rod moves in linear direction. If the piston moves for suction stroke, then creates vacuum in the cylinder. This vacuum causes suction valve to open and liquid is forcibly sucked by the suction pipe into the cylinder.

b) When the piston moves for discharge stroke, the liquid in the cylinder gets pressurized and makes the delivery valve to open and liquid discharges through delivery pipe.

c) Generally, the above process can be observed in a single acting reciprocating pump when there is only one delivery stroke per one revolution of crank. But when it comes to double acting reciprocating pump, there are two delivery strokes per one revolution of crank.

2.3 Uses of Reciprocating Pump

Reciprocating pump is mainly used for:

a) Oil drilling operations

b) Light oil pumping

c) Feeding small boilers condensate return

d) Chemical solution dosing

2.4 Causes of Volumetric Efficiency Loss

i. Common causes of loss in volumetric efficiency include worn valves, seats, liners, piston rings, or plungers, pockets of air or vapor in the inlet line or trapped above the inlet manifold, cylinder heads, loose belts or bolts in the pump inlet manifold. Routine maintenance and inspection in these areas can greatly increase output over time.

ii. Obstructions such as a safety relief valve partially held open or failing to maintain pressure, foreign objects preventing the pump inlet or discharge valve from closing or blocking liquid passage, or a vortex in the supply tank are some of the other more easily remedied factors reducing efficiency.

2.5 Efficiency Calculation

The actual reciprocating pump discharge is usually defined by the following expression:

Q_S= Actual discharge= Q_T - Q_L

Q_T= Theoretical Pump discharge

Q_L= Total Leakage

Volumetric Efficiency (η_{vol}) = Q_S / Q_T

3.0 Gear Pump

A gear Pump is a type of positive and fixed displacement rotary pump at a fixed speed regardless of change in pressure. In a gear pump, a fluid moves at a fixed volume using the interlocking gears and then transferring it as the gear rotates.

This pump delivers a pulse-free flow that is directly proportional to the rotational speed of the gears.

The gear pump is mainly used for developing high pressure.

3.1 Different parts & working of gear pump

i. Different parts of a gear pump are shown in Fig-2.6 in external gear pump, a kind.

ii. Driving gear shaft is driven by motor.

iii. The driving gear engages the other gear, called driven gear.

iv. Fluid on the inlet side flows into and is trapped between the rotating gear teeth and the housing.

v. The fluid is carried around the outside of the gears to the outlet side of the pump.

vi. The liquid must exit through the outlet port as the fluid cannot seep back along the path it came, nor between the engaged gear teeth (they create a seal).

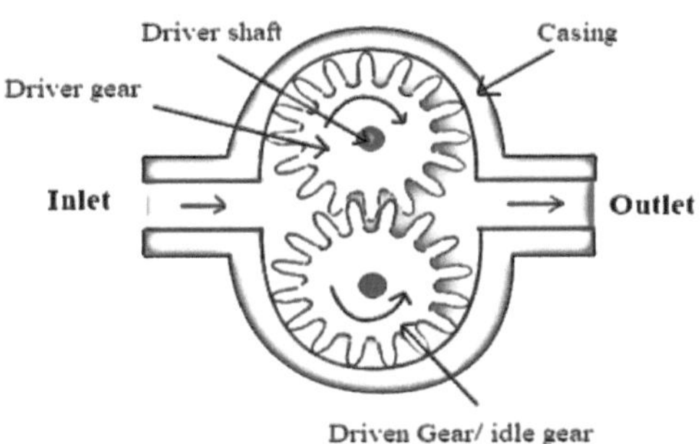

▲ **Fig-2.6: Gear Pump**

In some gear pumps, there are side plates usually made of brass which can be replaced or re-ground when the gap between the face of the gear and the end housing becomes too large.

3.2 Types of Gear Pump

Generally, gear pumps are of five types, Fig-2.7

a) External Gear Pump

b) Internal Gear Pump

c) Lobe Pump

d) Ge-rotor Pump (Generated Rotor Pump)

e) Screw Pump

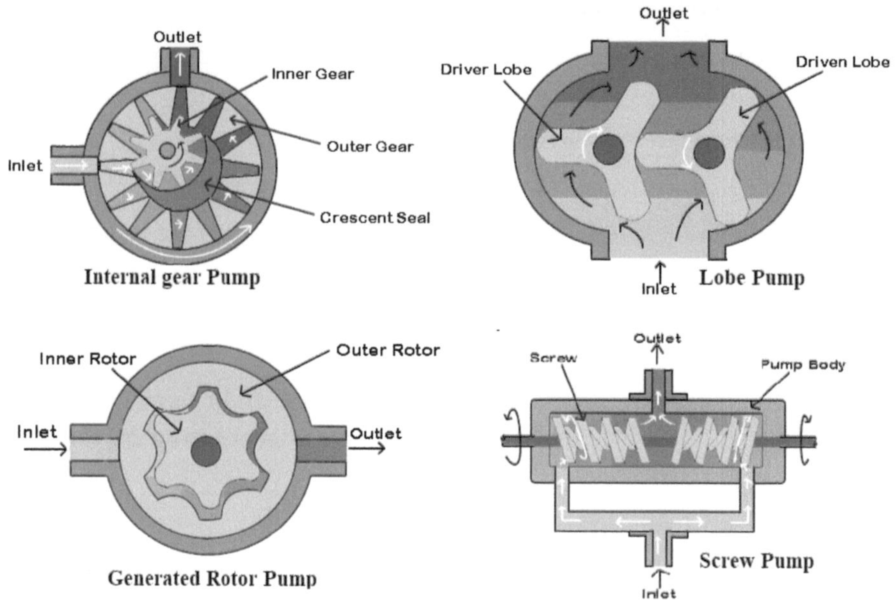

▲ **Fig- 2.7: Internal Gear, Lobe, Rotor & Screw pump**

3.3 Advantages & disadvantages of Gear Pump

i. Advantages

a) These pumps are very simple and compact with very few numbers of moving parts.

b) The maintenance cost of this type of pump is very low; It has a low cost.

c) Gear pump can be used for generating very high pressure up to 200 bar.

d) It can be used to pump highly viscous fluids like oils which cannot be pumped using centrifugal pumps.

e) There are very low chances of leakage while pumping high viscous liquid like oil in the gear pump. Hence, the efficiency of this pump increases when pumping highly viscous liquids.

f) Gear pumps can run in both directions. Hence, a single pump can be used for both loading and unloading purposes.

g) This pump is very less sensitive to contamination.

ii. **Disadvantages**

a) As meshing gears are used, abrasive fluids cannot be used in the gear pump.

b) These pumps are very noisy.

c) The size of gear pumps is limited, so it cannot be used for large bulk flow rates.

3.4 Power and efficiency measurement

Nomenclature:

a) Power input shaft (P_{in}), Watt

b) Power output to fluid system (P_{out}), Watt

c) Shaft rotational speed (ω), rad/s

d) Pressure increase between inlet and outlet (p), Pa

e) Flow rate through the pump (Q), m^3/s

f) Mechanical efficiency (η_m).

3.5 Pump Power Calculation

The pump takes power from a rotating shaft:

$$P_{in} = T \times \omega$$

A portion of this power is dissipated in the pump through **Coulomb Friction** and viscous dissipation. This is not easily quantified theoretically and is often determined experimentally.

Coulomb friction $F = \mu \times m \times g$. [When, m is the mass, μ- Viscosity and g-acceleration due to gravity.]

The friction acts to oppose the motion of the object.

This power is denoted at P_{loss}.

$$P_{loss} = P_{fric} + P_{vis} \text{(friction, viscous effects......)}$$

Some fluid seeps through the gap between the sides of the gears and the endplates This gap must be small in order to maintain the pressure increase across the pump. Increasing the gap diminishes the pumps ability to hold a pressure difference between the inlet and outlet. The gap is typically around 0.0125 mm.

The power which can then be derived from the fluid which comes out of the pump is:

$$P_{out} = (P \times Q) = P_{in} - P_{loss} = T \times \omega - P_{loss}$$

This can also be expressed using the efficiency:

$$P_{out} = \eta_p \times P_{in.}$$

3.6 Efficiency

$\eta_m = P_{out}/P_{in}$.

Mechanical efficiency is a function of:

i. The fluid viscosity

ii. Clearance between internal components

iii. Friction between mating components

iv. Other variables.

Typically, gear pumps have efficiencies of around 85%.

4.0 Centrifugal Fan

4.1 Working of Centrifugal fan

A centrifugal fan uses the centrifugal power supplied by the rotation of impellers to increase the kinetic energy of air/gases. When the impellers rotate, the air/gas particles in the impellers are thrown off from the impellers, then move into the fan casing. In a centrifugal fan, the air stream enters the impeller in an axial direction, acquires acceleration due to centrifugal force as it flows through the fan blades and gets discharged radially or axially from the fan housing.

4.2 Different parts and accessories

Important parts of centrifugal fans are, Fig-2.8:

i. Fan impeller- Impeller is a rotating device consisting of a number of fan blades mounted on a hub with a drive shaft that forces the air/gas in the desired direction with increased pressure.

ii. Fan Blades- Blades are fitted on the hub to form the impeller.

iii. Fan Housing- Housing encases the impeller, impeller hub, and diffuser in a fan system. improves the efficiency as it reduces noise and acts as a protection for moving parts like impellers, hubs etc.

iv. Drive shaft- It is a rotating device that holds the impeller & blades and determines the speed of the fan wheel through various drive mechanisms like direct, belt, or variable. Shaft design depends on the installed impeller weight and the velocity of the inlet gases. Fan can be driven by the drive in different as below:

a) **Direct driven-** In direct-driven fans, the fan speed is directly coupled with the fan motor.

b) **Belt driven-** In the belt-driven mechanism, the wheel shaft and the motor shaft are connected by the belt which is mounted on the sheave.

c) **Variable:** To get the desirable variable speed of the fans, variable-driven mechanisms use couplings to establish a connection between the motor shaft and the wheel shaft. Some of the fan systems are equipped with variable speed drives that can automatically control the fan speed at the desired level.

v. **Inlet and Outlet Ducts-** Ducts that are attached at the fan inlet and outlet as air flow path to and from the fan.

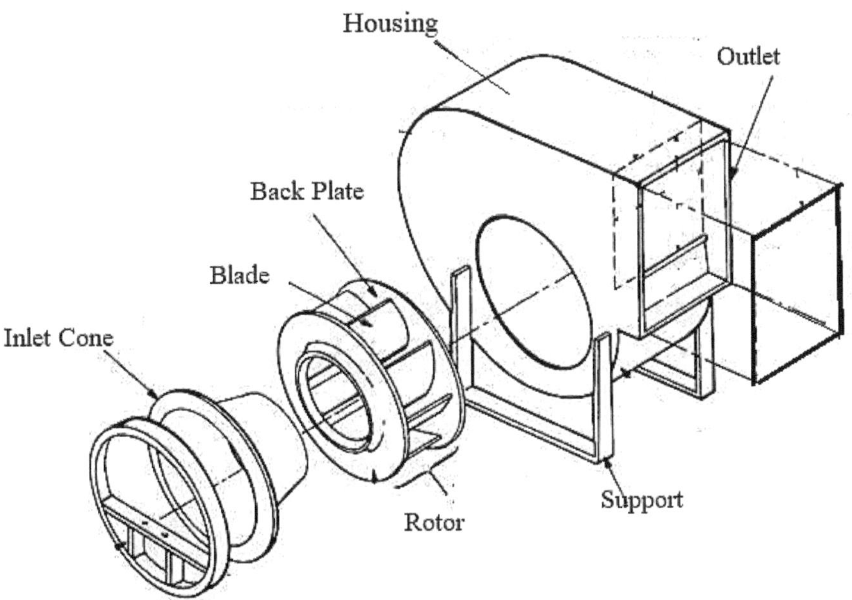

▲ **Fig-2.8: Components of a Centrifugal fan**

vi. **Dampers & Louvers-** Dampers and louvers are the plates or sheets inside the ducts which are used to control or block the airflow at the inlet or outlet of a fan.

4.3 Types of Centrifugal Fans

Centrifugal fans are of two type based on their discharge a) Radial b) Axial. Fans are also classified based shape of fan blade as Fig-2.9.

a) Radial

b) Forward curved

c) Backward curved

d) Aero foil

Axial fans are classified, Fig-2.9 as:

a) Vane axial

b) Tube axial

c) Propeller

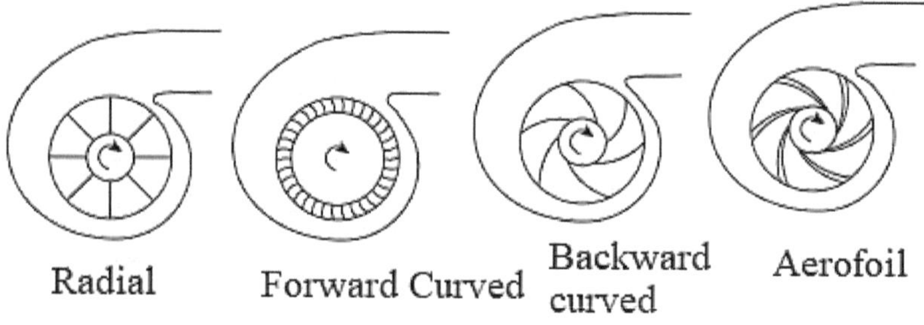

Radial centrifugal fans with different blade profile

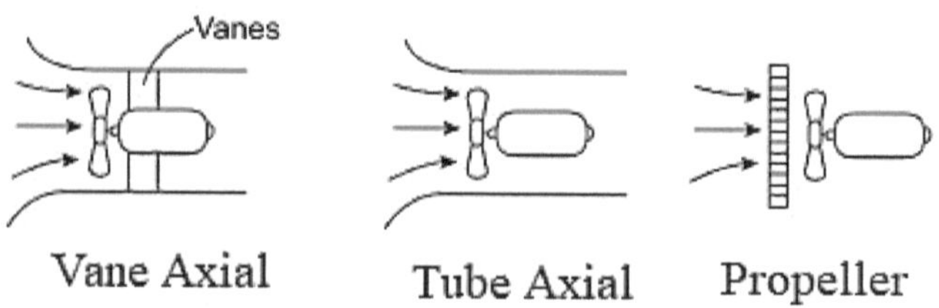

Axial centrifugal fans- different types

▲ **Fig-2.9: Blade profiles for Radial & Axial Centrifugal Fans**

4.4 Advantages & disadvantages of Centrifugal fans

I. Advantages

a) Efficient- Centrifugal exhaust fans are typically designed to be highly efficient and can handle large volumes of air.

b) It can handle higher static pressure than axial fans.

c) Easy to use- They are simple to install and require very little maintenance.

d) Noise levels- Centrifugal fan is quieter than other type of fan as it moves air at right angles to the intake.

e) Centrifugal fans are less likely to clog than axial fans. This is because the airflow created by a centrifugal fan is more turbulent, which means that debris is less likely to settle on the blades.

f) Centrifugal fans are better suited for use in applications when there is a need to move large volumes of air at high pressures.

II. Disadvantages

a) High initial cost: Centrifugal exhaust fans are more expensive than axial fans. This is especially true for high-end models that are designed to operate in harsh environments.

b) Installation: Installing a centrifugal exhaust fan requires ductwork, which can be time-consuming and expensive.

c) Maintenance: Regular cleaning and maintenance of the fan and ductwork is required to maintain its efficiency.

d) They also have a higher operating speed, which can make them less efficient.

4.5 Advantages & disadvantages of axial fan

i. Advantages of Axial fans

a) One advantage of axial fans is that they can move large volumes of air at relatively low pressures. This makes them ideal for applications when high airflow is more important than high pressure, such as in cooling systems.

b) Another advantage of axial fans is that they are relatively easy to maintain and repair. Because they have fewer moving parts than centrifugal fans, there are fewer opportunities for something to go wrong.

ii. Disadvantages of axial fans

There are several disadvantages of axial fans when compared to centrifugal fans, the most significant being that they are not as effective in moving air through ducts. This is because the airflow generated by an axial fan is not as focused and tends to disperse more, making it less efficient at moving air through specific areas.

4.6 Ideal Power Equation

The ideal power consumption for a centrifugal fan (without losses) can be expressed as:

i. $P_i = (dp \times q)/1000$ (for dp is in Pa, N/m^2)

ii. $P_i = (dp/10 \times q)/1000$ (for dp is in mm WC) [1mm WC= 10 Pa].

When,

a) P_i = Ideal power consumption (kW)

b) dP = Total pressure developed in the fan (Pa, N/m^2) or mm WC

c) q = Air volume flow rate delivered by the fan (m^3/s).

4.7 Fan Power & Efficiency

Fan efficiency is the ratio between power transferred to airflow and the power used by the fan.

The fan efficiency is generally independent of the air density and can be expressed as:

$\eta_{fan} = dp \times q / P \times 100\%$

When,

a) η_{fan} = Fan efficiency

b) dp = Total pressure developed (Pa)

c) q = Air volume delivered by the fan (m^3/s)

d) P = power used by the fan (W, N-m/s)

i. Power used by the fan can be expressed as: $P = (dp \times q) / (\eta_{fan} \times \eta_m)$

ii. Power used by the belt driven fan expressed as: $P = dp\, q / (\eta_{fan} \times \eta_b \times \eta_m)$

When,

a) η_b = belt efficiency

b) η_m = motor efficiency

4.8 Typical motor and belt efficiencies

i. **Motor Efficiency for** 1kW: 40%, 10 kW: 87%,100 kW: 92%

ii. **Belt Efficiency factor**

   a) Belt 1 kW: 0.78

   b) Belt 10 kW: 0.88

   c) Belt 100 kW: 0.93.

4.9 Fan and Installation Loss (System Loss)

The installation of a fan influences on the overall system efficiency as follows:

$dp_{system} = X_{system} p_d$

When,

a) dp_{system} = installation loss (Pa)

b) x_{system} = installation loss coefficient

c) p_d = dynamic pressure in the nominal intake and outlet of the fan (Pa).

4.10 Fan Loss and Temperature Increase

Near all of the energy lost in a fan heats up the airflow and the temperature increase can be expressed as dt = dp / 1000

When,

a) dt = temperature increase (K)

b) dp = increased pressure head (Pa).

4.11 Operating vs Reference

i. **Operating volume Flow vs. Reference Volume Flow**

   The volume flow through a fan is constant as long as the physical dimensions and the speed of the fan are not changed. The amount of air (mass) passing through the fan varies with the air density and the air density depends on the temperature (as long as the pressure is constant).

If the volume flows for a fan at a reference condition (typical NTP) - the operating volume flow for the fan at other operating temperatures can be calculated as:

1. $q_o / q_r = (273 + t_o) / (273 + t_r)$
2. $q_o = q_r (273 + t_o) / (273 + t_r)$.

When,

a) q_r = reference volume flow (m^3/s) - in general NTP conditions

b) q_o = operating volume flow (m^3/s)

c) t_r = reference temperature ($^\circ C$) - in general 20 $^\circ C$ NTP conditions

d) t_o = operating temperature ($^\circ C$)

ii. **Operating vs reference Pressure Head**

The ratio between developed pressure at different temperatures with respect to reference temperature can be expressed as:

1. $dp_o / dp_r = (273 + t_o) / (273 + t_r)$
2. $dp_o = dp_r (273 + t_o) / (273 + t_r)$

When,

a) dp_r = reference pressure developed (Pa) - in general at NTP conditions

b) dp_o = operating pressure developed (Pa)

With temperature higher than the reference temperature, the volume flow is higher and the required pressure developed in the fan is higher.

iii. **Operating Power vs reference power**

The ratio between power consumption at different temperatures can be expressed as:

1. $P_o / P_r = (273 + t_r) / (273 + t_o)$
2. $P_o = P_r (273 + t_r) / (273 + t_o)$

When,

a) P_r = Reference power consumption (W)

b) P_o = Operating power consumption (W)

5.0 Blower

An industrial blower is a device like fan that uses an electric motor, impellers, and aerofoils to effectively and efficiently improve airflow in a workstation. Centrifugal, axial, and positive displacement blowers are the three types of industrial blowers.

5.1 The differences among a fan and a blower & a Compressor

According to ASME, a fan is a device with a pressure ratio of up to 1.11. A blower has a pressure ratio between 1.11 and 1.2. On the other hand, the pressure ratio in a compressor is more than 1.2.

Discharge pressure of fan is 1136 mm WC, blower is maximum 1136-2026 mm WC and for compressor it is more than 2026 mm WC.

5.2 Types

i. Centrifugal blower

ii. Root blower-2 lobe root blower, 3 lobe root blower (Fig-2.10)

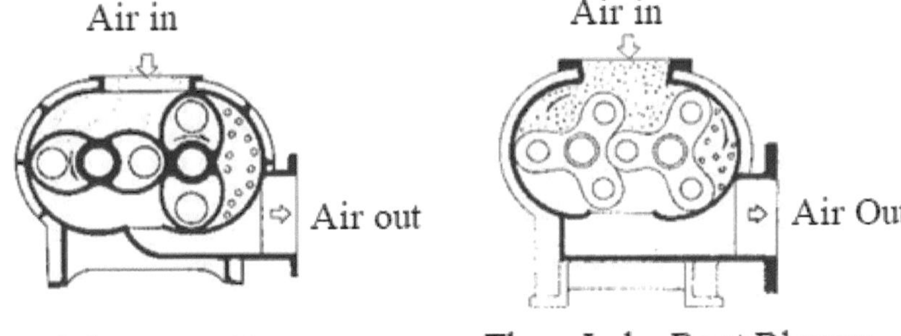

▲ **Fig-2.10: Lobe Root Blower**

5.3 Applications of Industrial Blowers

Blowers have varied applications across industries for handling air and gas. Here are some applications of industrial blower:

a) Dust collectors

b) Exhaust and HVAC

c) Filtration and suction

d) Exhaust extraction

e) Cooling and refrigeration

f) Fresh air supply and ventilation

g) Unidirectional air flow applications

h) Conveying systems

i) Ventilation, cooling, aspiration, and exhaust systems.

6.0 Compressor

6.1 Introduction

In industries, air compressors are vital equipment that delivers high-pressure air for multipurpose use.

To compress air, the internal mechanisms within the compressor move to push air through the chamber. There are two primary types of air displacement used for this purpose:

- **Positive displacement-** Most air compressors use this method, in which air is pulled into a chamber. There, the machine reduces the volume of the chamber to compress the air. Next, it is moved into a storage tank and saved for later use.
- **Dynamic displacement-** This method uses an impeller with rotating blades to bring air into the chamber. The energy created from the motion of the blades builds up air pressure in a shorter amount of time. Dynamic displacement can be used with turbo compressors because it works quickly and generates large volumes of air.

Air compressors are available in different types and shapes.

6.2 Types of Compressors

i. **Positive displacement**

   a) Reciprocating

   b) Rotary, rotary compressors can be sub-divided based on type of rotor as

      1. Vane
      2. Roots
      3. Screw

ii. **Dynamic**

   a) Centrifugal

   b) Axial

   c) Mixed

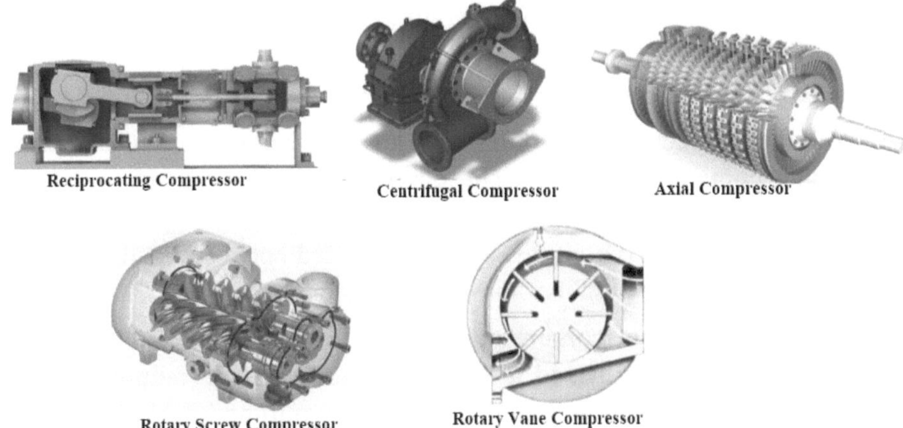

▲ **Fig-2.11: Different Types of Compressors**

6.3 Working of air compressor

i. Reciprocating Compressor

a) Working Air is sucked into the cylinder through the suction valve during its suction stroke. During the discharge stroke, the suction valve gets closed and the discharge valve opens through which high-pressure air is pushed out of a compressor and stored in air receiver.

Some compressors, called single-stage compressors, use only one piston. Others, called two-stage compressors, use two pistons and are able to pressurize more air. The reciprocating type of air compressor is one of the most common.

b) **Regulation of compressor**

Once the storage tank reaches its maximum air pressure, the compressor turns off. This process is called the duty cycle. The compressor will turn back on when the pressure drops below a set pressure.

A regulator attaches to the outlet for the receiver tank which receives an air pressure signal from tank pressure. Regular is set with loading and unloading pressure. As the pressure increases beyond the set pressure, the air from the regulator forces opens the suction valve to reduce airflow into the compressor. As discharge pressure reduces to loading pressure, the regulator closes the supply control air to the suction valve facilitating the loading of the compressor.

For many single-stage air compressors, the preset pressure limit is 8 bar for unloading and 6.5-7 bar for loading.

ii. **Rotary screw compressor**

Working-A rotary screw compressor has two screws inside the casing, turning continuously in opposite directions. The motion of the screws creates a vacuum that sucks in air. That air becomes trapped between the screw threads and is compressed as it is forced between them. Finally, it is sent through the output or into a containment tank.

Most rotary screw compressors are industrial-sized and lubricated with oil. The pressurized air is forced through the output and into the primary oil separator tank while still combined with the oil as a mist. The centrifugal force inside the tank causes most oil molecules to form into droplets and collect at the bottom as reusable oil. Air then enters a secondary separation filter when more oil is removed, purifying the air further. The oil-free air exits the system, where it is stored in a tank or used immediately in a connected pneumatic tool or machinery.

iii. **Rotary Vane Compressor**

Working- A rotary vane compressor or vacuum pump has a similar principle to a rotary screw. With a rotary vane, a motor is placed off-center inside a rounded cavity. compressor has blades with automatically adjusting arms. As the arms approach the air input, they are elongated, creating a large air cavity. As the motor spins, moving air with it, the arms approach the output and get smaller, creating a smaller space between the vanes and the round casing, which compresses the air.

iv. **Pressure regulation**

Control schemes for positive displacement compressors include:

a) **Stop/start:** This approach either provides power to the motor, or it does not, according to the application.

b) **Load/unload:** The compressor is powered continuously, with a slide valve that reduces the tank's capacity when a specific compression demand is met. This scheme is common in factory environments, and if it involves a stop timer, it is called a dual-control scheme.

c) **Modulation:** Modulation also uses a sliding valve to adjust pressure by throttling/closing the inlet valve, matching the compressor's capacity to the demand. These adjustments are less effective on rotary screw compressors than on other types. Even when set to zero capacity, the compressor would still consume about **70 percent of its full power load**. Still, modulation is applicable for operations in which frequently stopping the compressor is not an option.

d) **Variable displacement:** This control scheme adjusts the volume of air that is pulled into the compressor by recirculating from discharge. In rotary screw compressors, this method may be used alongside modulating inlet valves to **improve efficiency and pressure control accuracy**.

e) **Variable speed:** Variable speed is an efficient way to control a rotary compressor's capacity, though it may respond differently with different types of air compressors. It varies the speed of the motor, which affects the output.

6.4 Air Compressor Power Ratings and capacity in CFM

CFM-Compressor capacity is rated as cubic feet per minute (CFM). To consider these internal and external factors, manufacturers use standard cubic feet per minute (SCFM), which combines CFM with those outside factors of pressure and humidity. Capacity of compressor is called CFM FAD, which stands for **Free Air Delivery**.

6.5 Capacity assessment of a compressor

a) Isolate the compressor along with its individual receiver (cap-V) that is to be taken for a test

b) Open the water drain valve and drain out the water fully and empty the receiver and the pipeline.

c) Start the compressor and activate the stopwatch.

d) Note the time taken (T) in minute to attain the normal operational pressure P_2 (in the receiver) from initial pressure P_1. Po is the atmospheric pressure.

e) Calculate the capacity as per the formulae given.

f) FAD is to be corrected by a factor (273 + t1) / (273 + t2); t1-reference temp, 20 ^{0} C, t2- discharge temp, °C.

$Q = (P_2 - P_1)/P_0 x V/T$ Nm³/ min.

6.6 Air Compressor Isothermal efficiency

The reported value of efficiency is normally the isothermal efficiency. Isothermal efficiency is calculated as follows:

Isothermal Efficiency = Actual measured input power / Isothermal Power

Isothermal power (kW) = P1 x Q1 x ln r/36.7

When, P1 = Absolute intake pressure kg/ cm^2; Q1 = Free air delivered m^3/hr; and r = Pressure ratio P2/P1.

This equation can be converted as:

Kg/cm^2x10^4 x 9.8 Pa [pressure unit conversion to Pa]

m^3/hr = 1/3600 (m^3/s)

Power = (Pa x m^3/s)/1000 kW

The calculation of isothermal power does not include the power needed to overcome friction and generally gives an efficiency that is lower than adiabatic efficiency.

6.7 Air Compressor Volumetric Efficiency

Volumetric efficiency = Free air delivered m^3/min / Compressor displacement

a) Compressor Displacement = Π x D^2/4 x L x S x χ x n

b) When, D = Cylinder bore, meter; L = Cylinder stroke, meter; S = Compressor speed rpm; χ = 1 for single acting and 2 for double acting cylinders; and n = No. of cylinders

6.8 Calculation of Total leakage in percentage

Leakage (%) = [(T x 100) / (T + t)]. When, T = on-load time, and t = off-load time.

Leakage is expressed in terms of the percentage of compressor capacity lost. The percentage lost to leakage should be less than 10 percent in a well maintained system.

7.0 Industrial Mixer & Stirrer

7.1 Introduction

Tanks holding liquid are often equipped with an agitator. The agitator can have many functions, like improving the heat transfer, homogenizing different liquid, preventing sedimentation.

7.2 Different parts of a stirrer (Fig-2.10)

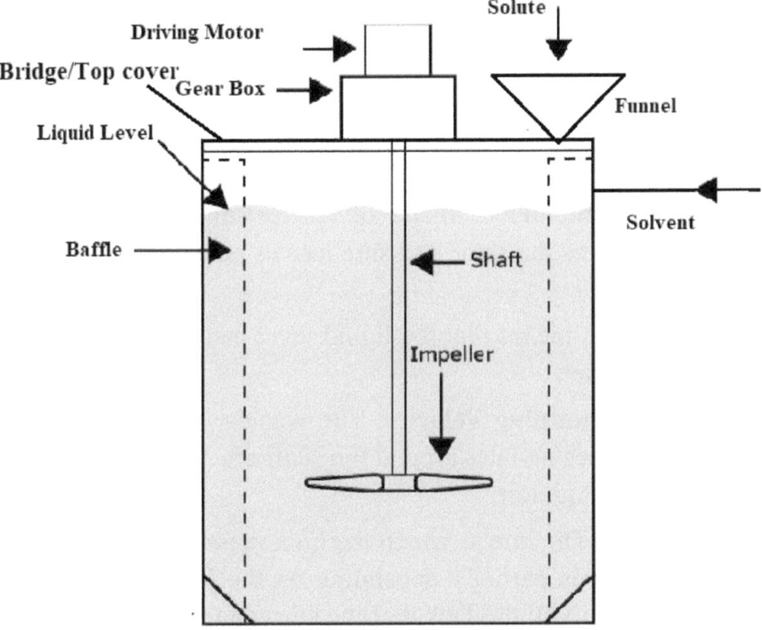

▲ **Fig-2.12: Stirrer**

a) Mixing Tank- Contains the materials for mixing. Tanks can be constructed from a variety of materials such as specialty metals or plastics to best suit the mixing application. Tank linings or coatings are often an economic choice for reactive fluids or gasses. Tanks may be open at the top or sealed depending on process requirements.

b) Bridge/ Top cover-Supports the motor, gearbox, shaft and impeller above the mixing tank.

c) Baffles -Flat metal plates attached to the wall of the mixing tank, used to prevent swirling and vortex formation in the tank. Mixing tanks contains three or four baffles.

d) Motor-Provides the power for the mixing system.

e) Gearbox-Selected to provide the optimal rotational speed of the impeller at the required torque for the mixing application.

f) Shaft -Positions the impeller in the correct location in the tank and transmits the power from the drive system to the impeller.

g) Impeller-Circulates fluids, solids, and/or gasses within the tank. Impellers come in a variety of types to suit different mixing applications. Many applications can use more than one impeller mounted to the same shaft and often different types of impellers on the same shaft to effect different process results.

7.3 Definition related to a stirrer

i. **Equivalent Diameter**: Diameter of reactor/solution preparation vessel diameter that gives the same pressure loss as an equivalent diameter of a rectangular vessel.

   In other words, it means that the liquid level inside the vessel is equal to the vessel diameter.

ii. **Bulk Velocity/Pumping Velocity**: The velocity at which the bulk mass in the vessel mixes/agitates around the shaft after gaining the momentum provided by the Agitator.

iii. **Pumping Rate**: The rate at which the bulk mass gets pumped inside the stirred tank, this is partially depending on the Hot Spots of the Stirred Tank and the utility in the Jacket of the stirred tank.

iv. **Power Number**: This number signifies the effect of power consumed over the agitation, in other basic words it is ratio of resistance force to inertial force, it is dimensionless.

v. **Reynolds Number**: This number defines the fluid flow, so here it indicates the flow is whether laminar or turbulent.

7.4 Scale of Agitation

Scale of Agitation gives the Intensity of Agitation or Mixing, generally the Scale of Agitation is in range of 1 to 10.

Scale of Agitation Range Values and their Significance over Agitation:

a) If Scale of Agitation is 1, then the effect of agitation is ***Mild***, and this is used when the operation is a crystallization, that too if it's a supersaturated Crystallization.

b) If Scale of Agitation is 3, then the effect of agitation is **Normal**, this is commonly used for the Dryers.

c) If Scale of Agitation is 6, then the effect of agitation is **Vigorous**, this means the mass that need to be stirred is somewhat denser but the bulk density will be not much.

d) If Scale of Agitation is 10, then the effect of agitation is ***Violent***, this is applicable for the mass when there is need for extraction of product in between layers, and also the settling time during this extraction should be higher than regular cases.

7.5 Tank agitator power calculation

Steps for Power calculation for typical blade turbine impellers:

a) To calculate the power number (Np) of the agitator using standard power number ($N_{standard}$).

b) To calculate the power number at actual process conditions (N_{actual})

c) To calculate the motor power required for the agitator

d) To select the actual size of the motor.

One of the most common agitator design for holding tanks or for mixing liquids is a turbine impeller. The following design are standard:

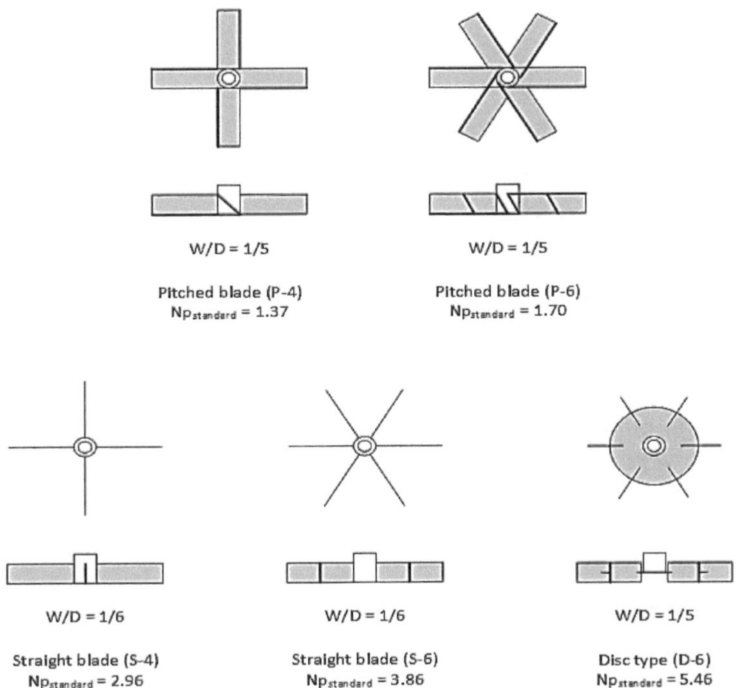

▲ **Fig-2.13:** Values of turbulent power number Np for various impeller geometries (W/D is the actual blade-width-to-impeller-diameter-ratio)

When,

a) $Np_{standard}$ = standard power number

b) W = width of the blades of the impeller (m)

D = diameter of the impeller (m)

One of the key design parameter to calculate when implementing such an agitator is the size (power) of the motor which drives the agitator.

Step-1: To calculate the power number Np of the agitator (using standard Power number)

The power number in turbulent conditions is tabulated for standard design in the table above. The Engineer must select the type of agitator and then may correct the standard power number according to the actual size of the agitator in the tank.

a) If (W/D) standard = 1/5 on table above, Np = $\mathbf{Np_{standard}\ [(W/D)/(1/5)]^{1.25}}$

b) If (W/D) standard = 1/6 on table above, Np = $\mathbf{Np_{standard}\ [(W/D)/(1/6)]}$

When,

a) Np = power number in the geometry considered

b) $Np_{standard}$ = standard power number

c) W = width of the blades of the impeller (m)

d) D = diameter of the impeller (m)

Step 2: Calculate the power number at actual process conditions (actual power number: (N_{actual}) using viscosity power factor & Reynolds number.

Step 2.1: Calculate the Reynolds number

The Reynolds number for an agitator can be calculated with the following formula: $\mathbf{N_{RE} = D^2.N.\rho / \mu}$

When,

a) N_{RE} = impeller Reynolds number

b) D = impeller diameter (m)

c) N = agitator speed (r/s)

d) ρ = liquid density (kg/m^3)

e) μ = liquid viscosity (Pa.s)

Step 2.2: Calculate the turbulent power number considering the actual viscosity

The Reynolds number allows to calculate a viscosity power factor by using the graph below:

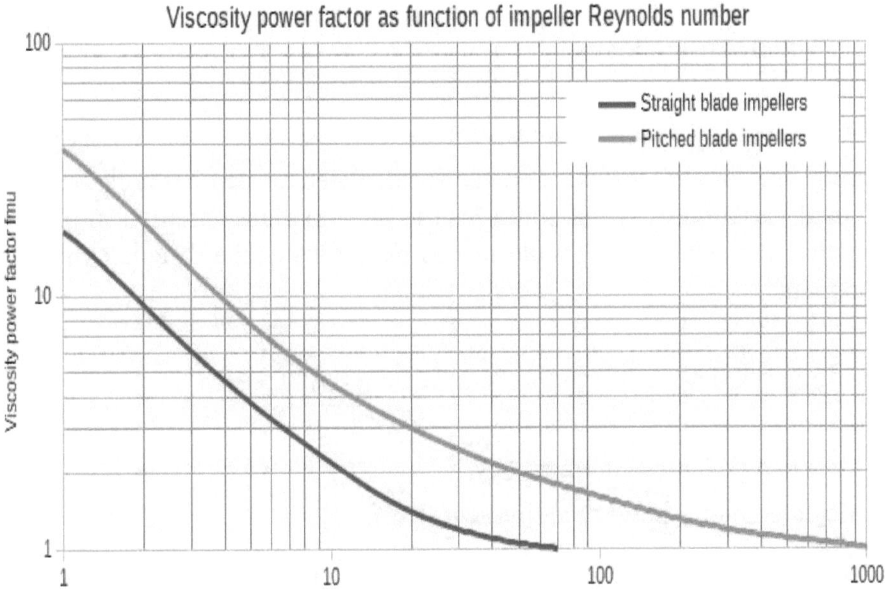

▲ **Graph 1: fμ as a function of $N_{Re} = D^2.N.\rho/\mu$**

The actual turbulent power number can then be calculated as: $Np_{actual} = f\mu \times Np$

When,

a) $N_{p\ actual}$ = actual turbulent power number (-)
b) fμ = viscosity power factor from graph above
c) Np = power number in the geometry considered see step - 1

Step 3: Calculate the motor power required for the agitator

Now, that the value of the power number is known, it is possible to go back to the definition of the power number to calculate the required power:

$Np_{actual} = P / (\rho.N^3.D^5)$; $P = Np_{actual}$ x $(\rho.N^3.D^5)$

$\rho.N^3.D^5 = kg/m^3 \times 1/s^3 \times m^5 = (kg \times m/s^2 \times m)/s = J/s = Watt$

When,

a) P = motor power required (Watt)

b) Np_{actual} = actual turbulent power number

c) D = impeller diameter (m)

d) N = agitator speed (r/s)

e) ρ = liquid density (kg/m^3)

Step 4: Select the actual size of the motor

Selecting a motor with the exact power as calculated on step 3 is not advisable as a small variation of speed could lead to an important variation in power. In practice, the actual load should not be over 85% of the calculated power requirement.

Thus: $P_{actual} = P/0.85$

When,

P_{actual} = actual power requirement (kW)

P = calculated motor power requirement (kW).

Example:

Given:

a) Type of Agitator: P-4

b) Diameter of impeller (D): 1.47 m

c) Width of blades of impeller (W): 300 mm=0.3 m

d) Rotational speed of impeller: 84 rpm=1.4 rps

e) Density of fluid: 1150 kg/m^3

f) Viscosity of fluid: 12 Pa-s

Calculation:

i. As per chart (Np standard): 1.37

ii. W/D ratio=0.3/1.47=0.2

iii. Power number $Np = Np_{standard} \times [(W/D)/(1/5)]^{1.25} = 1.37 \times (0.204/0.2)^{1.25} = 1.404$.

iv. Now, Reynold number=$D^2 xN \times \rho / \mu = 1.47^2 x1.4x1150/12=289$.

v. Viscosity power factor ($f\,\mu$, from graph) = 1.2

vi. Actual Power number $Np_{actual} = Np \times f\,\mu = 1.404 \times 1.2 = 1.7$.

vii. Power required for the agitator (P) = $Np_{actual} \times \rho \times N^3 \times D^5 = 1.7 \times 1150 \times 1.4^3 \times 1.47^5 = 36822$ Watt =36.8kW

viii. Motor capacity= 36.8/0.9=41 kW [motor efficiency=90%].

Questions:

1. What are different types of centrifugal pumps, list differences between the types?
2. What are hydraulic power and shaft power of a pump?
3. What are the different losses in a centrifugal pump?
4. What is meant by positive displacement pump, list the pumps in this category.
5. What are the different parts of a reciprocating pump?
6. List out the advantages & disadvantages of gear pump.
7. What are differences between a fan, blower and a compressor?
8. What are the different types of centrifugal fan?
9. List some of advantages & disadvantages of centrifugal fan.
10. What is the equation to measure fan power?
11. Discuss the steps to calculate power calculation for tank agitator.

SECTION 03

FLUID FLOW & SOLID MATERIAL TRANSPORTATION

1.0 Fluid flow

1.1 Introduction:

Transportation of fluid from one place to another occurs through an enclosed system, called a pipe or channel. Fluid Flow is a part of fluid mechanics and deals with fluid dynamics and is subjected to unbalanced forces. This motion continues as long as unbalanced forces are applied and suffers losses in energy.

Loss in head in flow through pipes: Two types of energy loss in fluid flow through a pipe network as described below.

1.2 Types of flow & Reynolds number

There are mainly two kinds of flow, such as:

i. Laminar flow - Laminar flow is the continuous movement of flowing fluid in streamlines.

ii. Turbulent flow - Turbulent flow is characterized by erratic movement of fluid particles in the flow.

There are three fluid flow regimes: laminar, turbulent, and transition region. The conditions that lead to each type of flow behaviour are system-specific. The type of flow is determined by a specific number, called Reynold number which is defined as –

Reynold No (N_{RE}): ρ d v/μ

When, ρ= Density kg/ m^3), d= diameter of flow path (m), v= Velocity of fluid (m/s), μ - Viscosity (Pa-s).

It is a dimensionless number: (kg/m^3) x (m) x (m/s) ÷ [(kg x m/s^2)/m^2]x s [Pa-s=N/m^2 x s]

a) N_{RE}<2000: Laminar Flow

b) 4000> N_{RE}>2000: Transition Flow

c) N_{RE}>4000: Turbulent Flow

1.3 Bernoulli's equation and continuity equation

i. **Bernoulli's equation**

In a fluid flow path, it can be stated as:

Pressure Energy + Velocity Energy + Potential energy =constant

$P_1 + 1/2\rho V_1^2 + \rho g h_1 = P_2 + 1/2\ \rho V_2^2 + \rho g h_2$ [Energy equation in Kg-m/s²]

Or,

Pressure head + Velocity head+ Potential head = Constant

$P_1/\rho g + V_1^2/2g + h_1 = P^2/\rho g + V_2^2/\rho g + h_2$ [Head equation in m]

[When, P=Pressure, V=Velocity, h=static head at two points 1&2].

ii. **Continuity Equation**

It can be expressed as:

A x V=constant

$A_1V_1 = A_2V_2$ [A=Area, V= velocity at two points 1&2].

1.4 Major losses

Major losses are associated with frictional energy loss caused by the viscous effects of the medium and roughness of the pipe wall and flow velocity.

$H_L = f\ L/d\ (V^2/2g)$ or, $H_L = (4\ C_f\ L/d)\ (V^2 / 2g)$ (m)

When, f= Friction factor, L= Length of Pipe (m), V= velocity (m/s), d= diameter (m), C_f= Fanning friction Factor, and g= Acceleration due to gravity.

Friction factor, f= 64/ Re_n for laminar flow.

When, Re= Reynold number (Rn= ρ d V/μ; ρ = density (kg/ m³), d= Diameter (m), V= Velocity (m/s), μ = Viscosity (Pa-s).

1.5 Minor losses

Minor Losses in the pipe flow system due to various piping components such as valves, fittings, elbows, contractions, enlargement, tees, bends, and exits.

$H_L = K\ V^2/2g$ (m)

When, K is the minor loss coefficient and H_L indicates the minor loss.

K for various piping components or fittings are mentioned in the following table.

▼ **Table 1: Types of Fittings vs loss coefficient**

Type of Piping Components or Fittings	Minor loss coefficient, K
Tee, Flanged, Dividing Line Flow	0.2
Tee, Threaded, Dividing Line Flow	0.9
Tee, Flanged, Dividing Branched Flow	1.0
Tee, Threaded, Dividing Branch Flow	2.0
Union, Threaded	0.08
Elbow, Flanged Regular 90°	0.3
Elbow, Threaded Regular 90°	1.5
Elbow, Threaded Regular 45°	0.4
Elbow, Flanged Long Radius 90°	0.2
Elbow, Threaded Long Radius 90°	0.7
Elbow, Flanged Long Radius 45°	0.2
Return Bend, Flanged 180°	0.2
Return Bend, Threaded 180°	1.5
Globe Valve, Fully Open	10
Angle Valve, Fully Open	2
Gate Valve, Fully Open	0.15
Gate Valve, 1/4 Closed	0.26
Gate Valve, 1/2 Closed	2.1
Gate Valve, 3/4 Closed	17
Swing Check Valve, Forward Flow	2
Ball Valve, Fully Open	0.05
Ball Valve, 1/3 Closed	5.5
Ball Valve, 2/3 Closed	200
Diaphragm Valve, Open	2.3
Diaphragm Valve, Half Open	4.3
Diaphragm Valve, 1/4 Open	21
Water meter	7

1.6 Fluid flow through the duct system

Loss in duct system:

Head loss/ unit Length: $H_L /L = f V^2/2gD$ or, $H_L/L = (4 C_f V^2) / (2gD)$

When,

D= the hydraulic diameter of the duct(m), D = 2(a x b)/ (a + b) [a &b are dimensions of a rectangular duct]. V= the mean flow velocity (m/s).

1.7 Fluid Flow Through Packed Bed

Flow through a packed bed can be described by the *Ergun Equation.*

$$\Delta p /L=150\ \mu\ (u_0/ d_p^{\ 2})\ \{(1-\varepsilon)^2 / \varepsilon^3\} + 1.75\ \rho\ (u_0^{\ 2}/ d_p)\ \{(1- \varepsilon) / \varepsilon^3\}$$

When,

a) $\Delta p /L$= pressure drop/unit length (Pa/ m)

b) μ = Fluid viscosity (Pa-s)

c) u_0= Superficial velocity (m^2)

d) d_p = particle diameter (m)

e) ε = void factor

f) ρ = Fluid density (kg/m^3

The ***Ergun equation*** reveals that the pressure drop along the length of the packed bed gives some fluid velocity. It also tells us that the pressure drop depends on the packing size, length of bed, fluid viscosity, and fluid density.

Pressure drop in a packed bed is due to:

i. Pressure drop in a packed bed due to viscosity is proportional to viscosity (Pa- s), velocity (m/s), and inversely proportional to the square of particle diameter.

   [Pa -s x m / s x1/m^2 = Pa/m]

ii. Pressure drop due to density is proportional to density, square of velocity, and inversely proportional to particle diameter.

   Unit derivation:

   $\Delta p /L$ =Kg / m^3 x m^2/s^2x 1/m = [(kg x m/s^2) x1/m^2] x1/m=N/m^2 x 1/m = Pa /m

Example:

i. Viscosity of water at 20 °C = 10^{-3} Pa- s

ii. Superficial velocity = 0.1 m/s

iii. Particle dia.= 5 mm = 5/1000 m

iv. Void Factor= 0.3

v. Packet bed height = 2m

vi. Pressure drop Pa/m due to viscosity = 150 x 10^{-3} x (0.1 x1000/5) x ($0.7^2/0.3^3$) =0.00054 bar

vii. Pressure drop Pa/m due to density (Pa/m) = 1.75 x 1000 x (0.5^2 x1000/5) x ($0.7/0.3^2$) = 0.27 bar

viii. Total pressure drop = 0.2705 bar x 2 = 0.55 bar.

ix. Excess Pressure drop = Actual Pressure drop- Design Pressure drop (m-head)

x. Excess power consumption = Fluid Flow (kg/s) x g x m-head/ (1000x Pump Efficiency) kW.

2.0 Solid Material Flow & Belt Conveyor

2.1 Introduction

Material flow is an important process in the manufacturing or production industry where material moves from one equipment or system to another equipment or system to accomplish different processes. Solid materials are conveyed from one point to the processing equipment using a conveyor or bucket elevator. Power consumption during conveying is vital for monitoring the industry. This section covers several conveying processes for the transportation of material and energy consumption therein.

2.2 Types of Conveying

Commonly types of conveying systems are:

i. Mechanical Conveying system

   Mechanical conveying systems use:

   a) Belt/ slat/pan conveyor

   b) Screw conveyor

ii. Fluid conveying system

In a fluid conveying system, water or air is used as conveying medium. Common systems are:

a) Hydraulic conveying

b) Pneumatic conveying

2.3 Mechanical conveying system

i. **Belt conveyor**

A conveyor belt works by using two pulleys, driven by a motor, that loop over a long stretch of thick, durable material, called belt. When motors in the pulleys operate at the same speed in the same direction, the belt moves between the two. Rollers are provided under the belt to support it during carrying material.

Different parts of a belt conveyor system are shown in fig-3.2.

ii. **Slat conveyor**

Such conveyor uses a slat and chain system to move components along an assembly line. They are often used when production operations are performed with the parts located on the conveyor. Steel panels are attached to the chain with special chain attachments, and the chain is driven by an electric motor and gearbox. It is used for conveying biomass material **Fig-3.1.**

iii. **Deep Pan Chain Conveyor**

Deep Pan Conveyor is a kind of trough formed by pan continuously fixed in series on the endless chain or belt. The chain is subjected to the traction of the conveyor, and the trough is used as a carrier for conveying materials to form a trough-chain conveyor belt. The conveyor belt bypasses the head sprocket and the tail redirection wheel, and the sprocket meshes to drive the chain, and the conveyor belt is dragged to continuously circulate along the fixed track along the roller at the bottom of the trough. The material is loaded at the receiving point, and is taken along the belt to be removed to realize the distance transportation of the material **Fig-3.1**. It has been widely used in metallurgy, coal, chemical, machinery manufacturing and other industrial.

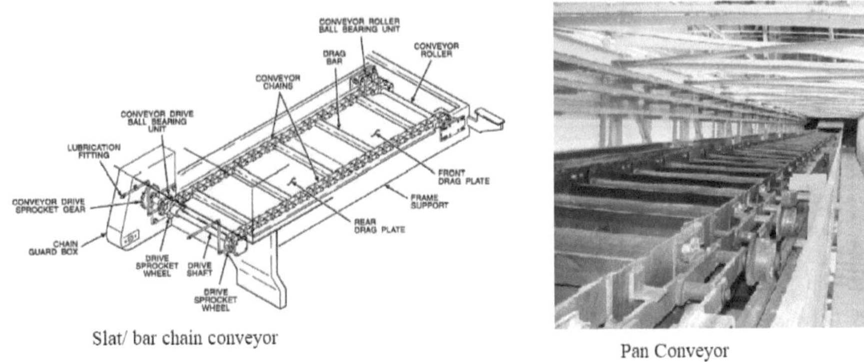

Slat/ bar chain conveyor

Pan Conveyor

▲ **Fig-3.1: Chain Conveyor and Pan Conveyor**

2.4 Energy Consumption in Belt Conveyor

The energy consumption in the typical belt conveyor system can be divided into three parts:

i. The energy needed to run the empty conveyor

ii. The energy needed to move the material horizontally over a certain distance.

iii. The energy needed to lift the material a certain height.

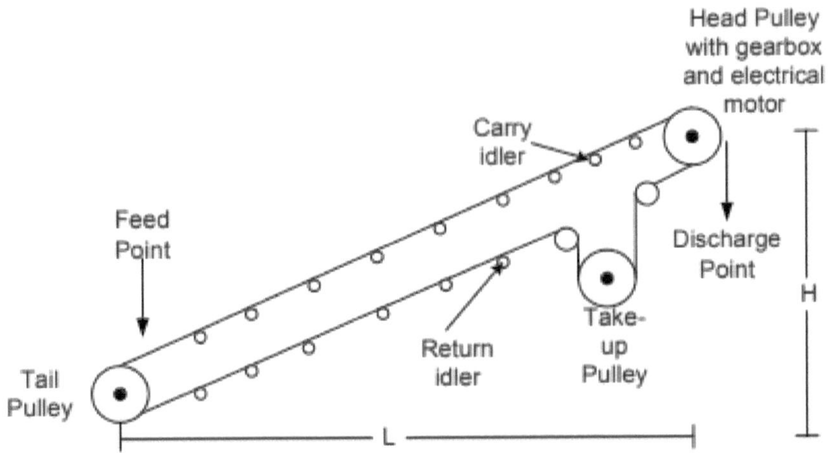

▲ **Fig-3.2: Conveyor Components**

2.5 Power required to Run the Empty Conveyor

In order to run the empty conveyor (no load condition), energy is needed to move the different parts of the conveyor and to overcome friction in the conveyor system.

The empty conveyor friction force can be calculated as (F1)

= Friction Factor x Normal force (mg) = C x [Q x (L + Lo) x g]

When,

a) F1 = empty conveyor friction force (N)
b) g = gravitational acceleration = 9.8 m/s^2
c) C = friction factor
d) Q = Mass of belt per meter length (kg /m)
e) L = the Centre-to-Centre distance or the horizontal projection of this distance for incline or decline belts. (m)
f) Lo = compensation length (m)

The power to overcome this friction force is ($P_{\text{empty conveyor}}$) = F1 x S/1000 kW

Belt speed (S) = Diameter pulley (D) x 3.14 x Rotations per minute (N)/60 = π D x N/60 m/s

When:

a) D= Diameter of pulley
b) $P_{\text{empty conveyor}}$ = power to run the empty conveyor(kW)
c) S = belt speed. (m/s)

2.6 Power to Move Material Horizontally

A loaded conveyor belt experiences an additional friction force due to the load on the belt.

This friction force can be calculated as:

(F2) = g C (L +Lo) x T /S.

When:

i. F2 = load friction force (N)
ii. T = Material transfer rate in (kg/s)

The power to transfer material horizontally can be obtained by the equation:

$P_h = F2 \times S/1000$ kW$= g \times C \times (L+ Lo) \times T/1000$ kW.

In cases when, the conveyor is skirted and the load is conveyed between skirt boards, the friction force is:

$(F_S) = 0.2 \times g \times d^2 L \times \rho$.

When:

i. F_S = skirt friction force (N)

ii. d = load depth (m)

iii. ρ = material density (kg/m^3)

iv. 0.2 = constant combination of average material repose angle and coefficient of friction.

The power needed to overcome this skirt friction is:

$P_s = F_s \times S/1000$ kW.

2.7 Energy to Elevate or Lower Material

The vertical component of force along the incline to lift or lower the load can be calculated as follows:

$F_L = g \times T \times H/s$.

When:

i. F_L = component of force along the incline (N)

ii. H = the net change in elevation (m)

The power can be calculated as:

$P_L = g \times T \times H/s \times s/1000$ kW

$= g \times T \times H/1000$ kW.

2.8 Total Energy for the Belt Conveyor System

The total energy consumption in a conveyor system is the sum of the components of energy consumptions and can be represented as follows: $P_{total} = P_{empty\ conveyor} + P_h + P_L$

i. **Transfer Rate**

The relationship between belt speed, per unit belt mass and the transfer rate in kg/s can be obtained by converting the belt speed and per unit belt mass into the transfer rate. From the following equation the transfer rate can then be expressed as the following: $T = PU_{belt\ mass} \times S$

When:

a) T = transfer rate of the material [kg/s]

b) $PU_{belt\ Mass}$ = per unit belt mass [kg/m]

c) S = Conveyor speed [m/s]

d) T = PU x S = kg/m x m/s = kg/s.

This equation is helpful for measuring and monitoring flow rate through belt weight measuring system.

ii. **Power Supply**

The power flow to the motor can be determined $Pm = \sqrt{3}VICos\varphi$

The motor power can then be obtained by measuring:

a) Motor voltage

b) Current

c) Power factor.

This value does not take the motor and mechanical drive unit efficiencies into account. In order to determine the power absorbed by the conveyor these efficiencies are taken into account and the following equation can be used: $Pc = \eta_m \times \eta_d \times P_m$

When:

a) Pc = conveyor power (kW)

b) P_m = motor power (kW)

c) η_m = motor efficiency

d) η_d = mechanical drive unit efficiency

2.9 Deviation in power consumption

Deviation in power is defined as:

$$\Delta P = (P_{measured} - P_{design}) / P_{measured} \times 100\%$$

When:

a) $\%\Delta P$ = Percentage power deviation between measured power and design Power

b) $P_{measured}$ = measured power (kW)

c) P_{design} = Design power (kW)

2.10 Losses in Belt Conveyor & remedies

i. **Losses in belt conveyor**

Following are the assemblies that cause a loss in power while in operation:

a) Belt cleaners
b) Frozen idlers
c) Accumulation on return idlers
d) Damaged idler rubber
e) Pulley rubber damage.

ii. **Remedies**

a) Stated losses are due to bad maintenance practice, can be minimized by following SMP.
b) Today there are highly efficient motors available when efficiencies of up to 96% are available against the usual type of efficiency between 89 and 92%.
c) Mechanical losses in the gearboxes and conveyor components contribute to a total loss of about 10%.

2.11 Effect on Power consumption due to different conditions

i. **Conveyor Loading**

The conveyor becomes more efficient when transfer capacity is closer to the designed parameters. There might be scope for improvement on total conveyor efficiency in cases when conveyors are running empty or only lightly loaded compared to design values.

The possible solution would be:

a) Installation of a variable speed drive
b) A buffer to regulate the feed onto the belt
c) A controller to control the operations

d) To stop the belt if it is unnecessary to have it running.

e) When there is a huge difference in conveyor loading and designed values it might be beneficial to replace the conveyor.

ii. **Power Factor**

Power factor correction may have an impact on the reactive energy cost and or the maximum demand cost charged in KVA.

3.0 Bucket Elevator

A bucket elevator conveys bulk material vertically. They are considered similar to conveyor belts, with the greatest difference being that bucket elevators move material using buckets attached to a rotating belt or chain. The buckets work to pick up material, move it to the desired endpoint, discharge material, and finally, return to the starting point to pick up a new load, **Fig-3.3**

Bucket elevator configurations are designed based on the application, material, required horsepower, and elevator height.

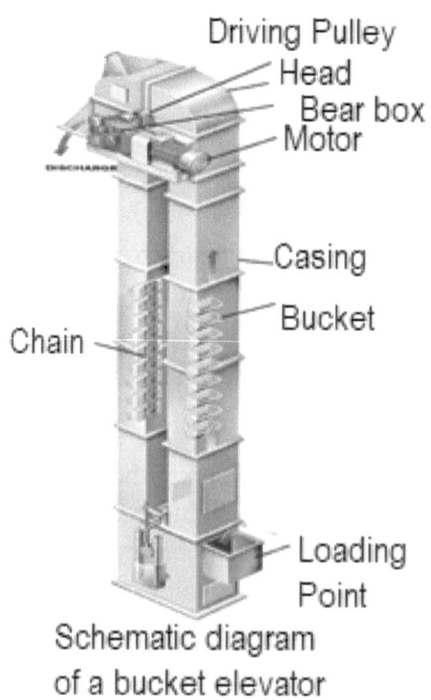

▲ **Fig-3.3: Bucket Elevator**

3.1 Application of Bucket Elevator

The following are some specific application industries:

i. Grain processing plant: The bucket elevator can be used to raise grains, food and other materials from one position to another position for transportation or processing.

ii. The coal mine industry: The bucket elevator machine is often used for vertical transportation of ore and other materials, which can be increased from the ground to the ground, or vertical conversion between different positions on the ground.

iii. Cement Plant: Bucket elevators can be used to increase raw materials such as limestone, clay, additives etc. from the ground to the top of the Pre-heater.

iv. Power plant: Feeding of coal from ground. to the grinding machine and other equipment.

3.2 Elevator speed/ Belt speed

Equation: Diameter pulley (D) x 3.14 x Rotations per minute (N)/60

Elevator Speed: $V = \pi D \times N/60$ m/s.

When,

a) D= Diameter pulley (M)

b) N= Rotations per minute (RPM)

c) V= elevator speed (m/s).

3.3 Derivation of Capacity of a bucket elevator

Capacity of bucket elevator: $Q = n \times V \times B \times v$ kg/s.

When,

a) Q = capacity in kg /s

b) n = numbers of buckets per meter

c) V = bucket volume in m^3

d) B = Bulk Density of the material kg/ m^3

e) v = Belt speed in m/s.

3.4 Power consumption by BE

Equation: P= Q x H x 9.81 /1000 kW.

When,

a) P = power in Kw
b) Q = capacity in kg/s
c) H = conveying height in meters
d) g = 9.81 m/sec^2

3.5 System Friction and power consumption

Friction includes the following variables:

i. Cup Digging
ii. Belt slip on the head pulley
iii. Chain slip on sprockets
iv. Bearing friction
v. Drive Inefficiencies

Power consumption with system friction: P= Q x H x 9.81 /1000 kW x C

Factor "C" is an estimate of the friction in the system and is required to accurately determine the power requirements of a Bucket Elevator. Commonly, Friction factor (C) = 1.15.

▼ **Table-2: Bulk density of commonly used material in industries**

Material	Bulk Density (Kg/ m^3)	Material	Bulk Density (Kg/ m^3)	Material	Bulk Density (Kg/ m^3)
Earth	1,600	Peat	410	Aluminum	2,800
Pit coal	860	Anthracite	1,700	Asbestos	2,800
Fly ash	1,000	Salt	1,100	Bauxite	2,550
Sand	1,600	Blast furnace slag	2,800	Boiler slag	1,000
Sawdust	600	Bronze	8,800	Brown coal	780
Sugar	1,600	Cement	1,600	Lime, caustic	1,300
Lime stone	2,800	Charcoal	400	Clinkers	2,000
Marble	2,700	Concrete	2,400	Ore, crude	2,200

4.0 Screw Conveyor

4.1 General description & different parts

i. **Description-**Screw conveyors, or auger conveyors, are industrial equipment used in transporting bulk quantities of granular solids (e.g., powder, grains, granules), semi-solids, liquids, and even non-flowing materials from one point to another. Screw conveyors primarily consist of a rotating screw shaft **Fig-3.4,** that is installed within a trough. As the screw shaft rotates, the material moves linearly. They can be designed to provide horizontal, vertical, and inclined travel paths.

ii. **Different Parts** -The components of the screw conveyor include the conveyor screw, trough, hanger bearings, couplings, internal collar, end lugs, and electric motor.

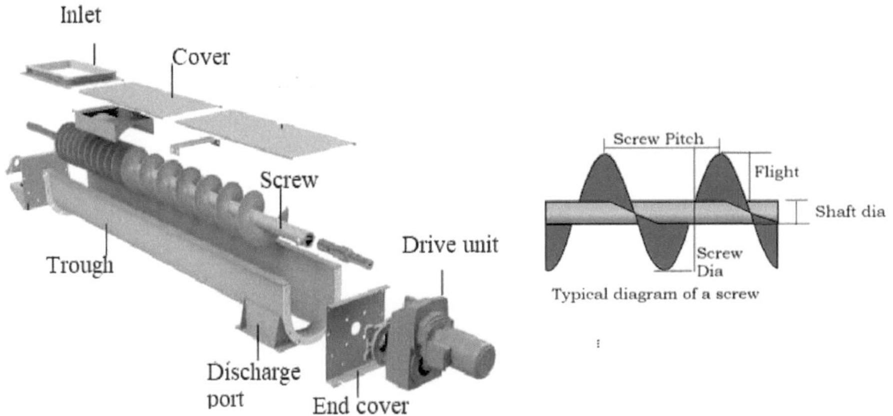

▲ **Fig 3.4: Different parts of a screw conveyor**

4.2 Types

The types of screw conveyors (based on the pitch and flight design) are the standard pitch screw conveyors, variable pitch screw conveyors, half-pitch screw conveyors, long pitch screw conveyors, short pitch screw conveyors, double flight screw conveyors, tapered flight screw conveyors, and mass flow screw conveyors.

Screw conveyors may be right-handed or left-handed. The handedness determines the material flow direction when the conveyor screw is rotated in a specific direction.

The main types of troughs are the U-shaped trough, rectangular trough, and tubular trough. Tubular troughs are ideal for inclined screw conveyors.

The types of screw conveyors (based on flow path) are the horizontal screw conveyor, inclined screw conveyor, and vertical screw conveyor.

4.3 Uses

A screw conveyor efficiently transports bulk quantities of materials. The materials may be granular solids, semi-solids, liquids, and non-flowing solids.

4.4 General formula for capacity

The capacity of a screw conveyor with a standard screw flight can be estimated the following way:

Equation: $Q = \pi/4 x D^2 x P x N x K x \rho x C x 60$

When,

a) Q = screw capacity in kg/h
b) D = screw diameter in m
c) P = screw pitch in m
d) N = screw speed in rpm
e) K= loading ratio
f) ρ = material loose density in kg/m^3
g) C = inclination correction factor

The CEMA association gives the capacity of a screw conveyor as:

Equation: $Q = \pi/4 \text{ x } (Ds^2 - Dp^2) \text{ x } P \text{ x } K \text{ x } \rho \text{ x } 60$

When,

a) Q= capacity kg/hr/rpm
b) D_s = Diameter of the screw flight in m
c) D_p = Diameter of the pipe -shaft supporting the screw flight in m
d) P = pitch of the screw in m
e) K = percent trough loading

It is the capacity over 1 rpm. If the dimensions are known, in the case of an existing screw, the screw capacity / hr can then be found by multiplying by the rpm at which the screw is used.

4.5 Capacity calculation for horizontal conveyor

Example: for capacity calculation for screw conveyor at 3500kg/hr.

i. Step 1: To define the requirement; The design of the screw must reach a capacity equal or greater than this value.

ii. Step 2: To calculate the capacity of the screw conveyor

a) Assume a diameter D

b) Define the screw pitch (depends on the characteristics of the product to be conveyed, among other parameters) according to the diameter of the screw as per the below chart.

Pitch	Pitch Length (m)
Standard	S=D
Short	S=2/3 D
Half	S=D/2
Long	S=1.5D

c) Estimate the loading ratio K of the screw according to the flow properties of the solid to be conveyed

Material	Minimum Loading ratio	Maximum loading ratio
Not free flowing	0.12	0.15
Average flowability	0.25	0.30
Free flowing	0.4	0.45

4.6 Capacity calculation for inclined screw conveyor

i. Define if the screw conveyor is flat (which is always preferable) or has to be inclined. Determine the corresponding correction factor as per below table.

ii. Adjust the screw speed so that the capacity of the screw is higher than the requirement.

Inclination in degree	Correction factor C
0	1
5	0.9
10	0.8
15	0.7
20	0.65

Example:

a) The diameter assumed is 0.1 m

b) No specific duty for the screw, the pitch is chosen standard, S=0.1 m

c) Sugar is free flowing, a loading of 0.45 is selected

d) The screw is installed without inclination, C=1

e) Sugar density is 800 kg/m^3

Equation: $Q= \pi/4 x D^2 x P x N x K x \rho x C x 60 = 0.785 x 0.1^2 x 0.1 \ x1 \ x \ 0.45 \ x 800 x 1 x 60 = 17 kg/hr$.

The calculation gives 17 kg/h for 1 rpm. Adjusting the speed, 207 rpm are required to reach a capacity of 3500 kg/h.

Step 3: Compare the calculated capacity to the max screw speed

Some reference max screw speed is given in the table below:

▼ **Table-3**

Screw dia (m)	15%	30%A	30%B	45%
0.1	69	139	69	190
0.15	66	132	66	182
0.23	62	122	62	170
0.25	60	118	60	165
0.30	58	111	58	157
0.36	56	104	56	148
0.41	53	97	53	140
0.46	50	90	50	131
0.51	47	82	47	122
0.61	42	68	42	105

a) If the calculated speed at step 2 is < than the max speed for the screw diameter selected, the design is satisfactory.

b) If the calculated speed at step 2 is > than the max speed for the screw diameter selected, the design is not suitable and the calculation must have rechecked by changing the diameter.

c) Note that when handling powders susceptible to lead to a dust explosion, the max speed should result a conveyor tip speed < 1 m/s.

Example:

The diameter selected was 0.1 m for which the max screw speed at 45% is advised to be 190 rpm.

The calculated speed is too high; the calculation must be done by changing a parameter.

Larger diameter can be selected, D=0.15 m.

The calculation is checked to be a speed of 62 rpm is calculated. It is less than previous one and seems to be satisfactory.

5.0 Pneumatic Conveying System

5.1 Introduction

This system works by moving materials through enclosed, airtight pipelines through a combination of airflow or another type of gas at a pressure. Pneumatic Conveying Systems uses air pressure to move materials from one location to another. System is operated in a timer mode or probe mode.

5.2 Main Components of a Pneumatic Conveying System

Pneumatic conveying components include various piping components, valves, fittings, and subsystems.

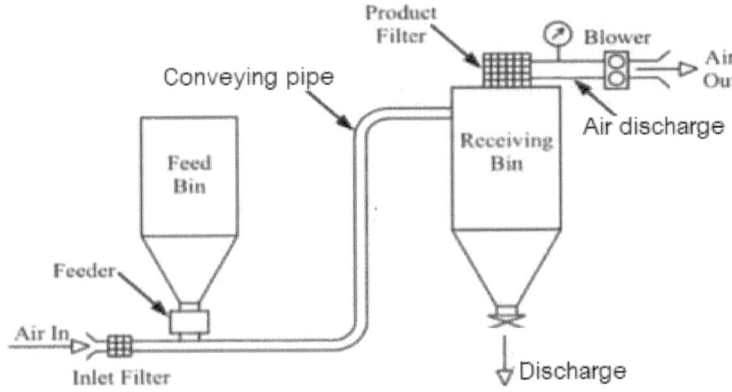

▲ **Fig-3.5: Pneumatic Conveying System**

The main components are shown in Fig-3.5:

a) Compressor/ compressed air: It generates the flow and pressure of air in the system

b) Feeder: Rotary valves and venturi type ejectors are used for this purpose

c) Pipes and fittings: These serve as the conveyor line through which the bulk materials are conveyed.

d) Material-gas separator: This is a sub-system that separates the solid particles from the air when conveying materials and filling the silos. Bag filter reverse jet cleaning system is used for this purpose.

e) Silos or storage: It does not participate in the actual conveyance of materials, but it is used to store the conveyed materials.

5.3 Working of Pneumatic conveying system

There are two main types of pneumatic conveying systems a) lean phase and b) dense phase. Each is used for different applications and includes different functions.

i. **Lean Phase:** The most commonly used pneumatic conveying system is a lean phase system used for granules and powders. Lean phase system conveying can be subdivided as:

   a) **Dilute Phase**-Dilute phase conveying systems use gas velocities of greater than 17-18 m/s with lower vacuum or conveying pressure below 0.1 bar. In dilute phase conveying systems, material particles are wholly suspended within the air or gas stream while the solids loading ratio is significantly lower. This type of system is ideal for conveying non-abrasive materials with low bulk density. Common applications for dilute phase systems include:

      1. Non-abrasives like flour
      2. Products that are difficult to break
      3. Light bulk-density materials

   b) **Medium Phase**-In medium phase pneumatic conveying systems, the gas or air velocity typically exceeds 17-18 m/s with a vacuum or pressure of up to 0.35 bar. Particles traveling through the pipelines remain suspended throughout, with a low concentration or solids loading ratio. Exhausters and roots blowers are often used to push air or other gas through the system.

ii. **Dense Phase:** This system is operated at lower velocity to transfer products, which helps prevent impact damage to the product. Dense phase systems use pulses of air to push the product through, followed by another pulse of both the air and product.

The overall density, size of the particles, and the run or length help determining the ratio of air to product when using such system. Common applications for dense phase systems include:

a) Abrasives including sugar

b) Materials with heavy bulk densities

c) Fly ash handling in boiler system

5.4 Advantages and disadvantages of Pneumatic conveying system

i. **Advantages**

a) While conveying the material, the path may involve many turns and lifts. In such cases, other types of conveying become costly.

b) The pneumatic conveyor is economical compared to other types of conveyors.

c) Friction losses are small because there is no movement of mechanical parts.

ii. **Disadvantages**

a) The pneumatic conveyor needs more power than other types of conveyor systems.

b) Erosion of the internal surface of the vessel and attrition of solids may take place.

c) Uneven contact of gas and solids is observed.

5.5 Specific Power Consumption in Pneumatic Conveying

The specific power consumption is defined as power consumed per kilogram of material conveyed. It can be calculated by using the following expression:

Equation:

$$P = M_s \times 202 \times V_{air} \times Ln\,(P_{in}/P_{out})\ \text{kW}$$

$$P_{spec} = 1/M_s \times 202 \times V_{air} \times Ln\,(P_{in}/P_{out}).$$

When:

a) P= Power consumption

b) P_{spec} = specific power consumption kw/(kg/s)

c) M_s = Solid mass flow rate (kg/s)

d) V_{air} = Volumetric flow rate of free air (m^3/s)

e) P_{in} = Pressure at inlet of pipe (pa)

f) P_{out} = Pressure at outlet of pipe (pa).

Specific power consumption increases with increase in inlet Froude number for fly ash as conveying material. This indicates that specific power consumption reduces when conveying occurs at a higher solids mass flow rate while needing a minimum amount of inertia force or driving force. So, there exists an optimum value of specific power consumption at a minimum inlet Froude number for each conveying material.

5.6 Effects on Specific Power Consumption

The specific power consumption is influenced, among others, by the inlet Froude Number, the inlet air kinetic energy, as well as the air mass flow rate.

i. **Inlet Froude Number:** The Froude number for a given powder material is defined as inertia forces (or driving forces) needed for a certain gravitational force. It is the ratio of inertia force to gravitational force. The inlet gas Froude number (Fr_i) can be expressed as: $Fr_{inlet} = V_{air} / \sqrt{(g \times D)}$

   When:

   a) Fr_{inlet} = inlet Froude Number

   b) V_{air} = inlet air velocity (M/s)

   c) g = Acceleration due to gravity m/s^2

   d) D= Diameter of Pipe (M)

ii. **Inlet Air Kinetic Energy:** The inlet air kinetic energy per unit volume (Ei) can be expressed as: $E_{inlet} = \frac{1}{2} \times \rho_{air} \times V_{air}^2$

   When:

   a) E_{inlet} = Inlet air KE / unit Volume (J/ m^3)

   b) ρ_{air} = Inlet air Density (kg/ m^3)

   c) V_{air} = Velocity of air (m/s)

   Minimum specific power consumption occurs when inlet air requires a minimum amount of kinetic energy for the beginning of conveying process.

iii. **Air Mass Flow Rate**

a) Specific power consumption increases with air mass flow rate at constant solids mass flow rates.

b) High specific power consumption occurs for conveying with low solids loading ratios.

c) The minimum or optimum point of conveying occurs for the conveying using low air mass flow rate, high solids loading ratio and low specific power consumption.

5.7 Power Consumption Coefficient

The most important parameters influencing the performance of a pneumatic conveying system are:

i. Solids mass flow rate

ii. Air mass flow rate

iii. Total pressure drop.

These three factors can be taken into consideration for developing a non-dimensional parameter, called the power consumption coefficient (η). The minimum value of power consumption coefficient indicates the requirement of a lower total pressure drop for a higher solids mass flow rate.

The power consumption coefficient is calculated by:

$$\eta = (\Delta P \times V_{air}) / (m_s \times g \times L) = \Delta P / (\Phi \times \rho_{air} \times g \times L)$$

When:

i. η = Power consumption coefficient

ii. ΔP = Total pressure drop (Pa)

iii. V_{air} = Inlet volumetric air flow rate (m^3/s)

iv. L= Total length of pipe line (Meter)

v. Φ = Solid loading ratio (Solid flow/Air flow)

5.8 Optimum Consumption Coefficient

The optimum or minimum value of power consumption coefficient and specific power consumption varies with type of conveying material as given in Table 4. These values are based on total length of pipeline.

▼ Table-4 Optimum values of power consumption coefficient

Conveying Material	Pressure drop KPa (bar)	Air flow rate (kg/s)	Solid mass flow rate (kg/s)	Solid loading ratio Solid flow/ air flow	Optimum Specific Power consumption [kw/(kg/s)]	Optimum power consumption coefficient
Fly ash	247(2.47)	0.0257	1.263	49.1	4.31	0.732
Alumina	277 (2.77)	0.0361	2.06	57.1	3.96	0.651

Questions:

1. What is Reynold's number? How N_{RE} is important in fluid flow?
2. Explain Bernoulli's equation in terms of head & energy.
3. What are the major and minor losses for fluid flow- explain?
4. Explain Ergun Equation showing that pressure drop in packed bed is due to viscosity & density of fluid.
5. What are the different types of mechanical conveying system? Explain different components of belt conveyor.
6. What are the losses in belt conveying system?
7. How to calculate capacity & power consumption of a bucket elevator?
8. Explain Lean phase & dense phase of pneumatic conveying system.
9. What are the factors that affect pneumatic conveying?

SECTION 04

SIZE REDUCTION & SCREENING PROCESS

1.0 Size Reduction:

1.1 Introduction

Raw materials are generally present in large size and cannot be used directly in any process. So, it is required to be converted into small size particle or powder. Size reduction is the operation carried out for reducing the size of bigger particles into smaller one of desired size and shape with the help of external forces. Conversion of large size material into small size particle or powder form using energy is called as size reduction that increases surface area of material.

1.2 Mechanism of Size Reduction

Size reduction is done through following processes:

i. **Cutting -** Material is cut down in small pieces with sharp blade, knife. It involves application of force. In industry, cutter mill is used to cut larger material to reduce in smaller size.

ii. **Compression** - Size reduction is done by crushing material by applying pressure. Roller mill is used in industry to reduce size by compression mechanism.

iii. **Impact -** When the material is stationary and hit by object moving at high speed, impact acts. Hammer mill works on this principle.

iv. **Attrition -** It occurs when there is collision between two body surfaces having high kinetic energy or high velocity impacted on particle. Roller mill works on this principle.

v. **Combination of attrition and impact** –Ball mill works on both these mechanisms, impact and attrition.

1.3 Factor affecting size reduction

a) **Hardness-** It is a surface property of the material like a strength. The harder the material the more difficult it is to reduce in size. It is easy to break soft particle as compared to hard material.

b) **Toughness-** Generally fibrous material having higher moisture content are tough in nature. Soft but Tough material causes problem during size reduction compared to hard but brittle substances.

c) **Stickiness-** Sticky material adheres to the grinding surface or in sieve surface which hinders size reduction. To overcome this, material is required to dry completely.

d) **Material Structure-** Material having special structure causes difficulty in size reduction. Mineral having lines of weakness produce flakes like particle on size reduction.

e) **Moisture Content-** Materials do not flow well if it contains moisture. Presence of moisture causes some changes in property of material like hardness, toughness or stickiness which affect size reduction. Materials having 5% moisture in dry grinding and 50% moisture in wet grinding does not pose any problem.

1.4 Equipment used in size reduction

The principal types of size reduction machines are:

a) Crushers

b) Grinders

c) Ultrafine grinders

d) Cutting machines.

1.5 Working of size reduction equipment

Different types of size reduction equipment are used in industry depending on the input material size and its characteristic and requirement of output size. The equipment works on the principle of impact, attrition, compression or combination of any two principles. Some of the commonly used size reduction equipment are described below.

1.6 Hammer Mill

i. **Working Principle:** The hammer mill, **Fig-4.1**, operates on the principle of impact grinding, when the size reduction is achieved by repeatedly striking particles against a hard surface. The hammer mill has a rotor, inside a drum, that is fitted with hammers. The hammers strike the particles and cause them to be broken down into smaller pieces.

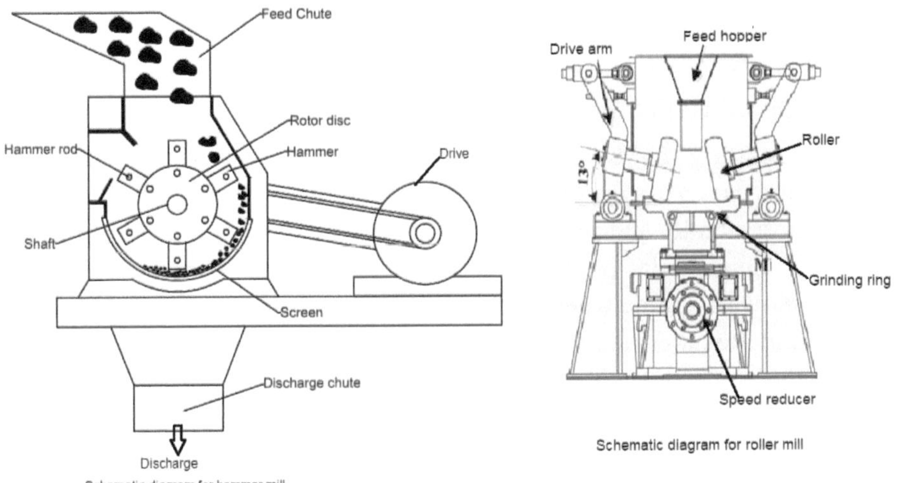

▲ **Fig-4.1: Shows Hammer mill & Vertical Roller Mill**

The mill also includes a feed hopper, which is used to feed the material to be ground and a discharge chute, used to remove the ground material.

During operation, the material to be ground is fed into the mill through feed hopper. As the rotor rotates, the hammers strike the particles, breaking them down into smaller pieces against the screen bar which acts as anvil. The ground material is then discharged through the discharge chute through screen bar.

ii. **Advantages of Hammer Mill**

a) High efficiency in size reduction

b) Able to handle a wide range of particle sizes

c) Versatile and can be used for a variety of materials

d) Simple and easy to operate

e) Low cost of maintenance

iii. **Disadvantages of Hammer Mill**

a) Not suitable for materials that are heat sensitive

b) Can generate a lot of dust and noise

c) Can cause wear and tear on the hammers and other parts of the mill

d) Not suitable for materials that are sticky or have high moisture content

1.7 Roller Mill

Roller mills are of two types-a) Horizontal roller mill b) Vertical roller mill.

i. **Horizontal Roller Mill**

a) **Working of a Horizontal Roller Mill-** It is a form of compression mill that uses a single, double, or triple cylindrical heavy wheel mounted horizontally and rotated about their long axis either in opposing pairs or against flat plates, to crush or grind various materials. **Fig-4.2** shows a double wheel roller mill. One of the rollers is run by a motor and the others are rotated by friction as the material is drawn through the gap between the rollers.

Roller mills use the process of compression and attrition in milling of solids. The rollers rotate at different speeds and the material is sheared as it passes through the gap.

To obtain the desired particles size, the following should be controlled;

1. The gap between the rollers
2. The speed differential between the rolls
3. Feed rate

b) **Advantages of Horizontal Roller Mills**

1. Produce more uniform product
2. Generates less heat (0 – 3 °C) unlike hammer mill that generates up to 10 °C
3. Less moisture loss
4. Better work environment due to low machine noises level
5. It is energy efficient
6. Reduce working hazard due to dust generation from the milling material.

c) **Disadvantages of Roller Mills**

1. Not effective on fiber or two-dimensional products. Works best on easy to grind materials like corn, cereal grains, etc.
2. At design stage additional inputs are required since the rolls need to be adjusted to suit grind requirements.

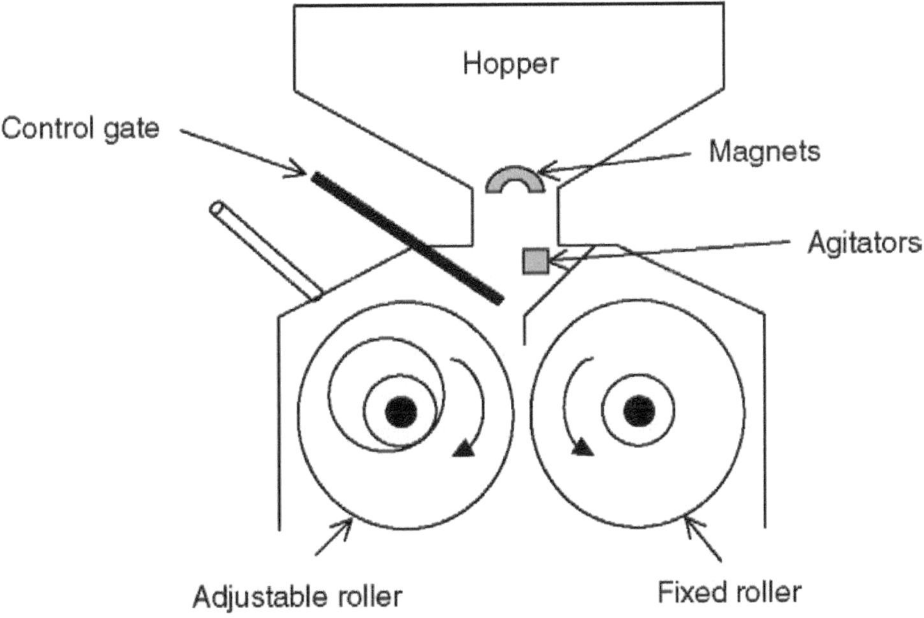

▲ **Fig-4.2: Two Roller Horizontal Roller Mill**

ii. **Vertical Roller Mill**

a) **Working principle of Vertical Roller Mill (VRM) -** In VRM, **Fig-4.1**, material layer is maintained between the grinding roller and the grinding disc. Rollers exert enough grinding pressure, so that the material is rolled and crushed on grinding disc. Air is supplied through the nozzles, fitted near the disc, to take way the ground material from the grinding area to classifier. Oversize materials from classifier fall back to the grinding area and the fines (product) escapes from the mill.

Factors for grinding in VRM are- grinding pressure by roller, the material grindability, moisture, wind speed. Roller pressure depends on hydraulic system pressure and the grinding roller weight which can be adjusted in the operation.

b) **Advantages of VRM**

1. More efficient
2. More precise
3. Safer &low wear rate
4. Lesser production costs
5. Small coverage area
6. Low noise generation.

c) **Disadvantages of VRM**

1. Limited work-piece Size
2. Limited machining Capabilities
3. High Initial Cost.

1.8 Ball Mill

a) **Working**: The ball mill operates on the principle of impact and attrition, **Fig-4.3**. The size reduction is achieved by the impact of the balls on the particles as well as by the attrition between the balls. The mill uses a combination of both impact and attrition forces to achieve the desired particle size.

A ball mill typically consists of a cylindrical drum partially filled with balls, typically made of steel or ceramic. The drum is mounted on a horizontal axis and is rotated by a motor. The mill also includes a feed hopper, which is used to introduce the material to be ground, and a discharge chute, which is used to remove the ground material.

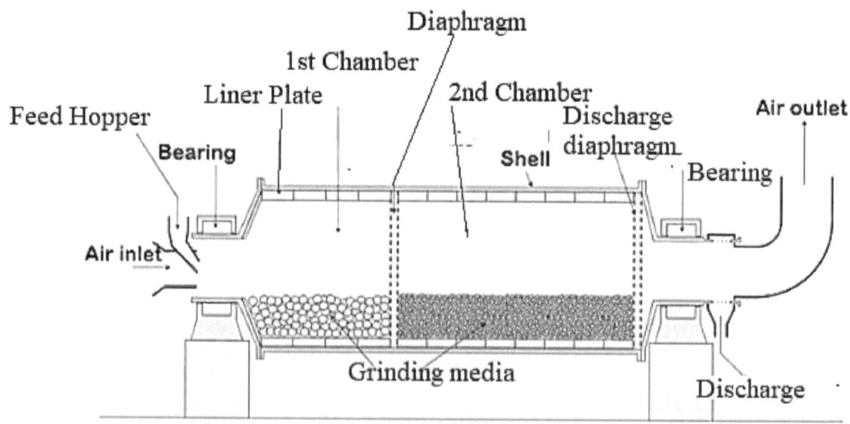

▲ **Fig-4.3: Schematic diagram of a ball mill**

The ball mill works by introducing the material to be ground into the feed hopper. The material is then pulled into the mill by the rotation of the drum. As the drum rotates, the balls inside the drum strike the particles, breaking them down into smaller pieces. The ground material is then discharged through the discharge chute. As the size reduction takes place in stages, diaphragms are placed inside the drum.

a) **Advantages of Ball Mill**

1. High efficiency in size reduction
2. Able to handle a wide range of particle sizes
3. Versatile and can be used for a variety of materials
4. Simple and easy to operate
5. Low cost of maintenance.

b) **Disadvantages of Ball Mill**

1. Not suitable for materials that are heat sensitive
2. Noisy operation
3. Wear and tear on the balls and other parts of the mill
4. Not suitable for materials that are sticky or contains high moisture.

1.9 Energy Consumption for Size Reduction

Following laws are followed for determination of energy consumption in size reduction processes.

i. **Rittinger's law**

Equation for energy consumption:

(J/kg or Watt/kg per sec) = K (1/product diameter-1/feed diameter)

$= K (1/D_p - 1/D_f)$.

When, K- Rittinger's constant, D_p= Product size, D_f=Feed Size.

Application of the law:

a) When surface area created is significant i.e. fine grinding
b) Particle size less than 0.05mm
c) Energy input is not very high.

ii. **Kick's Law:** The Kick's law states that the energy required to reduce the size of particles is proportional to the ratio of the initial size of a typical dimension (the diameter of the particles) to the final size of that dimension irrespective of the initial size.

Energy Consumption (J/kg or Watt/kg per sec) = K Ln (feed dia./ product dia.) = K Ln (D_f/D_p)

Application of Kicks law:

a) More accurate than Rittinger's law for coarse crushing.

b) Applicable for feed size greater than 50mm.

iii. **Bond's law:** Rittinger's law and Kick's law have their own limitations over a particular range of size reduction. Bond's law is a realistic way of estimating power consumption for crushing and grinding.

Bond's law states that: Work required to form particles of size Dp from a very like size feed is proportional to the square root of the surface to volume ratio (s_p/v_p=6/ϕ x D_p; Where, ϕ = sphericity) of the product.

As per equation: Specific power consumption:

Energy required/Ton (P/M) is proportional √ (s_p/v_p) = K_b/√Dp;

When, K_b= bond's constant; K_b= Wi x √Dp, Wi=work index.

If P/M(kWh/ton) is required to reduce a large feed to such a size that 80% of the product passes through 100-micron screen, then:

a) K_b = Wi √(100) =10 Wi; if Dp is in μm

b) K_b = Wi √(100x10^{-3}) =0.316 Wi if Dp is in mm

Work index: Wi, is the gross energy requirement in kilo Watt hour per ton of feed (kWh/ton of feed) to reduce a very large particle to such a size that 80% of the product will pass through a 100 micrometer or 0.01 mm screen.

Energy consumption:

W (kWh/ Ton) =10 x Wi x [1/ √ (product dia.)-1/√ (feed dia.)]

= 10 x Wi x [1/ √(D_p)-1/√(D_f)], dimensions are in μm

= 0.316x Wi x [1/ √(D_p)-1/√(D_f)], dimensions are in mm

80% feed & product pass through feed Screen sizes- D_f and D_p respectively.

If feed size is very large, energy consumption: **(W)= 10 x Wi x [1/ √(D_p)].**

When,

a) Wi =work index

b) D_p = product size

c) D_f = feed size.

1.10 Grinding Machine Technical Efficiency

The technical efficiency of a mill refers to the ratio between the content (%) of the qualified particle size in the product obtained after grinding and the content (%) of the original feed that is greater than the qualified particle size.

Efficiency = Product qty. of qualified size/Feed quantity having size more than qualified product size.

1.11 Procedure to calculate grinding efficiency using Bond Work Index

i. **Calculate the Actual Operating Bond Work Index of the grinding circuit.**

a) Estimate the F80 and P80.

b) Calculate the work or specific energy input from the size reduction equipment power and circuit tonnage.

Ignoring auxiliary equipment power, Specific energy unit, W=P/T

When,

1. W is the specific work input (kWh/t)
2. P is the equipment power (kW)
3. T is the circuit tonnage (metric t/h).
4. F80 & P80 are the feed size & product size 80% passing through input screen & product screen respectively.

c) Calculate the Actual Operating Bond Work Index.

$Wi_{act} = Wx10/(1/\sqrt{D_p}-1/\sqrt{D_f})$

When,

1. Wi_{act} is the Actual Operating Bond Work Index (kWh/t),
2. W is the specific energy input (kWh/t),
3. D_p is the 80% passing size of product (μm) through product screen
4. D_f is the 80% passing size of feed (μm) through feed screen.

d) Calculate the Standard Circuit Bond Work Index (Wi_{std}) from standard chart (table-1) and compare.

e) Calculate the circuit Wi Efficiency Ratio:

Wi Efficiency Ratio = Wi_{std} / Wi_{act}.

ii. **Inference**

a) If the Wi Efficiency Ratio is 1.0 or 100%, the circuit is performing with the same efficiency as the Bond Standard Circuit i.e. the circuit is using the same energy per ton as the design energy predicted with no correction factors.

b) If the Wi Efficiency Ratio is greater than 1.0 or 100%, the circuit is performing at an energy efficiency that exceeds the Bond Standard Circuit.

c) If the Wi Efficiency Ratio is less than 1.0 or 100%, the circuit is performing at an energy efficiency that is lower than the Bond Standard Circuit.

Example:

Power draw of mill(s) at pinion(s)- (kW) 3,150

Circuit dry tonnage -(metric t/h) 450

Circuit P80 -(µm) 212

Circuit F80- (µm) 2,500

Standard Test ball mill Wi of circuit feed ore- (kWh/t) 16.1

Calculation

a) Specific energy calculation (W)= 3150/450 kWh/ton=7

b) Wi_{act} =W/[10 x (1/√212-1/√2500)]=7/[10 x(1/14.5-1/50)]=14.58 kWh/t

c) Wi efficiency ratio = Wi_{std}/Wi_{act}=16/14.58=1.1=110%

d) Bond standard energy factor= Wi_{act}/ Wi_{std}=14.58/16=0.91=91%.

▼ **Table-1: Work Indexes of few materials for dry crushing**

Work indexes for dry crushing†

Material	Specific gravity	Work index, W_i
Bauxite	2.20	8.78
Cement clinker	3.15	13.45
Cement raw material	2.67	10.51
Clay	2.51	6.30
Coal	1.4	13.00
Coke	1.31	15.13
Granite	2.66	15.13
Gravel	2.66	16.06
Gypsum rock	2.69	6.73
Iron ore (hematite)	3.53	12.84
Limestone	2.66	12.74
Phosphate rock	2.74	9.92
Quartz	2.65	13.57
Shale	2.63	15.87
Slate	2.57	14.30
Trap rock	2.87	19.32

† For dry grinding, multiply by 4/3.

1.12 Improvement of Grinding Efficiency in ball mill

i. **Cares advised in ball mill**

Commonly following problems are faced in material processing plant when operating the ball mill.

1. Low grinding efficiency
2. Low processing capacity
3. High energy consumption
4. Unstable product fineness.

Following are some measures to improve the grinding efficiency of ball mill.

a) **Change the original grindability-** Grindability is determined by material hardness, toughness, dissociation and structural defects. Higher the grindability index easier to break and vice versa and the wear of lining plate and steel ball is lower, and the energy consumption is also lower. Therefore, the property of raw ore directly affects the productivity of the ball mill.

1. Some chemical reagent, called grinding aid (inorganic & organic) can be added to change the grinding effect and promote the grinding efficiency during the grinding stage;
2. Change the grindability of the raw ore, such as heating the minerals, change the mechanical properties of the ore, reduce the hardness of the ore.

b) **More crushing and less grinding to reduce the feed size into mill-** The larger feed size, the more work that the ball mill needs to do on the material. To achieve the specified grinding fineness, the workload of ball mill will be increased inevitably, then, the energy consumption and power consumption will be increased accordingly.

Hence, feed size should be as per design to get better output as per qualified size.

c) **Filling rate of steel ball-**At a fixed milling speed, larger filling rate, bigger grinding area have a stronger grinding effect with a large power consumption. If the filling rate is too high, the mill motion is to be reduced, then the impact effect on large size materials is reduced. Conversely, the smaller filling rate has the weaker grinding effect. The specific value should be decided by the mineral processing test (normally 40-50%) of mill volume.

d) **Reasonable size and proportion of steel ball-**The contact between steel ball in the ball mill and material is point-to-point. If the diameter of steel ball is too large, the crushing force is also large, so the material is crushed not the interface among different materials. Also, too large diameter of the steel ball results in less steel balls in numbers, low crushing probability, serious over-crushing phenomenon and uneven product particle size.

But if the steel ball is too small, the crushing force on the ore is small, and accordingly the grinding efficiency is low. Therefore, it is very important to select accurate size of the steel ball and its proportion for better grinding efficiency.

e) **Refill steel ball accurately-**The grinding action between steel ball and material causes the wear of steel balls, which can affect the grinding process and causes the fineness change of grinding products. Therefore, only by adopting reasonable steel ball replacement system, ball mill can be kept at a stable operation.

f) **Appropriate grinding Density-**Low grinding density, fast pulp flow, the material is not easy to stick around the steel ball, so the impact and grinding effect of steel ball on materials is weak resulting unqualified product size.

High grinding density, the material is easy to stick around the steel ball, so the impact and grinding effect of steel ball on materials is good, but the pulp flows slow, which is not conducive to improve the processing capacity of the ball mill.

So, maintain grinding density is an important parameter to have a better efficiency. This can be done by:

1. Control the feed rate in ball mill,
2. Control the air supply of the ball mill,
3. Adjust the classifying effect,
4. Control the particle size composition and moisture.

g) **Improve the classifying Efficiency-** Higher the classifying efficiency means that qualified size particles get discharged timely and efficiently, while low classifying efficiency means that most qualified particles are not discharged and sent back to the ball mill for re-grinding resulting over-grinding. Using two-stage classifying equipment can be provided to improve the classifying efficiency.

h) **Auto-control of the grinding System-**There are many variable parameters during the grinding operation, and one change inevitably leads to the change of many factors one after another. Generally, the manual operation control may cause unstable production, while the automatic control can keep the grinding and classifying in a stable and suitable state, thus improving the grinding efficiency.

1.13 Cares in Vertical Roller Mill to Increase Efficiency

i. **Material Layer of the Mill-** A too thick material layer reduces the grinding efficiency of the vertical roller mill. When the pressure difference of the mill reaches the limit, the material layer will collapse and affect the operation of the main motor and the discharge system.

   A too thin material layer causes the mill to generate a stronger driving force and damage the rollers, millstone, and hydraulic system of the mill.

ii. **The Vibration of the Mill-**The strong vibration of the mill body may cause mechanical failure and has a bad effect on the product quality.

   Causes for high vibration in mill body are- the grinding pressure, material layer thickness, air volume and temperature, accumulator pressure, wear condition of the roller and the grinding plate, etc.

iii. **The Grinding Pressure of the Mill-**The grinding pressure of the mill should be adjusted according to the fed quantity, the particle size, and the grindability of the material. It must be well controlled to maintain the thickness of the material layer and to reduce the vibration of the mill body.

   Increasing the grinding pressure improves the grinding capacity of the mill, but it ceases when the grinding pressure reaches a certain critical point. If the set pressure of the hydraulic cylinder is too high, it will only increase the driving force and accelerate the wear of parts, but not improve the grinding capacity.

   Too low grinding pressure will lead to the increase of the material layer thickness, power consumption, pressure difference in the mill, and vibration of the mill.

iv. **The Gas Temperature at the Discharging Port-**If the gas temperature at the discharging port of the mill is too low, the fluidity of the material will become poor, causing difficult in transportation of products. However, at higher gas temperature exceeds 130 °C, causes damage to equipment.

v. **The Air Volume in the Mill-**In vertical roller mills, the air volume in the mill is determined by the material feed rate. If the air volume is too large, the pressure difference in the mill and power consumption decrease, the thickness of the material layer will become thin, and the vibration of the mill body may be satisfactory.

If the air volume of the system is too small, the thickness of the material layer, the pressure difference in the mill and the power consumption increases and vibration increases.

2.0 Screening

2.1 Screening Fundamentals

Mechanical screening is the probability of a particle of a given size passing through an opening of a given size as described below.

a) As the size of the particle approaches the size of the opening, the probability of passing that particle through the screen decreases.

b) As the particle size decreases and the screen opening increases, the probability of passing the particle increases.

c) Each opening in a screen presents an additional chance for a given particle to pass through; hence, the more openings a particle is presented with, the higher the probability of passing that particle.

2.2 Screening Process

When a given distribution of material is fed to a screening machine, there are three main screening steps need to accomplish on a single screen deck.

i. **Bed stratification:** The activation and agitation of the material through the screen motion or the movement of the material.

ii. **Crowded screening:** In this section the smaller particles are in contact with the screen media and begin to pass in volume.

iii. **Separated screening:** In this section, the majority of the undersize material is already removed and the near size material has the opportunity to pass.

2.3 Screening efficiency and different factors

Screening efficiency is affected due to following factors:

i. **Particle Distribution-**The particle distribution is the sizes of the particles in any feed stream.

a) The percentage of material at each size determines the capacity of a given machine.

b) The more material required to pass through a screen, the larger the required screen area for the duty.

c) The top-size of the material will also have a direct impact on the material's screening.

ii. **Moisture-** Surface moisture is the moisture that is coating the exterior of a particle. Surface moisture has the greatest impact on the screening performance of a material.

iii. **Particle Shape**-The shape of particles both in the feed stream and the desired products affects the screening process.

iv. **Density**-Density of the material decides the load and bulk screening capacity of machine. Maintaining feed volume and effective bed depth can handle the screen effectively.

v. **Feed Rate**-Regular, consistent feed is critical to the proper operation of a screen. Irregular feed results in the passage of oversize material when the screen load is running at its minimum, or results in carryover of fines when the load is at its maximum.

vi. **Bed Depth**-The depth of the material on the screen affects the ability of the screen to stratify the material and allow the fines to pass through the deck.

vii. **Screen Openings**- Screen opening is described as a) nominal size b) the effective size of the mesh. The effective size is the size of the product that the mesh produces. During screening the nominal mesh size will produce an effective product size smaller than the nominal size.

2.4 Basis of Industrial Screen Design

i. **Considerations**

a) A 20 ° inclined screen is typically used for pre- screening and sizing applications.

b) Decreasing the inclination slows the movement across the table thereby decreasing capacity but improving efficiency.

c) Above 12 mm, dry screening is preferred. Below 12 mm, wet screening using a low-pressure water spray is preferred with a water flow rate of 0.8- 1.4 $m^3/t/h$. Actual water requirements dependent on application.

ii. **Screen Motion:** The processes of particle stratification and probability are caused by the vibration of the screen.

a) On inclined screens, the vibration is caused by a circular motion in a vertical plane of 3-12 mm amplitude at 700 to 1000 cycles/min. The vibration lifts the particles up, thereby, causing stratification. The incline will cascade the particles down the slope and increase the probability of particle passage through the screen.

b) Horizontal screens transfer the material using a straight line motion at an angle of approximately 45 degrees to the horizontal. This motion lifts the particles up from the screen surface and moves the particles toward the discharge end.

iii. **Screen Feed Rate Factors:** For a given feed rate, the width of the screen is selected to control the bed depth and achieve the optimum stratification.

Rule of thumb:

a) For particles having bulk density of 1600 kg/ m^3 or greater, the bed depth at discharge should never be greater than 4 times the aperture size in the screen.

b) For particles less than 1600 kg/ m^3, the bed depth is limited to 3 times the opening size.

iv. **Other factors controlling stratification include:**

a) Material travel rate (length/time) which is a function of:

1. material specifications
2. screening media specifications
3. depth of bed
4. stroke characteristics
5. slope of screen.

b) Stroke characteristics

1. amplitude
2. direction of rotation
3. type of motion
4. frequency.

c) Surface moisture – high surface moisture hinders stratification.

v. **Screening Probability**

Upon stratification, the particles having a size less than the smaller screen aperture pass through the screen to the underflow.

Obviously, particles having a size significantly smaller than the aperture size pass through easily.

However, the probability P of a particle passing through the screen in one trial decreases as the particle size approaches the aperture size, i.e,

$P = [(a-x)/(a+b)]^{2,}$ When,

a = screen opening size
x = particle diameter
b = diameter of wire.

The probability of a particle being retained on a screen during a single trial Q is: Q = 1-P

2.5 Perfect Screening vs. Reality

100% screening efficiency is not commercially practical since a screen of infinite length would be required.

A "perfect" separation is typically defined from a sieve analysis in which the sample is vibrated on a sieve for a period of 1 to 3 minutes. This is equivalent to approximately 25-50-meter-long screen. A screen length of 8 meter is the largest manufactured screen.

The "commercial perfect" screening practice is typically based on efficiency values in the order of 90% to 95%, indicating that 5% to 10% of the undersize particles report to the screen overflow.

2.6 Screen Blinding Effects

Blinding is typically caused by particles having a 50% chance of passing through the screen. The critical particle size range that causes blinding is: **0.5 < X/a < 1.5**

Therefore, sizing of material having a large portion of material in the size class should be avoided.

High moisture and clay in the feed causes binding.

2.7 Mass Flow Rate Through a Screen

The mass flow rate at which particles flow through the screen varies as a function of distance from the feed point. During the passage of flow, the phenomena that happen are-Stratification, screen saturation & probability of passing through.

2.8 Feed Rate vs. Efficiency

Screens are designed to treat a given mass flow rate at which point a maximum screening efficiency value is obtained as below:

a) At very low feed flow rates, screening efficiency increases with the feed rate. This is due to the fact that a sufficient amount of coarse particles is required to prevent excessive bouncing caused by screen vibration.

b) At the point "a" Fig-4.4, screening efficiency achieves an optimum value.

c) Feed rate greater than the optimum value result in a decline in screening efficiency due to an increasing bed thickness and the inability of the undersize particles to expose to the underflow stream, i.e., mass flow rate through the screen is limited.

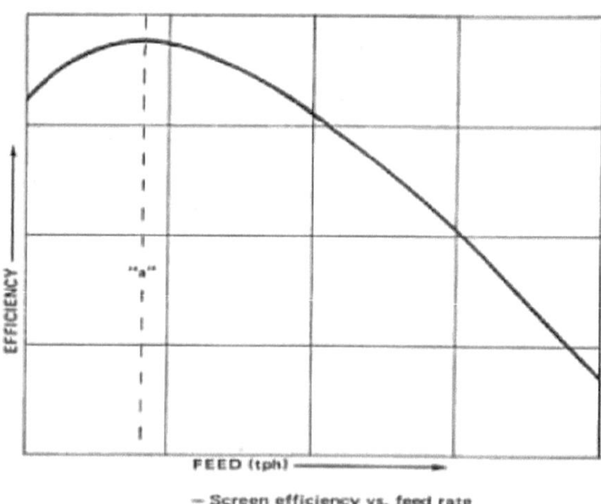

▲ **Fig- 4.4: Chart showing feed rate vs efficiency**

2.9 Screening Efficiency

Several definitions of screening efficiency exist, of which selected indexes as described below.

Screening efficiency

i. Screening efficiency is defined as the ratio of the material that should pass through a given opening and the material that actually passed through that opening.

   a) Screening efficiency in terms of the content of undersize in the overflow b when the overflow is considered the product.

   E = 100 – b

   Efficiency =% (of undersize in Feed) passes through the screen/%of undersize in feed

   b) Based on over size, efficiency equation:

   E = [% (or tph) of feed which is oversize /%(or tph) of feed which passes over] x 100

   For example, assume a sieve analysis of the screen overflow revealed that 9% of the screen overflow is undersize material. According to the equation, the screen efficiency is 100 – 9 = 91%.

   According to the determined ranges, screen efficiency should be in the range of 90%-95% with respect to the above formula.

ii. **Two product formula**

   a) **Two product formula - Oversize**

   Screening efficiency can also be determined by using the two-product formula.

   According to the two product formula, the recovery of oversize material to the screen overflow can be calculated using the following expression:

   $$R_O = O_O / F_f = o(f-u) / f(o-u)$$

   When,

   Ro- Recovery oversize, O_o- oversize output quantity, F_f- Feed quantity,

   o, u &f are the contents of the oversize material in the screen overflow, underflow and feed streams, respectively.

b) **Two product formula - Undersize**

   Likewise, the corresponding recovery of undersize material to the screen underflow is:

   $$Ru = (1-u)(o-f) / (1-f)(o-u)$$

   Thus, the overall screening efficiency $E = R_o \times R_u$

   $$E = [o(f-u) / f(o-u)^2] \times [(1-u)(o-f) / (1-f)]$$

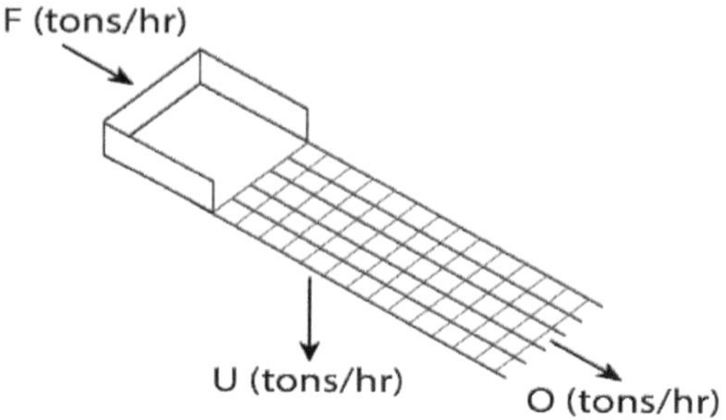

▲ **Fig-4.5: Screen**

2.10 Factors affecting Screening efficiency

Screening efficiency is affected by the factors listed below:

a) Wear, abrasion and tearing in the screen baffle.

b) Feeding too much or too little material onto the screen.

c) Large changes in grade.

d) Humidity level of the screened material.

e) Efficiency of the machine used prior to screening.

Questions:

1. List and briefly discuss the processes for size reduction.
2. Explain working, advantages & disadvantages of a hammer mill.
3. Discuss in brief about working of Vertical Roller Mill (VRM) and advantages over other mills.
4. Explain Rittinger's law, kick's law & Bond's law of grinding.
5. How to calculate grinding efficiency using Bond's work Index?
6. What cares to be taken for ball mill operation?
7. What are the stages for screening operation?
8. What are the factors that affect screening efficiency?
9. How to calculate screening efficiencies for two-product?

SECTION

05 BOILER, FURNACE AND HEATER

1.0 Boiler- its working and performance assessment

A boiler is an important and vital heat-generating equipment that can supply heat in the form of high-pressure and at high-temperature steam. This steam is used as a constant source of heat required for the process industry, also high-pressure steam is used in power generation for thermal power plants.

1.1 Boiler definition

As per the Indian Boilers Act, 1923/1950- A boiler is any closed vessel exceeding 22.75 litres in capacity having a pressure of minimum 1 bar(g) which is used expressly for generating steam under pressure and includes any mounting or other fitting attached to such vessel, which is wholly or partly under pressure when is shut off.

1.2 Different Types of Boilers

There are two types:

a) Fire tube boilers

b) Water tube boilers

1.3 Fire Tube Boiler

In such a boiler, Fig: 5-1, hot flue gases flow inside tubes that are submerged in water within a shell.

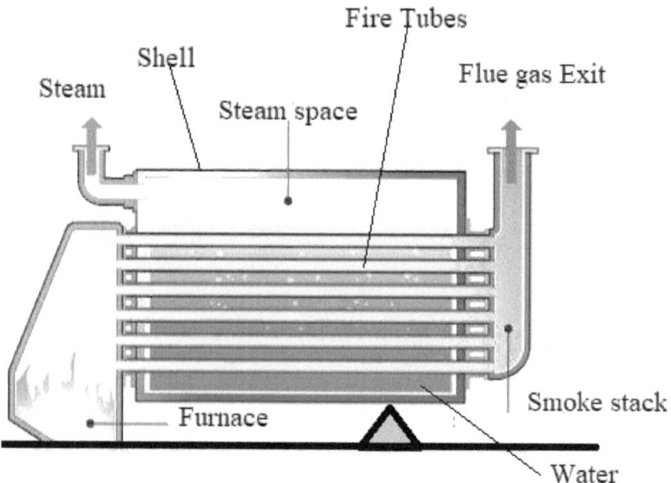

▲ **Fig-5.1: Fire tube boiler**

i. **Features of fire tube boiler**
   a) Pressures up to about 10 bar
   b) Produce up to 14 tonnes of steam/hr
   c) Can meet wide and sudden load fluctuations because of large water volumes
   d) Usually rated in HP

ii. **Types of fire tube boiler**
   a) **Wet back** –It has a water wall at the back of the boiler in the area when combustion gases reverse direction to enter tubes and exit from the front.
   b) **Dry back** -Refractory is used at the back, instead of a water wall. Internal maintenance is simplified, but refractory replacement is expensive, and overheating and cracking of tube ends at the entrance to return gas passages cause problems.

iii. **Advantages of fire tube boiler**
   a) Lower initial cost
   b) Few controls
   c) Easy to operate.

iv. **Disadvantages of fire tube boiler**

a) Shell exposed to heat, increasing the risk of explosion

b) Large water volume, resulting in poor circulation

c) Limited steam pressure and evaporation.

1.4 Water Tube Boiler

Fig: 5.2 shows a schematic diagram of water tube boiler. In water tube boiler, water flows through tubes that are surrounded by hot combustion gases in a shell.

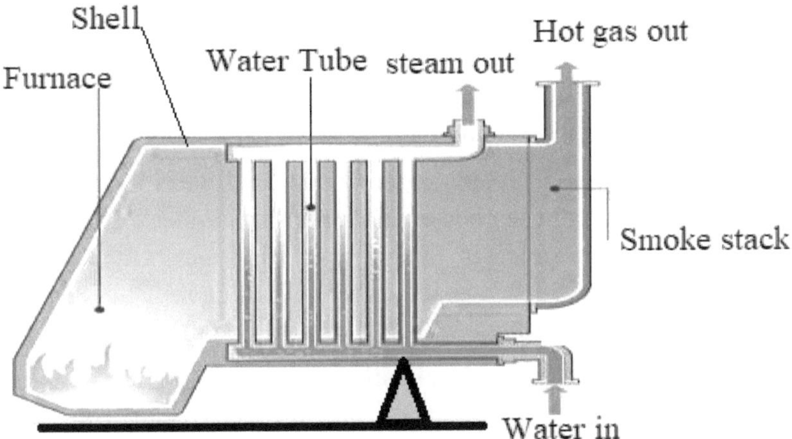

▲ **Fig-5.2: Water tube boiler**

i. **Features**

a) Usually rated in tons of steam/hr

b) Used for High Pressure steam

c) High capacity.

ii. **Advantages of water tube boiler**

a) Rapid heat transmission

b) Fast reaction to steam demand

c) High efficiency

d) Safer than fire tube boilers.

iii. **Disadvantages of water tube boiler**

a) More control than fire tube boiler

b) Higher initial cost

c) More complicated to operate.

1.5 Parts of a boiler

i. **Main Basic Parts:** Water tube boiler has a wide range of use from medium to heavy industry and for power generation. Hence, onward discussion will be based on water tube boiler, **Fig-5.3.**

a) **Fuel feeding system**-This system is meant for feeding fuel in the combustion chamber and it is different for different types of fuel. For solid and higher sized fuel, grate & spreader is used and for medium size <6mm, fuel is fed using air pressure. For liquid & gaseous fuel, burner is used to burn the fuel.

b) **Combustion space**- This space is enclosed with water filled tubes. Fuel is burnt either in bed or in suspension. Water changes its phase from liquid to steam in this region. For super-heated steam, SHs are also placed in this space either in bed or in hanging condition. Maximum heat, around 50%, gets transferred in this area,

c) **Convection Section**-Combustion gas, called flue gas, exiting from combustion space enters into the convection zone having a temperature of 450-600 °C. Heat in flue gas is recovered in this section by placing heat recovery equipment such as Primary super-heater, Economizer, Air pre-heater. Flue gas temperature after last heat exchanger remains at 140 °C.

d) **Stack**- The flue gas is discharged in the atmosphere through chimney, called stack, after passing through gas cleaning zone. ESP, bag filters are used to remove ash particles from gas and FGD & SCR are presently used to remove SOx & NOx.

e) **Air fans**-To supply air for combustion, fans are provided in boiler plant- they are called FD & Secondary air (SA) fan. To take out flue gas from the furnace, for balance draft system, Induced draft (ID) fan is provided. Another fan called primary air (PA) fan is provided to convey fuel into the combustion space.

f) **Attemperator**-It is a de-super heater placed in a steam path to control final steam temperature.

g) **Feed pump-** To supply water into the boiler, a dedicated pump is provided called, Boiler Feed Pump (BFP).

h) **Controls and Accessories-** For stable and safe operation controls systems are provided in boiler to control the parameters such as- drum level, furnace pressure, main steam pressure& temperature.

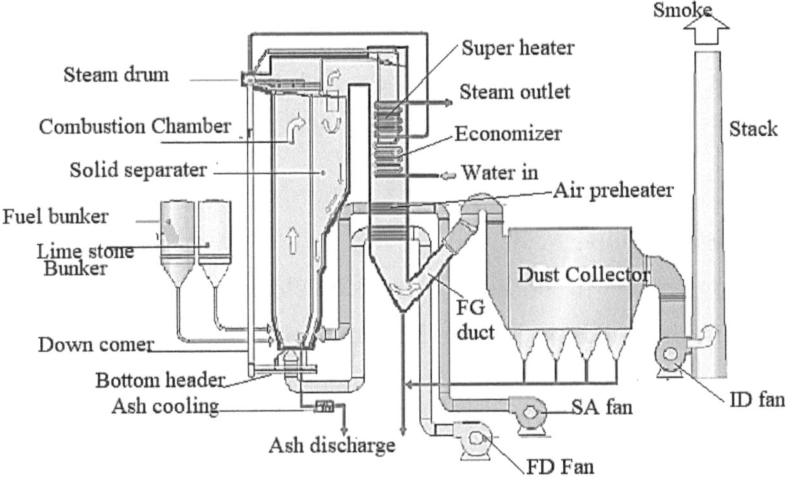

Schematic diagram for a water tube boiler showing different parts and systems

▲ **Fig-5.3: Water tube boiler with parts and systems**

1.6 Systems of a boiler

Following are the systems included in a boiler plant.

i. **Fuel system-** This system consists of storage of fuel, called bunker for fossil fuel fired boiler and fuel day tank for liquid fuel fired boiler. Fuel is fed into the combustion space with a regulation as per demand. For coal fired boiler this regulation is done through coal feeder and for liquid & gaseous fuel it is done through control valve/ damper.

ii. **Air system-** Air system supplies air required for coal feeding or for combustion. It is preheated in air pre-heater before use.

iii. **Feed water system-** Continuous water is supplied to boiler by BFP taking water from feed tank to boiler drum after preheating in economizer. Drum supplies water to evaporating zone either by natural circulation or by forced circulation.

iv. **Flue gas System**-Flue gas coming out from combustion section passes through different sections such as Primary SH, Economizer, APH, ESP/ bag filter and finally exits to atmosphere through chimney with the help of ID fan.

v. **Ash disposal system**- Fossil fuel discharges ash after combustion and collected as bottom ash and fly ash. The collected ash is removed from collected hopper either by hydraulically (wet disposal) or pneumatically (dry discharge).

vi. **Pollution control system**- Flue gas contains ash particles while getting discharge from boiler. The dust laden gas is cleaned in ESP or bag filters.

1.7 Operation of boiler

Typical start up process for boiler as described below:

i. Check all valves and dampers in their start up positions.

ii. Open the sight gauge and water level transmitter to take in service. The local water level safety system is to be in service.

iii. Open the drum and super heater vent valves.

iv. Start filling with water, the quality of water should be as per the recommendation of the boiler manufacturer. Keep BFP running.

v. Inject boiler water treatment chemicals including oxygen scavenger chemicals, so that the chemicals are added to the fill water.

vi. Check the boiler protection system.

vii. Once water is full to the operating level, take a trial for all auxiliary systems.

viii. After satisfactory checks, start fans as per sequence i.e. ID, FD, SA and maintain furnace pressure.

ix. Start the fuel system and fire the boiler.

x. Carefully bring the pressure up to 2 bar with the vent valves open, then close all the vents.

xi. The boiler start-up curve should be strictly followed. The standard warm-up curve for a typical boiler is not to increase the boiler water temperature over 60-65 °C per hour. Check manufacturers' guidelines.

xii. After closing the vents and slowly bringing the boiler up to operating pressure as per start-up curve, keeping start-up vent open.

xiii. Charge steam line as per operating pressure

xiv. Collect a boiler water sample and test for the proper chemical concentrations. Adjust as needed.

1.8 Cares of a boiler Operation &Maintenance

i. **Daily**

a) Observe operating pressures, temperatures and general conditions. Determine the cause of any unusual noises or conditions and make necessary corrections.

b) Observe flame/ fire conditions and adjust burner or air if flue gas at chimney exit is smoky.

c) Examine the safety relief valve and its drain for signs of leakage or malfunction. Repairs must be undertaken as soon as possible at the first indication of any defects, leakage, or malfunction.

d) Operate drum blow down, gauge glass flushing as per recommendation.

e) Keep the boiler room clean and free of combustible materials.

f) Maintain a boiler room log and record the various routines and tests that are performed. (Log repairs, as well as safety relief valve servicing and or replacement).

g) Observe the condition of Fans and feed pump for normal operation.

ii. **Monthly**

a) Operate the safety relief valve manually by means of the try lever.

b) Test flame detection devices as per manufactures instructions.

c) Test low-water fuel cut-off system.

iii. **Annually**

a) Thorough cleaning of pressure and non-pressure parts.

b) Do visual checking and measure the pressure part thickness.

c) Have the boiler, its controls and safety devices thoroughly examined.

d) Do hydro test and get it checked by competent authority.

e) Check insulation and repair if any damage is observed.

1.9 Performance, Efficiency, and Energy Balance Assessment of Boiler

1.9.1 Evaluation of combustion air quantity

i. **Stoichiometric and actual air requirements**

Stoichiometry or theoretical air -The quantity of calculated air, required for combustion, based on the fuel analysis (ultimate analysis) is called stoichiometric air quantity. It is always less than what is actually used during boiler combustion.

Example:

Combustion equation:

$C+O_2=CO_2$; $H_2+O_2=2H_2O$; $S+O_2=SO_2$.

O_2 required per kg of fuel:

a) For combustion of carbon =8/3 x C/100 kg

b) For combustion of H=8 x H/100 kg

c) For S combustion= S/100 kg

Hence, total O_2 required (O) = 1/100 (8/3C +8H +S) kg/ kg of fuel [C, H &S are expressed in%].

a) Air contains 23% O_2 by weight; hence, the theoretical quantity of air required is O/23x100 kg/ kg of fuel.

b) If fuel contains Oxygen = O^a%, theoretical actual air quantity= [O-(O^a/100)] x1/23x100 kg/kg of fuel.

ii. **Excess air:** To have complete combustion, the quantity of theoretical air is not sufficient; rather a bit higher quantity of air is required. The excess quantity of air than the theoretical quantity is called the excess air.

▼ **Table - Excess Air for different types of fuel combustion**

Type of fuel and combustion	% of excess air
Stoker fired boiler	25-35
Wood fired boiler	35-45
Coal-fired AFBC boiler	20-25
Coal-fired CFBC boiler	20-25
Coal- fired PF boiler	20-25
Gas fired boiler	5-10
Oil fired boiler	10-15

Actual air quantity = Theoretical air quantity + Excess air quantity.

Example: The theoretical air (or stoichiometric air) quantity required for combustion of fuel in a boiler is 8kg/ kg of fuel. If excess air is 25%, then actual air quantity= 8 + 8x25%=10 kg/ kg of fuel.

iii. **Calculation of quantity of air supplied and excess air percentage from coal & and flue gas analysis**

Example:

Fuel Analysis: C-81%, H-4.5%, O-8%, balance is ash.

FG analysis (v/v): CO_2-8.3%, CO-1.4%, O_2-10%, N_2-80.3%.

Constituent	%	Proportional weight Kg/kg-mol of gas	Fraction Kg/ kg of gas	Carbon content in FG Kg/kg of FG
CO_2	8.3	8.3/100 x 44 =3.65	3.65/29.72=0.1228	12/44x0.1228=0.0334
CO	1.4	1.4/100 x 28 =0.39	0.39/29.72=0.0131	12/28x0.0131=0.0056
O_2	10	10/100 x32=3.2	3.2/29.72=0.1076	
N_2	80.3	80.3/100 x28=22.48	22.48/29.72=0.7563	
	Total	29.72		Total C=0.039

a) Hence, C=0.039 kg contains in 1 kg FG

0.81 Kg C presents in = 1/0.039x0.81 kg FG

FG quantity = 20.76 kg/kg of fuel (total dry FG quantity).

b) N_2 remains unaltered and it is considered that it comes from Air.

N_2 in FG= 20.76x0.7563 = 15.525 kg/ kg of fuel.

c) Hence, air supplied= 15.525/0.77 kg/ kg of fuel. = 20.16 kg/ kg of fuel [air contains 77% N_2]

d) Minimum air required =100/23 [8/3 C+8 x(H_2-O_2/8) + S] = 100/23 [8/3x0.81+8 x (0.045-0.08/8) + 0] =10.61 kg/ kg of fuel.

e) Excess air percentage= (20.16-10.61)/10.61= 95.5%.

iv. **Air quantity Using Thumb Rule**

As a thumb rule: 1kg air is required to burn 775 kcal from fuel.

For example, If GCV of fuel = 4000kcal/kg; approximately air consumption= 4000/775 kg air/ kg of fuel = 5.16 kg air/ kg of fuel.

If excess air = 20%; actual air consumption= 5.16x1.2=6.2 kg/kg of fuel.

v. **Excess air in combustion and its effect**

a) **Percentage of Excess Air with respect to Oxygen in Flue Gas**

Excess air (%) = O_2% in flue gas/ (21-O_2% in flue gas) x 100.

Example: if O_2 in FG is 4%, then excess air= 4/ (21-4) x 100=23.5%.

b) **Relationship between excess air & CO_2 at the exit of the boiler**

Excess air (%) = (Theoretical CO_2 -Running CO_2)/Running CO_2 x100.

Example: If theoretical CO2 is 15% and measured CO2 in flue gas is 11 %, then excess air = (15-11)/11x100 %=36.36%.

c) **Good effects of excess air**

1. Better combustion, less un-burnt in ash
2. Less CO generation
3. Better control of furnace & bed temperature.

d) **Bad effect of excess air**

1. Boiler efficiency may be affected badly
2. Fines carry over may be more in flue gas
3. More erosion in flue gas path.

So, it is recommended to maintain excess air as per design only.

1.9.2 Evaluation of flue gas quantity per Kg of fuel burning

i. **By Calculation Method (Flue gas analysis known)**

Equation using gas analysis:

a) Kg of gas/ kg of fuel = [11 CO_2 +8O_2 + 7(N_2+CO)] x C / [3(CO_2 +CO)] for coal having negligible Sulphur (S).

b) Kg of gas/ kg of fuel = [11 CO_2 +8O_2 + 7(N_2+CO) x (C+3/8 S)] / [3(CO_2 +CO)] + 5/8 S for coal having considerable amount of Sulphur (S).

When, C= Quantity of Carbon Burnt, S= Sulphur in fuel, CO & CO_2= Carbon Monoxide & Carbon dioxide in flue gas, N_2 & O_2 = Nitrogen & Oxygen in FG.

ii. **Flue Gas Quantity using standard practice**

Flue gas quantity = (Actual air quantity +1) kg/kg of fuel= 6.2+1= 7.2 kg/kg of fuel.

[Refer example 1.9.1 (iv)]

1.9.3 Loss calculation due to excess air & unburnt in ash

i. **Percentage Loss in Flue Gas in Every% of Excess Oxygen beyond design**

For every% of O_2 in flue gas, excess flue gas quantity is around 4%.

Hence, loss due to flue gas shall be increased by additional 4% (at the same condition).

If the loss due to flue gas at design O_2% is 5%, then due to every extra% of O_2 loss is increased to = 1. 04 x 5%=5.2%. Hence, additional loss= 5.2-5=0.2%.

ii. **Unburnt loss in ash**

a) **Calculated Unburnt per cent in ash**

Calculated un-burnt in ash is based on combustion efficiency of fuel combustion system, GCV of fuel and ash content in fuel.

Example: Let, fuel GCV is 4000 kcal/kg & ash is 40%, is being burnt in AFBC boiler.

Let us consider, AFBC boiler combustion efficiency is 97%.

Heat loss due to inefficient combustion system (3%)

= 4000x (100-97)/100 kcal/kg of fuel =120 kcal /kg

= 120/8050 = 0.015 (U_b) (kg)/kg of fuel [When, (8050= CV of carbon)].

Un-burnt% = U_b / (U_b + ash) x 100 = 0.015/ (0.015+0.4) x 100=3.5%.

b) **Derivation of U_b quantity in refuse**

Let, 1 Kg coal is fed and feed coal contains: A% ash & C% carbon; Unburnt in refuse = U_b%

Ash balance: 1x A%= Refuse x (100-U_b)

Hence, Refuse quantity/ kg fuel= A%/(100-U_b)

Then, U_b quantity in refuse/ kg of fuel=A% x U_b /(100-U_b)]

Using above example: kg of U_b / kg of fuel= [3.5/(100-3.5)] x 40% =0.014.

This equation can be used for all areas like Bed ash, Economizer ash, APH, ESP Field-1,2,3…

In such case ash collection% of all the area is to be used in the equation as under.

U_b quantity /kg of fuel in a specific section = [(U_b / (100- U_b) x Ash% in fuel x% collection in the section].

c) **Loss in Efficiency due to unburnt in Ash**

Let U_b=12%, Ash% in coal= 20%, GCV of coal =5500 kcal/kg

Loss% = [(U_b /(100- U_b) x 8050 x Ash% in fuel]/4000)x 100=(12/88x8050x0.2)/5500=4%.

1.10 Evaluation of Boiler Efficiency by Direct Method

Thermal efficiency of boiler can be defined as:

i. **Heat output/ heat input x 100%**

Or, Q x (H_s- H_w)/ (q x CV) x100%

ii. **Heat output/ (heat output + losses) x 100%**

Or, Q x (H_s-H_w)/ [(Q x (H_s-H_w)] + H_L) x100%

When, Q= Quantity of steam generated, q= Quantity of fuel burnt, H_s= Enthalpy of steam,

H_w = Enthalpy of feed water at boiler inlet, H_L= Quantity of heat loss, CV= GCV of coal.

1.11 Calculation of Boiler Efficiency by Indirect Method or Loss Method

i. **Efficiency calculation by loss method**

Efficiency = 100- (% loss due to FG+ Moisture loss% + H- loss% +Unburnt loss% + loss% due to moisture in air +Sensible heat loss% through ash+ radiation loss)

Legend:

a) C_p= Specific heat of flue gas, kcal/ kg

b) C_s = Specific heat of steam at flue gas temperature and pressure.

c) T_f= Flue gas temperature

d) T_a= Ambient Temperature.

e) AAS= Actual air supply kg/kg of fuel

f) C_a= specific heat of ash at corresponding flue gas temperature

g) U_b= Unburnt%.

1. **Flue Gas Loss (%)**= Quantity of flue gas kg/ kg of fuel x C_p x $[T_f - T_a]$ / GCV of fuel x 100.
2. **Hydrogen loss (%)**= 9 x H x [584 + C_s $(T_f - T_a)$]/ GCV of fuel x100; When, H=% Hydrogen in fuel.
3. **Moisture (in fuel) loss (%)** =M x [584 + C_s $(T_f - T_a)$]/ GCV of fuel x 100. When, M=% moisture in fuel.
4. **Loss due to moisture in air (%)**=AAS x Humidity Factor x C_s $(T_f - T_a)$/ GCV of fuel **x 100.** Humidity factor= kg of moisture /kg of air.
5. **Sensible heat loss through ash (%)** = [% ash in fuel x100/(100-Ub)] x C_a x $(T_f - T_a)$/ GCV of fuel x 100.
6. **Unburnt Loss%** = [(U_b /(100- U_b) x 8050 x Ash% in fuel]/GCV x 100
7. **CO loss (%)** = [FG quantity/ kg of fuel x CO% /100 x 2410]/GCV x100.
8. **Radiation Loss (%)** = It is checked with the ABMA chart; it varies from 0.25-0.5% for bigger boilers.

Boiler Efficiency (%) = (100- All Losses%).

1.12 Classification of boiler losses

If the above-stated losses are classified, these can be classified as following categories:

a) **Loss due to inefficiency in combustion**

   The loss can be calculated considering losses due to:

   1. Unburnt in ash
   2. CO in Flue gas

b) **Loss due to inefficiency in heat transfer**

   The losses can be calculated considering losses due to: Sensible heat loss through Flue gas

c) **Uncontrolled Loss**

   1. Radiation loss
   2. Hydrogen Loss
   3. Moisture loss (moisture in fuel and moisture in the air)

Overall Thermal efficiency% = [(Combustion efficiency x heat transfer efficiency in fraction) x 100 – Uncontrolled loss%].

1.13 Heat Balance

Heat balance is the purpose of understanding how much heat is absorbed by which equipment or assemblies, and where heat is being lost. So that, input heat must be equal to output heat. The following items are considered while doing heat balance:

i. Heat absorbed by unit
ii. Heat loss due to dry gas
iii. Heat loss due to moisture
iv. Heat loss due to CO formation
v. Heat loss due to water formation from hydrogen combustion
vi. Heat loss due to unburnt in ash
vii. Heat loss due to radiation
viii. Heat loss due to sensible heat in drained ash
ix. Unaccounted loss (less than 1%)

1.14 Discussion on different benefits & losses on boiler operation

i. **Benefits on air heating:** Heating of combustion air can raise boiler efficiency about 1% for every 22°C in temperature increase.

   a) Let air consumption/ kg of fuel is 7 kg/kg of fuel having GCV of coal=4000 kcal/kg and ambient temperature =30 ^{0} C.

   b) Quantity of heat gain for every 10 ^{0} C rise of air above ambient Temperature= 0.25 x 10x7 = 17.5 kcal.

   c) % Increase in efficiency by saving heat through air preheater =14.5/4000x100=0.44%.

   d) Hence, every 22 °C it is around 1.0%.

ii. **Loss in Boiler for an increase in composition**

   a) **Moisture in Fuel-**Loss due to Moisture = Quantity of moisture in fuel (kg/kg of fuel) x (655- Ambient temperature)/ (GCV ARB of fuel); Enthalpy of steam at flue gas condition = 655 kcal/kg.

   Example: if moisture contains is 10% and fuel GCV= 4000 kcal/kg & Ambient temp=30^0 C.

   Then, loss due to moisture= [10/100x (655-30)/4000] x100% =1.56%.

   Hence, for every% increase in moisture, loss increases by =0.156% (at 4000 GCV coal).

b) **Flue Gas Temperature beyond Design-**Loss (%) due to flue gas = Quantity of dry Flue /kg of fuel x 0.25 x (FG exit temperature-ambient temp) / GCV of fuel.

Let, design flue gas temp is 140 ^{0} C, ambient temp= 40 ^{0}C.

For additional 10 ^{0} C increase in FG temp = (150-40) / (140-40) =110/100=1.1, that is 10% of FG loss.

If FG loss is 6%, then every extra 10 ^{0} C, increase in temperature additional loss is = 0.6%.

iii. **Boiler Blow Down**

a) **Percentage blow down in Boiler-**Blow Down% = (TDS in Feed water- TDS in steam)/ (TDS maintained in drum water- TDS in steam) x 100.

Example: let TDS in Feed water = 3 ppm, TDS in steam =1 ppm and TDS maintained in drum=100 ppm. Blow down% = (3-1)/99 x100= 2%.

b) **Boiler Blow Down Loss-**The loss can be calculated in following ways with an example:

Let boiler blow down is 5% for a boiler pressure at 67 ata & 485°C having capacity 40 t/h and FW enthalpy is 130 kcal/kg & boiler efficiency=80%.

Corresponding to 67 ata, blow down water enthalpy is at 285 kcal/ kg.

Loss due to blow down (%)= [5% x 40 x 285 Kcal] ÷ [40 x (807-130) /0.80 kcal] = 1.6%.

[Main steam enthalpy at 67 ata & 485 ^{0} C = 807 kcal/ kg].

c) **Flash Steam Generation in Blow Down Water-** Using above example flash steam generation can be calculated as under:

Let, flash steam is generated at 3 bar. Saturated steam enthalpy at 3 bar =651 kcal/ kg, enthalpy of condensate at atmospheric pressure= 99 kcal/kg.

Solution: Blow down 5% = 40x 5%= 2 t/h.

Flash steam generation= (drum water Enthalpy-Enthalpy of condensate) ÷ flash steam enthalpy x blow- down = (285-99)/651 x2=0.56 t/h.

iv. **Some important facts on boiler efficiency**

1. Air temp increased by 20°C =1% Fuel saving.
2. Flue gas temp reduced by 25°C = 1% Fuel saving.
3. Feed water temp increased by 5°C = 1% Fuel saving.
4. Oxygen% in FG at APH I/L increased by 1%=0.32% Fuel Loss
5. Every 1% reduction in excess air there is approximately 0.06% rise in efficiency.
6. Every 1% increase in moisture in fuel, boiler efficiency loss is 0.15%

Above facts can be checked with calculation considering standard parameter for fuel (GCV: 4000 kcal/ kg) and APH outlet flue gas temperature (140 ^{0} C) and ambient temperature is 40°C.

v. **Increase in air temperature**

a) Air required for combustion of 1 kg coal with 20% excess air = 4000/775 x 1.2=6.2 kg.

b) Additional heat input due to pre-heating of air = 6.2 x 0.30 x 20= 37.24 kcal/ kg of coal

c) Fuel saving =37.24/4000 x100 = 0.93% of fuel.

vi. **Flue gas temperature**

a) O_2% increase by 1% means an increase in the quantity of flue gas by= 1/(21-1)x100% = 5%

b) Hence, increased Flue gas quantity = 7.2 x 5%= 0.36 kg/ kg of coal.

c) Additional flue gas loss= 0.36 x0.3x (140-40) = 10.8 kcal/ kg of coal

d) Equivalent loss in fuel = 10.8/4000x100= 0.27%

vii. **Excess air reduction**

a) Reduction in excess air by 1%, reduction in FG quantity =7.2 x1% =0.072 kg Flue Gas /kg of coal

b) Reduction in the quantity of heat in FG taking out of boiler = 0.072 x 0.3 x (140-40) = 2.16 kcal

c) Decrease in loss i.e. increase in efficiency 2.16/4000 x100 = 0.054%

viii. **Moisture in coal**

i. 1% moisture increase in coal = 0.01 kg moisture/ kg of coal

ii. Extra Heat carried by moisture = 0.01x (655-40) = 6.15 kcal/ kg of fuel

iii. Extra loss i.e. loss in efficiency= 6.15/4000 x100= 0.15%.

2.0 Furnace- Working & Performance Assessment

A furnace is an equipment in which heat is generated and transferred to materials with the object of bringing about physical and chemical changes in them. The source of heat is usually the combustion of solid, liquid or gaseous fuel, or electrical energy. Electrical energy is applied through resistance heating (Joule heating) or inductive heating.

2.1 Types of Furnaces

Furnaces used in industries serve the dual purposes of providing heat and assisting in production. Industrial furnaces are mainly used for the purpose of annealing, melting, tempering, and carburizing of metals although they serve far more purposes and functions. Industrial furnaces can be divided between direct contact and indirect contact furnaces. Common Type of furnace are:

a) **Calcination Furnaces**- Calcination is a heat treatment process where feeds are heated to a point just below their melting temperature to produce thermal decomposition or to remove volatile substances. When ores are mined as carbonates or sulphates, the only way to extract the metal from the ore is to apply reduction; this is done in a calcination furnace.

   In the process of calcination, the ore is heated to a high temperature in the absence of air or oxygen; this removes moisture from the ore. In some processes volatile and oxides are removed from the ore.

b) **Tempering Furnaces**-A tempering furnace is meant for heat treatment of metal products to increase their durability and hardness. A tempering furnace imparts the beneficial properties of a metal and improves its mechanical characteristics.

   Tempering furnaces have ceramic and quartz heating elements that are lined with electrical coils to provide uniform heating of the chamber. Tempering takes place at temperatures between 300 °C and 750 °C.

c) **Annealing Furnaces**-Annealing is a heat treatment that relieves the internal stress of materials by heating them to their recrystallization temperature to make them ductile for further machining.

d) **Sintering Furnaces**- Sintering is a heat treatment process that agglomerate loose, fragmented material into a solid mass. The amount of heat provided during sintering is always slightly below the material's melting point. During sintering, the porous spaces in a work piece are minimized as the material is squeezed and shaped at high temperatures and pressures. The purpose of sintering is to improve the material's properties, such as thermal and electrical conductivity, strength, and transparency.

e) **Blast Furnaces**-A blast furnace is a cylindrical furnace that is used for smelting, the process of extracting metals from their ores. The furnace is loaded from the top with ore, fuel, and limestone. As the components move down the cylinder, a reaction takes place between them that produces molten metal and slag. At the bottom of the furnace are parallel pipes that push hot blast air up the cylinder to create the reaction between the materials.

f) **Electric Arc Furnaces**-Electric arc furnaces are used to produce carbon steel and alloy steel by recycling ferrous scrap. Scrap is melted and converted to steel by high-powered electric arcs that are formed by a cathode and one or more anodes. The scrap is loaded into a basket with limestone for slag formation, then charged in the furnace.

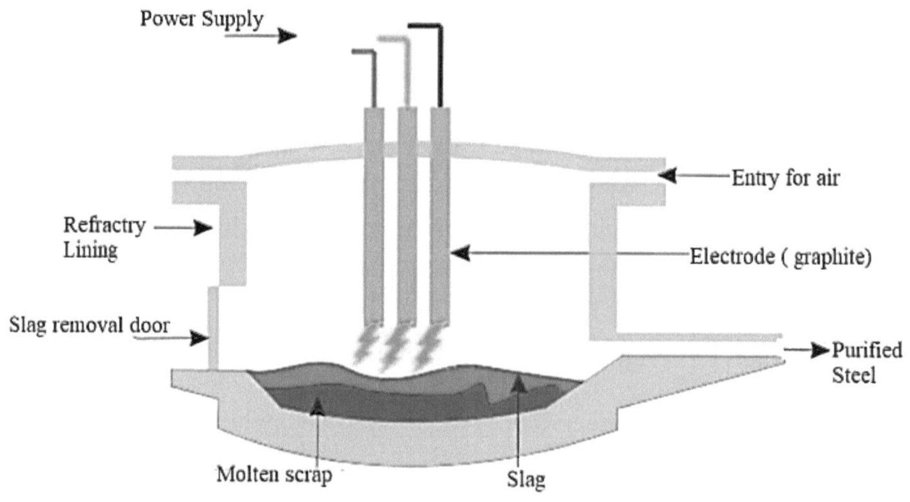

▲ **Fig-5.4(a): Electric furnace**

g) **Electric High Frequency Induction Furnaces**-Electric induction furnaces work on the same principles of transformers. The primary winding of an induction furnace is wound around the furnace and connected to an AC electrical supply. The charge inside the furnace acts as the secondary winding and uses induced current to heat up the charge. The primary coils are made of hollow tubes through which water circulates to keep the coils cooled to the appropriate temperature limits.

Heat is generated by eddy currents flowing concentrically, producing a high frequency supply of 500 Hz to 1000 Hz. A laminated core is used to protect the furnace's structure. Energy is transferred to the heated object through electromagnetic induction, so is the name.

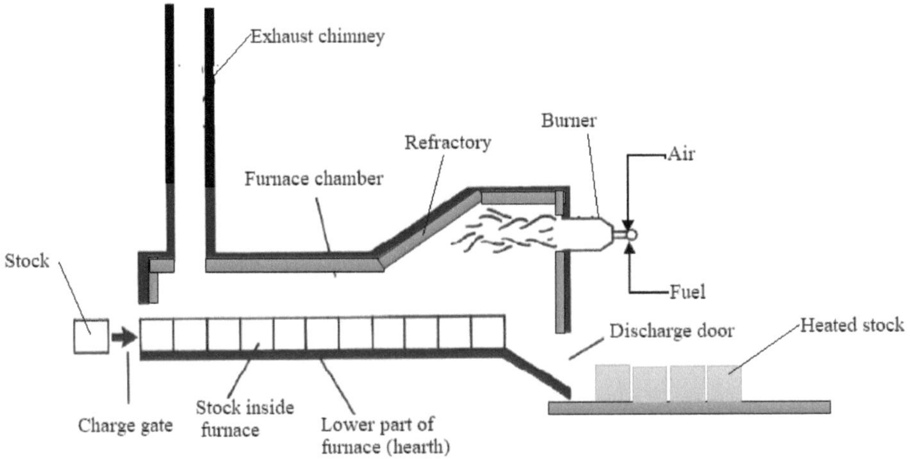

▲ **Fig-5.4(b): Furnace with different parts**

2.2 Different parts of a furnace

Following are the different parts of a furnace, fig-5.4(b).

a) Main heating space
b) Charge gate
c) Discharge gate
d) Refractory
e) Fuel burner
f) Exhaust chimney
g) Hearth

2.3 Heat losses in furnace

The losses in furnace described below for those where fuel are burnt as energy source.

For most heating equipment, a large amount of the heat supplied is wasted in the form of exhaust gases.

The furnace losses are, Fig-5.4:

a) Heat storage in the furnace structure

b) Losses from the furnace outside walls or structure

c) Heat transported out of the furnace by the load conveyors, fixtures, trays, etc.

d) Radiation losses from openings, hot exposed parts, etc.

e) Heat carried by the cold air infiltration into the furnace

f) Heat carried by the excess air used in the burners.

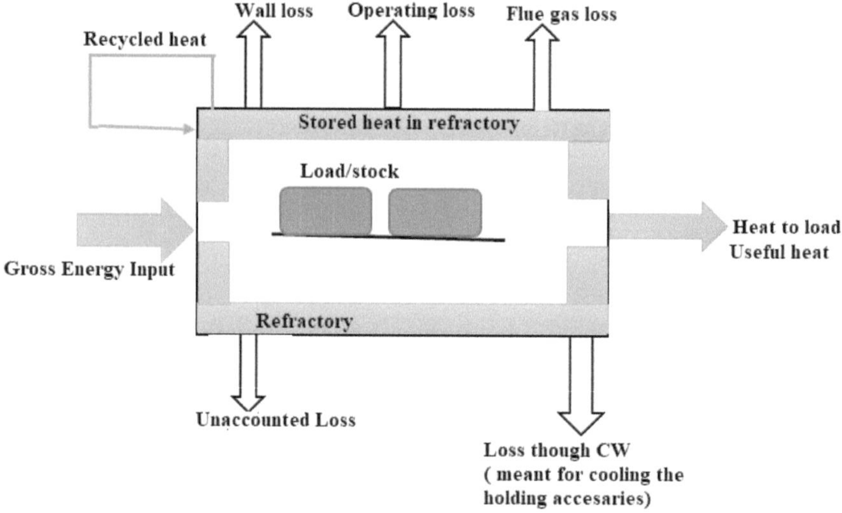

▲ **Fig-5.4: Furnace showing different losses**

2.4 Efficiency evaluation of furnace

i. Direct method

Thermal efficiency of process heating equipment, such as furnaces, ovens, heaters, and kilns is the ratio of heat delivered to a material and heat supplied to the heating equipment.

The efficiency of furnace can be evaluated as per below equation

Thermal efficiency of furnace (η):

= Heat content in stock/Total heat in fuel consumed for heating the stock

= [m x Cp (t_1 – t_2)]/ [q x GCV] x100%

When,

a) Q = Quantity of heat of stock in kCal [m x C_p x (t_1-t_2)]

b) q = quantity of fuel

c) GCV=Calorific value of fuel

d) m = Weight of the stock in kg

e) Cp = Mean specific heat of stock in kCal /kg°C

f) t_1 = Final temperature of stock desired, °C

g) t_2 = Initial temperature of the stock before it enters the furnace, °C.

ii. Furnace efficiency derivation by indirect method

By this method all the losses are evaluated and then deducted from input energy to find the efficiency of a furnace by indirect method. The derivation is explained using an example.

Example: Furnace Efficiency Calculation for a Furnace by indirect method

An oil-fired reheating furnace has an operating temperature of around 1340°C. Average fuel consumption is 400 litres/hour. The flue gas exit temperature is 750°C after air preheater. Air is preheated from ambient temperature of 40°C to 190°C through an air pre-heater. The furnace has 460 mm thick wall (x) on the billet extraction outlet side, which is 1 m high (D) and 1 m wide. The other data are as given below. Find out the efficiency of the furnace by both indirect and direct method.

a) Flue gas temperature after air preheater = 750°C

b) Ambient temperature = 40°C

c) Preheated air temperature = 190°C

d) Specific gravity of oil = 0.92

e) Moisture content=15%

f) Average fuel oil consumption = 400 Litres / hr = 400 x 0.92 =368 kg/hr

g) Calorific value of oil = 10000 k Cal /kg, H-content=11.2%

h) Average O2 percentage in flue gas = 12%

i) Weight of stock = 6000 kg/hr

j) Specific heat of Billet = 0.12 k Cal/kg/°C

k) Surface temperature of roof and side walls = 122 °C

l) Surface temperature other than heating and soaking zone = 85 °C

m) Specific heat of flue gas=0.24 K Cal/kg/°C

Solution

1. **Sensible Heat Loss in Flue Gas**

   Equation: Sensible heat loss = m x Cp x ΔT

   a) Excess air=12/(21-12)=133%

   b) Theoretical air required to burn 1 kg of oil = 14 kg

   c) Total air supplied = 14 x (1+1.33) =14 x 2.33 kg / kg of oil = 32.62 kg / kg of oil

   d) Weight of flue gas (m) = 32.62+1= 33.62 kg/kg of fuel

   e) Sensible heat loss = 33.62x0.24 x (750-40) kcal/kg of fuel=7161 kcal/kg of fuel

   Loss (%) =6206/10000x100=62.06.

2. **Loss Due to Evaporation of Moisture Present in Fuel**

   Equation: Loss (%)= M [584 + 0.45 (T_{FG} –T_{amb})]/GCV x100

   = [0.15x[584+0.46 x (750-40):10,000x100

   =1.36% [0.46-specific heat of steam at operating condition].

   When,

   a) M - kg of Moisture in 1 kg of fuel oil (0.15 kg/kg of fuel oil)

   b) T_{FG} - Flue Gas Temperature,°C

   c) T_{amb} - Ambient temperature,°C

   d) GCV - Gross Calorific Value of Fuel, K Cal/kg.

3. **Loss Due to Evaporation of Water Formed due to Hydrogen in Fuel**

   Equation: Loss% = 9 x H [584 + 0.45 (T_{FG} -T_{amb})]/GCVx100

   =9 x0.112x[584+0.45x(750-40)]/10000x100

   =9.1% [When, H – kg of H_2 in 1 kg of fuel oil (0.1123 kg/kg of fuel oil)].

4. **Heat Loss due to Openings:**

   If a furnace body has an opening on it, the heat in the furnace escapes to the outside as radiant heat. Heat loss due to openings can be calculated by:

   Equation: Heat loss= Blackbody radiation at operating temperature (kcal/cm²-hr) x emissivity (0.8) x radiation factor derived from shape factor(D/X) x opening area (cm²)

   Radiation loss is to be determined from curve, Fig- 5.5(a) &(b):

   a) Black body Radiant heat at 1340°C= 36 kcal/cm²-hr(fig:5.5a)

   b) Shape factor=Height(D)/ thickness (X) = 1x1/0.46=2.17; With respect to shape factor, radiation factor= 0.71 (fig: 5.5b)

   c) Opening area= 100x100 cm²

   Heat loss= 36x0.8x0.71x100x100 Kcal/hr=2,04,480 kcal/hr=204480/10000= 20.4 kg fuel/hr.

   Loss%=20.4/368x100=5.5%.

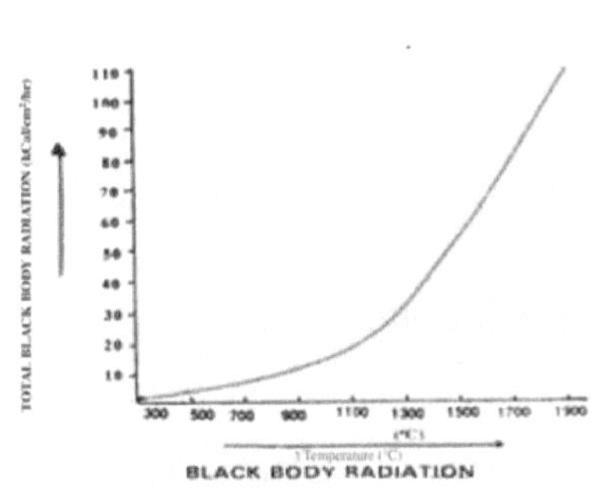

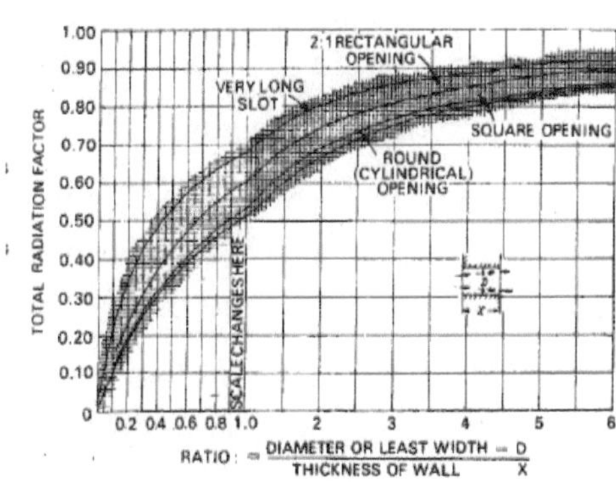

▲ **Fig-5.5 (a)&(b): Black body Radiation plot & Total Radiation Factor vs Ratio**

5. **Heat Loss through Furnace Skin:**

   A. **Heat loss through roof and sidewalls:**

   a) Total average surface temperature = 122°C

   b) Heat loss at 122°C (Refer Fig 5.6) = 1252 k Cal / m²- hr

   c) Total area of heating + soaking zone = 70.18 m²

d) Total heat loss = 1252 k Cal / m² / hr x 70.18 m²= = 87865 k Cal/hr

e) Equivalent oil loss (a) = 87865/10000=8.78 kg / hr

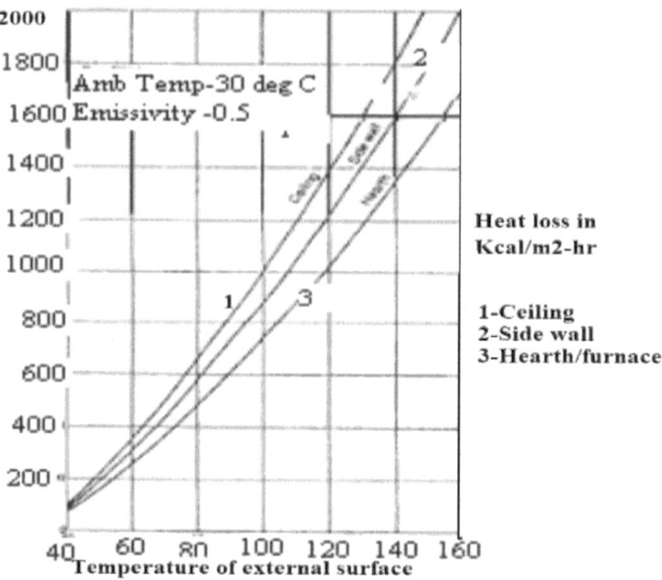

▲ **Fig-5.6: Heat Loss vs External Temperature Surface**

B. **Heat loss through wall skin, Fig-5.6.**

Total average surface temperature of area other than heating and soaking zone = 80°C

a) Heat loss at 80°C = 740 k Cal / m² / hr

b) Total area = 12.6 m²

c) Total heat loss = 740 x 12.6 = 9324 k Cal/hr

d) Equivalent oil loss (b) = 0.93 kg / hr

Total loss of fuel oil = a + b = 8.78 +0.93=9.71 kg/hr.

Total percentage loss = (9.71/368) x 100

= 2.64%.

6. **Unaccounted Loss:** These losses comprise of heat storage loss, loss of furnace gases around charging door and opening, heat loss by incomplete combustion, loss of heat by conduction through hearth, loss due to formation of scales.

7. **Evaluated Furnace Efficiency by Indirect Method:**

   a. Sensible Heat Loss in flue gas = 62.06%

   b. Loss due to evaporation of moisture in fuel = 1.36%

   c. Loss due to evaporation of water formed from H2 in fuel = 9.13%

   d. Heat loss due to openings = 5.56%

   e. Heat loss through skin = 2.64%.

   Total losses = 78.75%

   Furnace Efficiency = 100 – 78.75= **21.25%.**

8. **Furnace Efficiency Direct Method**

   Fuel input = 400 litres / hr = 368 kg/hr

   Heat input= 368x10,000 kcal/hr

   Heat output = m x C_p x ΔT= 6000 kg x 0.12 x (1340 – 40)=936,000 kcal/hr

   Efficiency = [936000 / (368 x 10,000)]x 100 = 25.43%.

   Losses = 74.6%.

2.4.1 Heat loss from surface of furnace body by calculation method

The quantity of heat release from surface of furnace body is the sum of natural convection and thermal radiation. This quantity can be calculated from surface temperatures of furnace.

i. Q1=Convection heat loss= a $x(t_1\text{-}t_2)^{5/4}$.

ii. Q2=Radiation loss=4.88 E $x[(T1/100)^4\text{-}(T2/100)^4]$. When, T1=$t_1$+273, T2=$t_2$+273

iii. Total heat loss (Q)=Q1+Q2

   When,

   a) Q- Quantity of heat released (k Cal/hr)

   b) a- factor regarding direction of the surface of natural convection: ceiling = 2.8, side walls = 2.2, hearth = 1.5.

   c) t_1 - temperature of external wall surface of the furnace (°C)

   d) t_2- temperature of air around the furnace (°C)

   e) E- emissivity of external wall surface of the furnace.

2.4.2 Derivation of efficiency for electric furnace

i. **Efficiency**

Energy supply in each phase/hr = V x I x pf kWatt-hr [V-voltage in KV, I-current in amp, pf-power factor]

Total Energy supply= 3x V x Ix pf kWatt-hr

Energy required to melt /ton of steel =Sensible heat + latent heat

= [1000x sp. heat (kcal/kg-°C) x [melting temp (°C)- ambient temp (°C)] + latent heat of steel Kcal/kg]/860 K Watt-hr.

Example: Melting one tonne of steel from an ambient temperature of 20°C. Specific heat of steel is (0.16 kcal/kg-°C), latent heat for melting of steel is (34.4 kcal/kg-°C). Melting point of steel = 1600°C. Actual Energy used to melt to 1600°C is 700 kWh/ton, Find efficiency.

a) Energy supplied= 700 kWh /ton of steel

b) Sensible Heat required = 1000x0.16x(1600-20) =2,52,800 kcal

c) Heat required for melting= 1000 x34.4 kcal=34,400 kcal

d) Total heat supplied= 252800+34400=287,200=287,200/860=334 kWh [1kWh=860 Kcal]

Efficiency= 334/700=47.7%

Typical thermal efficiencies for common industrial furnaces are given in Table-5.1

▼ **Table:5.1 Thermal efficiencies for common industrial furnaces**

Furnace Type	Typical thermal efficiencies (%)
1) Low Temperature furnaces	
a. 540 – 980 °C (Batch type)	20-30
b. 540 – 980 °C (Continous type)	15-25
c. Coil Anneal (Bell) radiant type	5-7
d. Strip Anneal Muffle	7-12
2) High temperature furnaces	
a. Pusher, Rotary	7-15
b. Batch forge	5-10
3) Continuous Kiln	
a. Hoffman	25-90
b. Tunnel	20-80
4) Ovens	
a. Indirect fired ovens (20°C-370°C)	35-40
b. Direct fired ovens (20°C-370°C)	35-40

ii. Specific Energy Consumption

Sp. Energy consumption: Quantity of fuel or energy consumed/ Quantity of material processed

Unit: Kg of fuel/kg of stock, Kcal/kg of stock or K Watt-hr/kg of stock

2.5 General Fuel Economy Measures in Furnaces

Corrective measures that will reduce loss in furnace operation are as below.

i. Complete combustion with minimum excess air
ii. Correct heat distribution
iii. Operating at the desired temperature
iv. Reducing heat losses from furnace openings
v. Maintaining correct amount of furnace draught
vi. Optimum capacity utilization
vii. Waste heat recovery from the flue gases
viii. Minimum refractory losses
ix. Use of Ceramic Coatings.

3.0 Fired Heater

A fired heater is an insulated enclosure that uses the heat generated by the combustion of fuels to heat fluids contained inside coils. Fired Heaters are also symbolically called as furnaces. They are mainly used in process industries such as refineries, petrochemical complexes, textile industries etc. to heat fluids up to the desired temperature. So, the main purpose of fired heaters is to raise the temperature of the process fluid that flows through the tubes. The heat energy is supplied by combusting fuels.

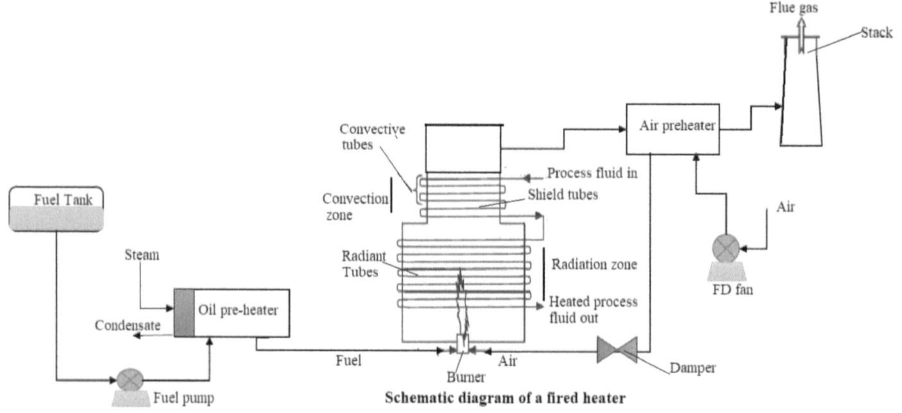

▲ **Fig-5.7: Forced draft fired heater**

3.1 Components of a Fired Heater

Fig-5.7 shows a schematic diagram for fired heater showing important equipment and auxiliaries.

A fired heater consists of:

a) Casing
b) Tubes
c) Return bends
d) Tube supports
e) Burners
f) APH/SAPH
g) ID & FD fans
h) Radiant, Shield, and Convection zone
i) Duct
j) Damper
k) Stack
l) Refractory
m) Louvers /Air registers

3.2 Types of Fired Heaters with Different Coil Arrangements

The type of heater is normally described based on structural configuration, radiant tube coil configuration and burner arrangement. Depending on the arrangement of tube banks and combustion chambers there are several types of fired heaters, **fig: 5.8,** that are used in industries. Some of the common types of fired heaters are:

a) Type A-Box heater with arbor coil

b) Type-B: Cylindrical with helical coil

c) Type-C: Cylindrical with horizontal tube

d) Type D-Box heater with vertical tube coil

e) Type E-Cylindrical heater with vertical coil

f) Type F-Box heater with horizontal tube coil.

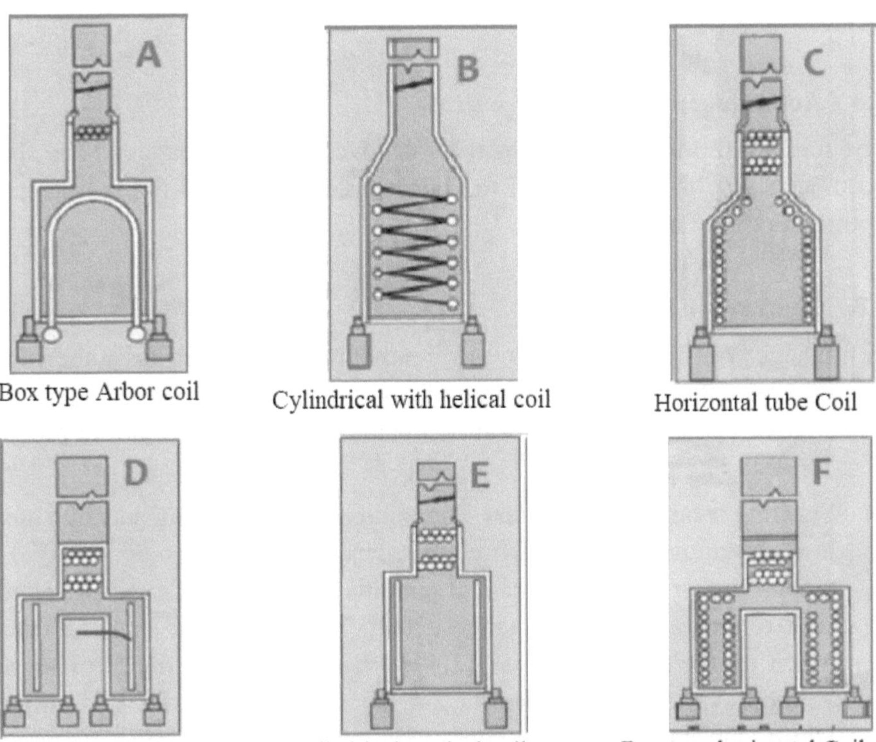

▲ **Fig-5.8: Types of Fire heaters**

3.3 Draft & Excess air Control Scheme

Draft and air are closely linked together & they should act together. The main objective should be achieving the optimum air level for the complete combustion of fuel.

a) Natural Draft fire heater- In this type of heater fuel gas or air is injected into the heater by using atmospheric pressure & the combusted gas is vented out through the stack. Natural Air flow into heater and exit of flue gas happens due to density differences as hot gases are having a lower density than the normal atmospheric air.

b) Force Draft fire heater- In this type of heater fuel gas or air is pushed into the heater by means of FD fan. The FD fan is installed before the furnace.

c) Induced Draft fire heater- In this type, the fired-heater ID fan is installed above the heater so that it can induce air through the combustion chamber into the burner. This fan causes a negative draft which pushes the burnt air out through the ventilation system.

3.4 Advantages of using Force draft

The forced draft system requires a lower level of excess oxygen. The flame becomes stable & small size of the burner is required. FD fan maintains an optimum ratio of air to fuel gas.

3.5 Features in fired heater

a) Bridge Wall Temperature- It is the temperature of the flue gas at the inlet to the convection section. The rate of heat transfer at the convection section is governed by the bridge wall temperature. It should be in the range of 760-900°C.

b) Snuffing steam in fire heaters-The main purpose of using snuffing steam is to extinguish unwanted fire (that can cause due to tube leakage) by preventing air ingress or prevent potential fuel from air exposure as well as it carries away heat to some extent. The amount of snuffing can be based on the requirement of 120-130 kg/hr per cubic meter of furnace volume. Normally LP steam at 2-4 bar steam is used for this purpose. During the start-up of the heater operation, snuffing steam is also used to remove combustible gas & excess air as well as create a negative draft.

c) Puffing-If the heater is not running properly, unburned fuel may accumulate in its combustion chamber. Upon starting the heating appliance, this oil can ignite and cause an explosion or misfire inside the furnace. This oil burner backfire, called a "puff back," forces soot or smoke through the heating exhaust system. It actually indicates a huge vibration of furnaces. If a burner is seriously out of fire, opening air control without reducing the firing rate can cause a hazardous situation called puffing. To prevent such a scenario first slow down the firing & then adjust the air louvers.

3.6 Start-up of fired heaters

i. Ensure that
   a) All the utilities are ready for operation.
   b) All instrument & safety devices are in operation.
   c) Fuel for the burner with sufficient operating pressure.
   d) Process fluid circulation is on.

ii. Purge combustible gas inside the furnace by snuffing steam.

iii. Maintain a negative draft of -5 to -15 mm WC in the radiant section by controlling fan damper or by snuffing steam for natural draft system controlling the louvers & the stack damper.

iv. Ignite the pilot burner and then the main burner.

v. Check the concentration of O_2 in flue gas and heater draft.

vi. The ramp of raising process fluid temperature at 30-50°C/hr to prevent over firing.

vii. Once the furnace has been brought up to a steady state, then switch the control mode from Manual to Auto mode.

3.7 Annual Maintenance

i. Tubes visual inspection prior to cleaning

ii. Inspection after cleaning

iii. Dimensional check-up (OD of a tube), thickness

iv. Visual inspection of header plug for any leakage

v. Inspection – tubes supports, hangers, etc.

vi. Inspection burner assemblies

vii. Inspection of refractory

viii. Inspection of explosion doors

ix. Dampers external, internal, operating linkages, etc.

3.8 Troubleshooting of fired heaters

Problem	Reason	Recommendation
High flue gas temperature	Fouling in convection section, Over-firing, tubes damaged	Clean convection section, Replace convection tubes, Reduce firing
High Fuel gas pressure	Burners are plugged	Clean burners
High-pressure drop in tubes	Coke formation High rate of vaporization	Decoking of tubes, Reduces the flow rate
Excess air	High furnace draft, Poor air-fuel mixing, Air leakage in the furnace	Reduce furnace draft Modify burners, Plug air leakage
Flame flashback	Low gas pressure	Raise fuel gas pressure
Burners go out	The gas mixture is too dilute	Reduce air.

3.9 Applications of Industrial Heater

Application of fired heater in industrial sectors are wide. Some of the key applications industries that utilize them in some form or the other are as below.

- Chemical Industries
- Petro chemical
- Oil & Gas Industry
- Fertilizer plant
- Food and beverage industry
- Steam super heating
- Textile Plants.

3.10 Efficiency assessment

i. Fuel Efficiency = Heat Absorbed/Total Heat from Fuel x 100

ii. Net Thermal Efficiency = ((Heat from Fuel + Air Heat + Fuel Heat)-(Losses due to moisture, H-burning& radiation + Stack Loss)) / (Heat From Fuel + Air Heat + Fuel Heat)

Most fired heaters have a thermal efficiency of around 85% – 90%.

Example: A Therminol heater's operating parameter are:

a) Fluid flow: 929 m^3/hr, density-0.71 kg/litre

b) Fluid inlet temperature-339°C, Outlet temp-374°C.

c) FO consumption-1538 kg/hr, GCV 10,500 kcal/kg

d) FG outlet temp-217°C

e) Sp. Heat of therminal is 0.62 kcal/kg-°C

f) O_2 in FG-1.7%

i. **Efficiency calculation by direct method**

a) Output =929x1000x0.71x0.62x(374-339)=14,313,103kcal/hr

b) Energy input= 1538x10500= 16,149,000 Kcal/hr

c) Efficiency= 14.313/16.149x100=88%

ii. **Indirect method**

a) Excess air=1.7/(21-1.7)=9%

b) Actual Air quantity=1538x14.1x1.09 kg/ hr =23,790 kg/hr

c) FG quantity= (23790+1538) kg/hr=25327 kg/hr

d) Loss through FG=25327x 0.26 x (217-40) kcal/hr=1,165,548.5 kcal/hr

e) Efficiency= (input-loss)/input= (16.149-1.068)/16.149=92.7%

f) Considering radiation loss, Moisture & H- loss = 5.4%

g) Actual efficiency= 92.7-5.4%=87.3%.

Question Review

1. What are the types of boiler based on fire & water flow?
2. What are the basic parts of a boiler?
3. Describe flue gas path & pollution control system in a boiler.
4. What are losses that are found out to derive boiler efficiency in indirect method?
5. How to deduce flash steam generation from boiler blow down?
6. Name at least five types of furnaces used in industry.
7. What are the losses in a furnace operation?
8. Write the equation to derive furnace efficiency by direct method.
9. What are different types of heater based on tube arrangement?
10. Define snuffing steam, puffing & bridge wall temperature.
11. Write in short for start-up of a heater.

SECTION

06 STEAM TURBINE, GAS TURBINE & HYDEL TURBINE

1.0 Steam Turbine

Steam turbine is a prime mover rotated by steam at high pressure and temperature. It is used to drive pump, turbine, fan etc. Its working, different parts involves and its efficiency etc. are discussed in this section.

1.1 Working Principle of Steam Turbines

A steam turbine works on principle of the "dynamic action of steam". Superheated steam passes through nozzles to generate high velocity which strikes the rotating blades that are fitted on a disc, mounted on a shaft. As that steam flows past a turbine's spinning blades, the steam expands and temperature drops. The energy of the steam is thus converted into kinetic energy in the rotating turbine's blades. Because steam turbines generate rotary motion, they are particularly suited for driving electrical generators for electrical power generation or driving a centrifugal pump or fan.

1.2 Function of Turbine blades- stages

Turbine blades are designed to control the speed, direction, and pressure of the steam as it passes through the turbine. For large-scale turbines, there are numbers of blades attached to the rotor, typically in different sets. Each set of blades helps to extract energy from the steam while keeping the pressure at optimal levels.

Blades are fitted in several rows to make it multi stage to reduce the pressure of the steam by very small increments during each stage. This, in turn, reduces the forces on them and significantly improves the overall output of the turbine.

1.3 Different parts of a steam turbine

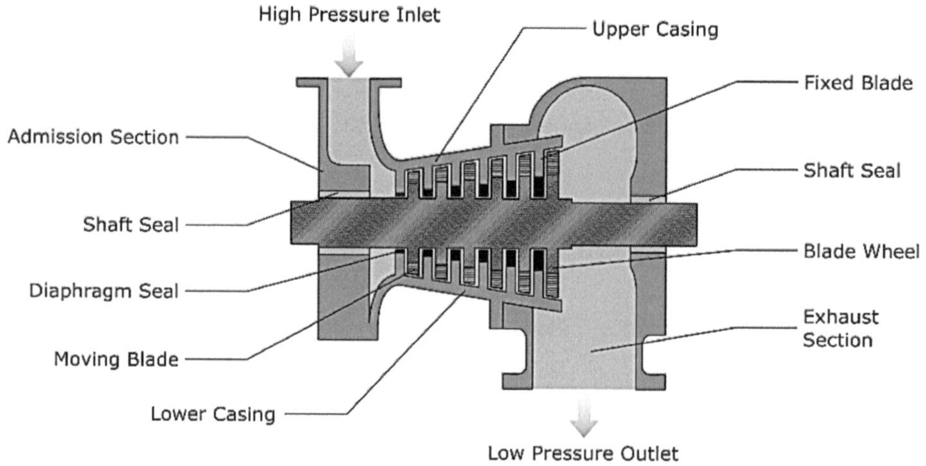

▲ **Fig-6.1: Schematic diagram of a turbine showing different parts**

Considering the design part there are various components of steam turbines like rotor, shaft, blades, casing, bearings, valves, nozzle, seals, trip system, governor system, casing etc. But only on 4 main parts of the steam turbine are discussed here, Fig-6.1.

i. **Rotor**- Rotor is the main rotating part of a steam turbine that is fitted inside a casing. The rotor consists of rows of moving blades and these blades rotate between the rows of fixed blades fitted on the casing. The high pressure steam flowing through the turbine passes through the fixed and moving blades alternatively.

   The mechanism is such that the fixed blade directs the steam at a right angle for entry into the moving blades. The utmost care is taken so that the rotor and its casing withstands thermal stress and the moving blades fitted with the rotor must securely withstand the high centrifugal forces.

ii. **Casing**- Casing encases the rotor and carries the fixed blades. Fixed blades act as nozzle or direct the flow path in the rotating blades; it also prevents steam leakage from the turbine. The ends of the casing, where the shaft of the rotor passes, are provided with sealing system to prevent steam leakages. Though leakage is not fully eliminated but can be controlled to a minimal amount.

iii. **Exhaust system-** Steam gets exhausted at a low pressure & temperature from turbine after doing necessary work on the rotor. The exhaust pressure is commonly at a vacuum or at a very low pressure. First kind of turbine is called condensing turbine and the other kind is called back pressure turbine.

iv. **Speed regulator-**The speed regulator regulates the overall turbine speed. The speed turbines are designed to operate at a rated speed called synchronous speed. At synchronous speed the efficiency of the turbine is maximum.

1.4 Support system of a turbine plant

The support systems are required to function turbine in a designed way and to prevent from any failure. It includes following subsystems.

i. **Lubrication system**: This system is for suppling lubricating oil for the bearings that supports the rotor also for absorbing the thrust force developed within the turbine. Main Oil pump (MOP), Auxiliary oil pump (AOP), Emergency oil pump (EOP) & Overhead oil tank (OHOT) are provided in lubrication system so that in no case lubrication oil flow in the bearing is interrupted.

ii. **Condensation & condensate system-** Exhaust steam is condensed in a water cooled or air cooled condenser to convert it to liquid water, called condensate. This liquid water is sent back to boiler through series of heat exchangers and heaters so that it can gain heat before it is fed to boiler.

iii. **Evacuation system**: To maintain a negative pressure in the exhaust, evacuation system is provided. Ejectors or vacuum pump continuously removes incondensable gases from condenser so as to maintain a negative pressure in the exhaust.

iv. **Gland sealing system-** This system is provided to prevent steam from escaping from turbine. Normally labyrinth sealing system is proved for the purpose.

v. **Governor-**Speed governor or governor is a speed-sensitive control system that is integrated with a steam turbine. Speed sensor, speed controller along with control valves control the speed of steam turbine. Speed sensor measures actual speed of turbine and sends to controller as feedback. In turn, controller generates a corrective signal against set point. The signal emanating from controller regulates the steam flow by operating the control valve hydraulically. The steam valve is linked to the servo motor system; the turbine speed is continuously compared with the reference speed that is set to the output signal of the servo motor. The speed is controlled by the quantity of inlet steam, extraction steam and exhaust steam of the turbine.

1.5 Types of steam turbine

Depending on their operating pressures, size, construction and lots of alternative parameters, there are two basic **varieties of steam turbines.**

a) Impulse Turbine

b) Reaction turbine

The main distinction in these turbines depends on the way the steam is expanded and action on blade while it passes through the turbine, **fig-6.2.**

i. **Impulse steam turbines:** In impulse turbine, thermal energy of steam is converted to a high speed steam through a fixed nozzle and strikes the blades mounted on the outer boundary of a rotor. While passing over the rotating blade, velocity changes and pressure remains fixed. The rotation of the turbine shaft is the due to change in momentum of steam. Examples of the impulse turbine are Brown-Curtis turbine, Curtis turbine and Rateau turbine. De Laval was the initial turbine having a single-blade wheel.

ii. **Reaction steam turbines:** In reaction turbine, as the steam passes over the blades, it expands in moving blades and velocity remains fixed. Fixed blade acts as nozzle and pressure drop takes place both in moving and fixed blades. In comparison to the impulse turbine, the pressure drop per stage is lower within the reaction turbine. Reaction turbines are more efficient. An example of this turbine is Parson's turbine.

A reaction turbine would need about double the quantity of blade rows than an impulse turbine for constant thermal energy conversion and this makes the reaction turbine much longer and heavier.

Typically, the degree of ***reaction and impulse*** from the blade root to its outer boundary takes place in turbine blades and based on such actions, modern steam turbines frequently use each impulse and reaction in the same unit, can be called as **impulse-reaction turbine**.

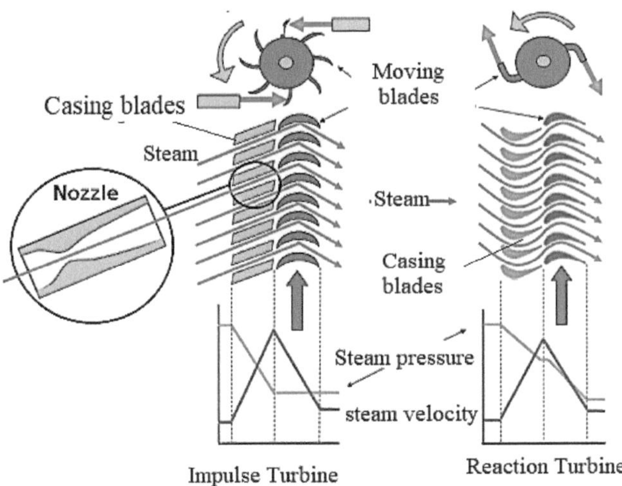

▲ Fig-6.2: Impulse & reaction turbine showing blade profile, variation of pressure & velocity

1.6 Highlighting features of impulse & reaction turbine

i. **Highlights of Impulse turbine**

   a) The blades mounted to the rotor are stricken by an impulsive force.

   b) Once the steam passes through the nozzles, it expands fully and its pressure remains constant.

   c) The blades are symmetrical in form.

   d) Speed is high in impulse turbine, since the speed of steam is high.

   e) The quantity of stages needed for manufacturing same power very less.

   f) High blade strength.

ii. **Highlights of Reaction turbine**

   a) The vector of reactive and impulsive force strikes the blades mounted to the rotor.

   b) The pressure cannot expand totally, only it passes through the nozzles and on the rotor blades, steam partly expands.

   c) The blades are asymmetrical in form.

   d) Because the steam speed is lower in reaction turbine, speed of turbine is also lower than impulse turbine.

e) To develop constant power, it needs additional stages.

f) The blade strength is lower compared to the impulse turbine.

1.7 Different configurations of steam turbine

Different configurations of steam turbine based on utility are shown in **fig: 6.3(a)& (b).**

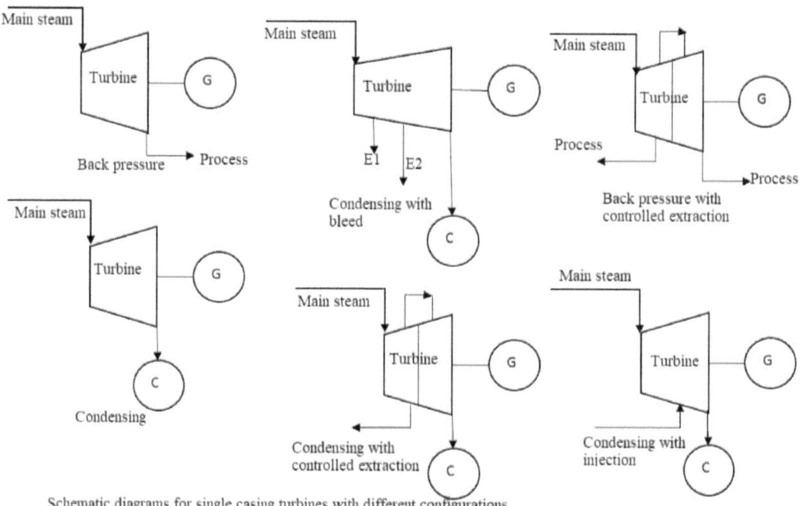

▲ **Fig-6.3(a): Single casing turbines with different configurations**

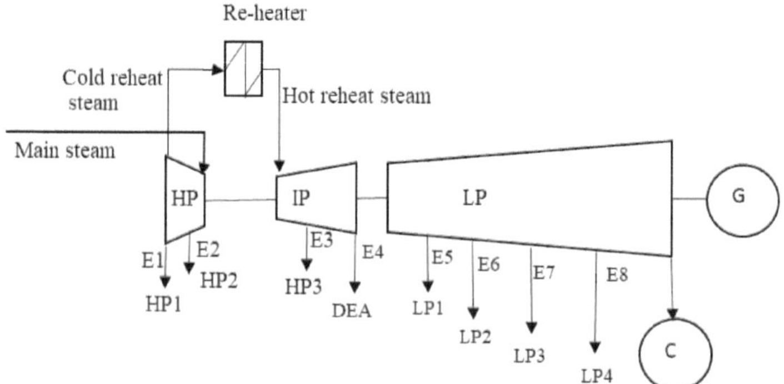

Schematic diagram showing Typical multi-casing Re-heat turbine configuration with several bleeds

▲ **Fig- 6.3 (b): Multi-casing Re-heat turbine configuration with several bleeds**

i. **Back pressure turbine-**Such configuration is called cogeneration plant, mainly used in process industry when both power and process steam at low pressure are required. Commonly back pressure is maintained at 4 bar.

ii. **Fully condensing turbine-** Low-capacity captive power plant (CPP) is configured for fully condensing turbine. When power interruption is not advisable in process and utility supply cannot be relied on. For uninterrupted supply of power, owner selects for such scheme.

iii. **Condensing with bleed-** This an energy efficient condensing turbine configuration. Feed water is preheated by bleed steam, extracted at different stages, called regenerative system, before feed water reaches to boiler. Though specific steam consumption for turbine increases partly, but overall power plant efficiency increases.

iv. **Condensing with controlled extraction-** This is a cogeneration scheme, when process steam requirement is at elevated pressure say, about 16/17 bar. Such scheme is very much flexible and is designed to take care of variation of both power and process steam requirement.

v. **Back pressure with controlled extraction-** Such cogeneration scheme is highly efficient. When process steam is required at different pressure levels say, 16/17 bar and 4 bar such scheme is adopted.

vi. **Condensing with Injection turbine-** This is a special type of configuration, when industry process generates steam and there is excess of steam after internal consumption. Turbine is started with its own source of steam, at high pressure and temperature, from power plant boiler. Steam from process is injected into the turbine in subsequent stage(s) to increase power generating capacity.

vii. **Multi-casing re-heat turbine-** Such scheme, **Fig-6.3(b),** is generally adopted in power utility sectors. To increase overall thermal efficiency of power plant, regenerative system includes bleeds from multi points of turbine. To increase heat input to turbine, steam is taken out from High pressure section (HP) and after reheating steam is allowed to re-enter into subsequent turbine section (IP). Overall thermal efficiency of such power plant remains about 45%.

1.8 Typical scheme of a thermal power plant

Important equipment and associated systems of a power plant are shown in **fig-6.4**. The function of each equipment and system are explained in brief as below.

i. **Boiler**- Required quantity of steam is generated in boiler using varied type of fuel starting from fossil fuel to bio-mass fuel. Additional super heater is placed in boiler, called re-heater, to reheat HP exit steam in bigger power plant.

ii. **Turbine & generator**- Turbine is the prime mover to rotate the generator rotor using steam energy. Generator magnetic field, which is rotating, is generated by supplying external source. Due to electro-magnetic induction, power is generated in stator.

iii. **Condenser and condenser cooling system**- It is provided for condensing turbine to condense the exhaust steam from turbine. Cooling water or air is used as coolant for condenser. Condenser using cooling water as coolant is called surface condenser where as, it is called Air Cooled Condenser (ACC) while using air as coolant.

iv. **Ejector system**- This system is provided to remove non-condensable gases generated in steam system. Motive steam is used in ejector or vacuum pump is provided to do the stated job, so as to maintain required vacuum in the system

v. **Condensate system**- Condensate generated in condenser is fed to deaerator after passing through series of heat exchangers & heaters such as ejector condenser & Vent steam condenser, drain cooler & LP heater.

vi. **FW system**- This system is comprised of feed tank (placed under deaerator), feed pump and series of HP heaters. Feed water is preheated in HP heater(s) before feeding in boiler to increase cycle efficiency.

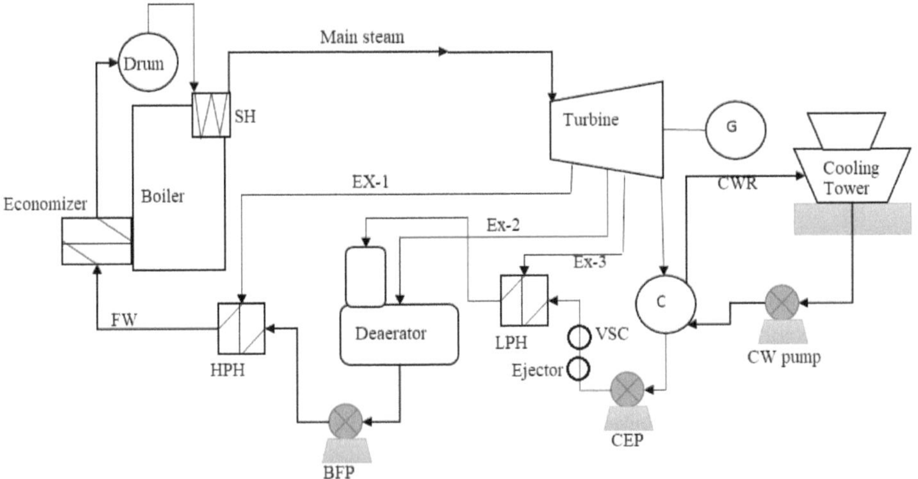

Schematic diagram for a thermal power plant with single casing turbine having three extractions

▲ **Fig-6.4: Thermal power plant with single casing turbine having three extractions**

1.9 Start-up of steam turbine

Following are the steps for Standard Operating Procedure (SOPs) of Steam Turbine which is connected to a generator.

I. **Pre-Heating Procedure of Steam Turbine**

a) CW system is in operation and charged in condenser and all other heat exchangers

b) Instrument air is available

c) Generator is ready in all respect

d) All interlock & protection system is in service

e) Lube oil circulation is on and turbine is on barring gear.

f) Condensate system is in service

g) Main steam line heating up completed

h) Normal vacuum is maintained

i) The casing drain of the turbine should be kept open.

II. **Start-Up Procedures of Steam Turbine:** Now most of the turbines are operated semi-automatically or automatically. Hence, during auto operation, turbine speed increases automatically as per sequence as described below.

a) Ready to start light is glowing
b) Initiate start command
c) See that turbine speed increases as per ramp up rate (500-600 rpm/ min for impulse turbine & 250-300 rpm/ min for reaction/ impulse-reaction turbine).
d) Holds at soaking speed for a pre-set time.
e) If there is a second soaking RPM, turbine holds again there.
f) Turbine ramps faster at critical speed zone(s) (double the normal ramping rate)
g) Finally, it reaches rated speed.
h) Close all casing drains that are connected to flash tank.
i) Check all parameters of turbine, if satisfactory, generator is ready for taking load.
j) Synchronize generator and load as per load increasing rate (normally 0.35-0.5 MW/ min).

III. **Operation monitoring**

a) Monitor the steam temperature & temperature.
b) Other parameters for oil & condensate system.
c) Temperature and vibration of all bearing points.
d) Abnormal sound, if any.
e) Monitor condenser vacuum system.
f) Monitor turbine ejector system works properly.
g) Keep normal the condenser level control system.
h) Monitor Turbine load or Steam Consumption.
i) Check the leakage from the stuffing box.
j) Adjust sealing system, if required.
k) Check the bearing lube oil flow and pressure.
l) Check lube oil, cooling water, pressure, and temperature.
m) Check control oil pressure & temperature.

IV. **Shutdown of Steam Turbine:** When a turbine operates at a rated speed and at full load, the following steps are to be adopted to shut down the turbine.

a) Reduce load of STG to 5-10%

b) Trip turbine directly by operating turbine trip switch.

c) See that, generator breaker opened, excitation system tripped and turbine speed is gradually decreasing.

d) Put on barring gear when turbine comes to a halt.

e) Keep LO system in operation for over 24 hrs, unless casing or main steam temperature drops below 60°C.

1.10 Performance and efficiency assessment for Turbine & thermal power plant

i. Heat rate & efficiency of Power plant

Performance of thermal power plant is assessed by following terms:

a) **STG Heat Rate (SHR)** - Quantity of heat required by STG to generate one unit of power is defined as STG heat rate. Unit for heat rate: Kcal/ kWh or, KJ/kWh.

b) **Plant Heat Rate (PHR)** - Quantity of heat required by the power plant to generate one unit of power. Unit for PHR is Kcal/ kWh or, KJ/kWh.

The equations for different Power plant configurations are differed as explained below.

ii. Heat rate & efficiency of STG

Notations:

a) Steam generation in Boiler-M

b) Main steam quantity –M_1

c) Main steam Enthalpy- H_1

d) Reheat steam quantity-M_2

e) RH steam enthalpy –H_2

f) Exhaust steam quantity- M_3

g) Exhaust steam enthalpy- H_3

h) Extraction/injection steam quantity-M_4

i) Extraction/injection steam enthalpy- H_4

j) Return condensate quantity-M_5

k) Return condensate enthalpy- H_5

l) Make up quantity-M_6

m) Make up water enthalpy- H_6

n) FW quantity- Mw

o) FW enthalpy- Hw

p) Boiler efficiency- ηb

q) Power generation – P

r) Quantity of fuel-Q

s) GCV of fuel-CV

a) **Heat rate for Condensing STG**

STG heat rate (SHR) = [Quantity of steam at turbine inlet x (enthalpy of steam –enthalpy of feed water)] ÷ power generation (motive steam quantity is also included).

$SHR=M_1 \times (H_1 - H_w)/P$.

b) **Reheat STG**

Heat rate for reheat STG

SHR= [(Quantity of steam at HP turbine inlet x (steam enthalpy– FW enthalpy) + Quantity of steam at IP turbine inlet x enthalpy)] ÷ power generation.

$SHR= [M_1 (H_1 - H_w) + M_2H_2]/P$.

c) **Heat rate for condensing cum extraction STG**

SHR=[Quantity of steam at turbine inlet x (enthalpy of steam –enthalpy of feed water)-quantity of extraction steam x enthalpy] ÷ power generation (motive steam quantity is also included).

$SHR= [M_1 (H_1 - H_w) - M_4 H_4] /P$.

d) **Heat rate for fully back pressure turbine**

SHR/THR= Quantity of steam at turbine inlet x (Enthalpy of inlet steam –enthalpy of back pressure steam) ÷ power generation. **(FW heating is not from turbine steam)**

$SHR= M_1 (H_1 - H_3) /P$.

e) **Heat rate for back pressure & controlled extraction STG**

SHR = (Quantity of steam at turbine inlet x enthalpy–Quantity of extraction steam x Enthalpy-Quantity of back pressure steam x enthalpy) ÷ power generation. **(FW heating is not from turbine steam)**

SHR= $(M_1H_1 - M_3H_3 - M_4H_4)/P$.

iii. **Plant Heat Rate (PHR) for condensing**

a) **PHR for condensing & reheat power plant**

PHR= SHR for condensing & reheat STG ÷ Boiler efficiency

PHR =SHR (condensing& reheat)/ η_b

iv. **Plant heat rate for Cogeneration plant**

a) **PHR condensing cum extraction**

PHR =SHR (condensing cum extraction)/ η_b

b) **PHR back pressure/back pressure cum extraction**

PHR= (fuel quantity x CV of fuel+ quantity of return condensate x enthalpy+ make up water x enthalpy) – (Quantity of extraction steam x Enthalpy+ quantity of exhaust steam x enthalpy)] ÷ power generation.

Plant Heat rate (PHR) = $(Q \times CV + M_5 \times H_5 + M_6 \times H_6) - (M_3 \times H_3 + M_4 \times H_4)]/P$.

v. **Efficiency evaluation**

a) Efficiency of STG = 860/Turbine Heat rate (THR) of STG

b) Efficiency of Power plant = 860/ PHR of power plant

c) Efficiency of Cogeneration plant = 860/ PHR of cogeneration plant

Example-1: A 43 MW power plant is operating at inlet steam at 89 bar & 515°C with a steam flow of 170 t/hr. FW at boiler inlet is 230°C, boiler efficiency is 87%. Derive the following performance values.

a) Heat rate & efficiency of turbine

b) Power plant heat rate and efficiency

Solution:

Enthalpy of steam at 89 bar & 515°C= 820 Kcal/kg, FW enthalpy=228 kcal/kg

STG heat rate=170 x (820-228)/43=2340 kcal/kWh

STG efficiency=860/2340=36.75%

Power plant heat rate= 2340/0.87=2690 Kcal/ kWh.

Power Plant efficiency=860/2690=32%.

Example-2: A 10 MW back pressure cogeneration plant configured with an inlet steam flow of 102 t/h at a pressure at 67 bar, 490°C, controlled extraction flow of 20 t/h at a pressure 12 bar & 310°C and back pressure at 4 bar and 220°C. Coal at 4400 kcal/kg is consumed at 20 t/h. Return condensate quantity is 90 t/h at 90°C and make up flow 15t/h at 30°C. Find -

a) Turbine heat rate & efficiency

b) Cogeneration heat rate & efficiency

Solution:

From steam table: Inlet steam enthalpy=808 kcal/kg, Extraction steam enthalpy =734 kcal/kg, Exhaust steam enthalpy=693 kcal/kg.

Turbine heat rate = 10^3x [102 x 808-20x734-693 x (102-20)]/10x10^3 =1090 Kcal/kWh

Turbine efficiency= 1/heat rate = 860/1090x100=78.8%

Cogeneration plant heat rate:

Make up quantity=15t/h, enthalpy 30 kcal/kg

Total heat input=Fuel heat+ Return condensate heat+ Make up water heat

=10^3 (20x4400+90x90+15x30) =10^3x96550 Kcal/hr

Total heat output= Through Controlled extraction + Through exhaust steam

=10^3 x (20x734+82x 693) =10^3 x 71506 Kcal/hr

Cogen heat rate= (96550-71506)/10=2504 Kcal/kWh

Cogen efficiency=860/2504x100=34.35%.

1.11 Factors affecting turbine efficiency

i. **Effect of inlet steam pressure:** Fig-6.5(a), (b) shows the effect of turbine inlet steam pressure on turbine performance. Figure indicates that increase in steam inlet pressure by 1 kg/cm^2 in condensing type turbine, steam consumption reduces in the turbine by about 0.3% and improves the turbine efficiency by about 0.1% respectively.

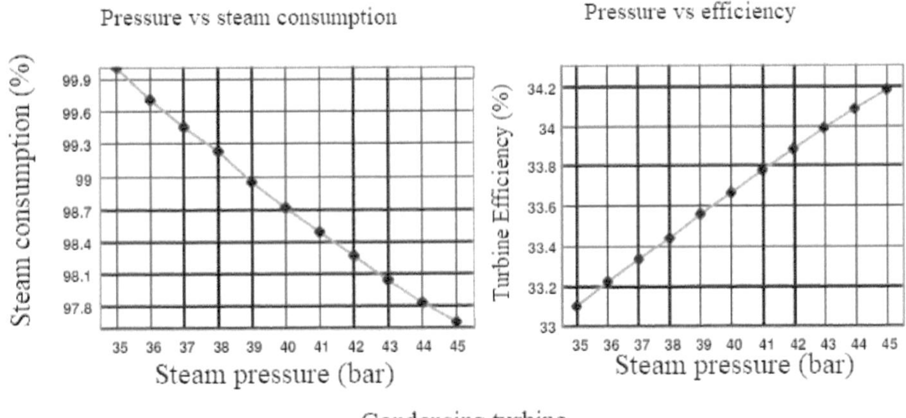

▲ **Fig-6.5(a), (b): Pressure vs steam consumption and pressure vs efficiency**

ii. **Effect of inlet steam temperature:** Enthalpy of steam is a function of temperature and pressure. At lower temperature, enthalpy is low, hence, work done by the turbine & its efficiency are also low. Fig-6.6 (a),(b) shows the effect of steam temperature on steam consumption & turbine efficiency keeping all other factors constant for the condensing type turbine. In other words, at higher steam inlet temperature, work done by the turbine is higher resulting higher output and reduction in steam consumption.

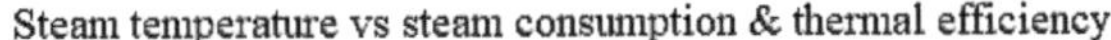

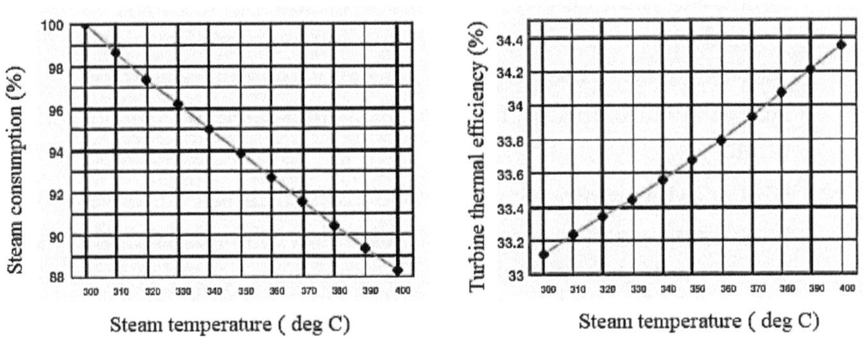

Typical effect of steam temperature for condensing turbine

▲ **Fig: 6.6(a), (b): steam temperature vs steam consumption &thermal efficiency**

iii. **Effect of exhaust pressure:** Fig-6.7 (a),(b) shows the effect of exhaust pressure on steam consumption and efficiency for a condensing turbine. Steam consumption increases by about 0.75% and efficiency decreases by 0.15% for increase in exhaust pressure by 0.01 ata. Few probable causes for reduction in exhaust pressure for condensing turbine are described below.

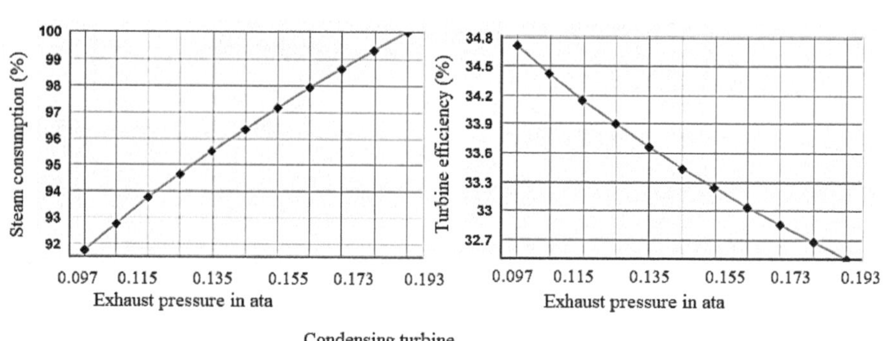

▲ **Fig-6.7(a), (b): Exhaust pressure vs steam consumption & efficiency**

a) **Vacuum ejector System-**Vacuum ejector system creates and maintains the vacuum in the surface condenser by removing the air/ inert gases. Removal of air/ inert gases is important, as accumulation of this hampers the performance of surface.

Motive steam condition shall be maintained as specified. Inter-after condenser shall be cleaned in the available opportunity, as they get choked due to foreign material coming with cooling water. Flange joints shall be tightened properly to avoid any ingress of air. Exhaust side of the turbine shall be properly steam sealed to avoid any ingress of air.

b) **Size of exhaust pipe-**In many condensing turbines it is observed that, the exhaust vacuum is much less than that of condenser. Mainly, it is due to the higher pressure drop in the exhaust pipeline from turbine exhaust to the condenser. In order to improve the vacuum at turbine exhaust so as to reduce steam consumption in the turbine, exhaust pipeline can be replaced with higher size. Hence, selection of exhaust pipe size is an important aspect for designing turbine system and it can be reviewed while stated problem is faced.

iv. **Part load operation:** During part load operation of STG, fixed losses become dominating resulting reduction in STG efficiency. The losses are: CV throttling, turbine internal loss, nozzle entry loss, fixed iron loss in generator etc. **Fig:6.8** shows efficiency variation at different load. Hence, maximum efficiency is achieved at rated capacity which is 80-90% of design capacity.

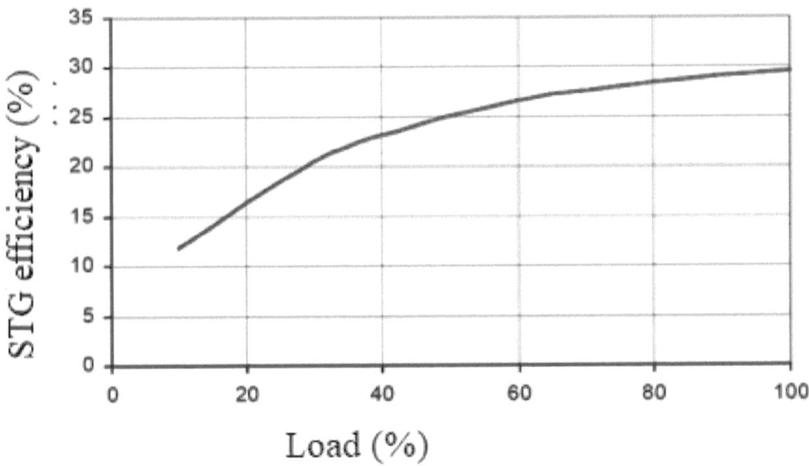

▲ **Fig-6.8: Diagram showing STG Load vs efficiency**

1.12 Losses in turbine

i. **Nozzle Friction Loss**-It is a very important loss for impulse turbine. When steam passes through the nozzles, friction loss and loss due to eddies occur. Friction occurs in the nozzle due to the factor of nozzle efficiency and it is the ratio of actual enthalpy drop to isentropic enthalpy drop.

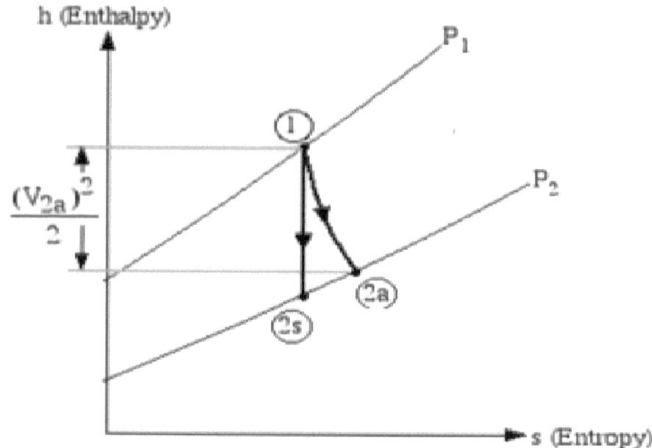

▲ **Fig-6.9: Velocity vs enthalpy drop**

From fig-6.9:

a) $h_1+V_1^2/2g=h_2+V_{2a}^2/2g$. Actual heat drop $(h_1-h_{2a}) = V_{2a}^2/2g$ (ignoring initial velocity, V_1)

b) $h_1+V_1^2/2g=h_2+v_{2s}^2/2g$.Isentropic heat drop $(h_1-h_{2s})= V_{2s}^2/2g$

$\eta_{nozzle} = (h_1-h_{2a})/ (h_1-h_{2s})= (V_{2a}^2/2g)/ (V_{2s}^2/2g)= V_{2a}^2/ V_{2s}^2$

Efficiency of nozzle can be expressed as:

η_{nozzle}=Actual KE/Isentropic KE=$\mathbf{(V_{2a})^2/(V_{2s})^2= (h_1-h_{2a})/(h_1-h_{2s})}$.

When,

a) h_1 & V_1 = Enthalpy & velocity of steam respectively at nozzle inlet

b) h_{2a} & V_{2a} = Actual enthalpy & velocity of steam respectively at nozzle exit

c) h_{2s} & V_{2s} = Isentropic Enthalpy & velocity of steam respectively at nozzle exit.

ii. **Blade Friction Loss**-This loss is crucial for both Impulse and Reaction turbines. Blade friction loss is due to the gliding action of steam over the blades and friction of the surface of the blades. The effect of turbine blades is considered as a blade velocity coefficient. The relative velocity of steam is reduced for this loss, Fig-6.10.

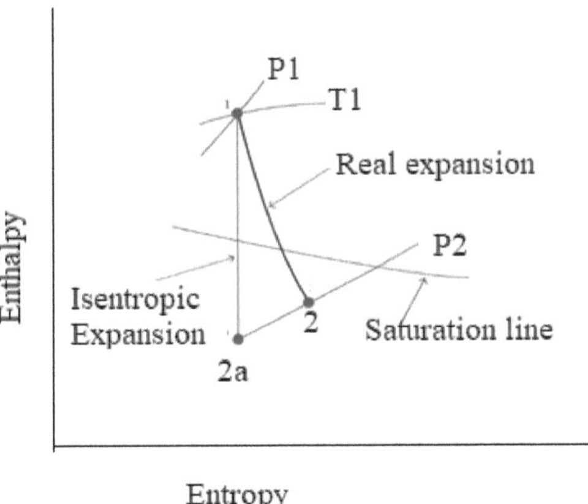

▲ **Fig-6.10: Enthalpy & Entropy curve**

Blade efficiency (η_b) =$(h_1-h_2)/(h_1-h_{2s})$x100%

Loss due to blade friction= (100- η_b)%.

iii. **Wheel Friction Loss**-When rotor rotates in steam space, steam produces some resistance on the turbine wheel, also called windage loss. As a result, it rotates at lower speed than expected. It is the loss in both Impulse and Reaction turbines.

The total blade frictional loss is about 10% of total turbine loss.

iv. **Losses due to mechanical friction**- Mechanical friction loss is due to the friction between the shaft and bearing. This loss may be reduced by proper lubrication of the moving parts of the turbine. This loss occurs both Impulse and Reaction turbine.

v. **Losses due to Leakage**-Leakage loss is different in both impulse and reaction turbines. In Impulse turbine, leakage loss occurs between the shaft & casing, nozzles and stationary diaphragms. For Reaction turbine, it may occur at the blade tips and shaft gland. This loss is due to the leakage of steam on each stage of the turbine. Total leakage loss is about 1 to 2% of total turbine loss.

vi. **Residual Velocity loss**- Residual velocity loss occurs when kinetic energy of steam leaves from the turbine wheel. Actually, steam leaves from the turbine with certain absolute velocity resulting loss in kinetic energy. Residual velocity loss can be reduced by multistage turbines. This loss is about 10 to 12% in a single stage turbine.

Let, inlet velocity of steam is V and the same quantity of steam is coming out at a velocity, v. Then, the energy absorbed by the turbine is $\frac{1}{2}mV^2 - \frac{1}{2}mv^2$. In other words, $\frac{1}{2}mv^2$ is the quantity out of the total energy is lost to the exhaust. This is residual velocity loss.

If it is expressed as a fraction $(\frac{1}{2}mv^2)/(\frac{1}{2}mV^2)$, then, residual velocity loss (%) = $v^2/V^2 \times 100$.

vii. **Loss in regulating valves**- Before entering the steam to the turbine, it passes through the boilers stop and regulating valve. Steam gets throttled in these regulating valves and as a result steam pressure will be less than the boiler pressure at the entry of turbine, fig-6.11.

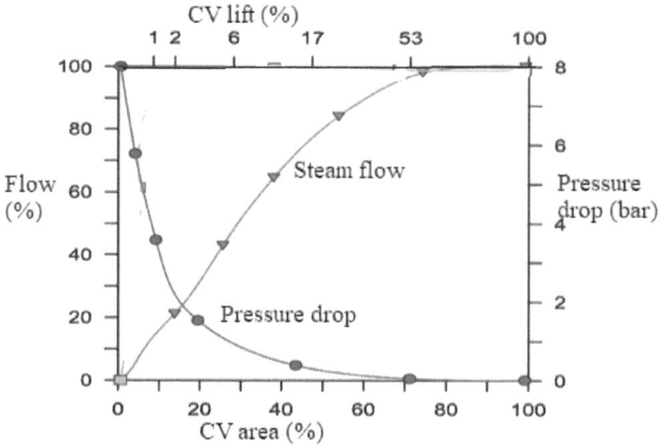

▲ **Fig-6.11: CV opening vs steam flow & pressure drop**

viii. **Loss due to wetness of steam**-This loss is due to the moisture present in the turbine. When steam passes through the lower stage of the turbine, it becomes wet. At the lower stage, the velocity of water and steam are different and do not form a homogeneous mixture resulting the velocity of water particle is less than that of steam. Water particles are dragged with the steam and some part of kinetic energy of steam is lost. This loss occurs both impulse and reaction turbine.

ix. **Radiation Loss**-This loss mainly takes place due to the temperature difference between turbine casing and its surrounding atmosphere. This loss can be minimized by the heavily insulated turbine. This loss is for both impulse and reaction turbine.

x. **Governing loss** -It is the loss in both impulse and reaction turbine and this loss is due to the throttling of the steam at the main stop valve of the governor.

xi. **Gland steam loss-** Gland seals, at the steam turbine inlet, is subjected to the highest steam pressure, causes leakage. This leakage loss causes the most notable reduction in turbine produced power; because the steam mass flow rate leaked through this gland-seal.

In running condition steam escapes through HP gland, called labyrinth. Initially steam is supplied externally to seal both HP & LP glands from preventing air entry into turbine. Once turbine is loaded, steam escapes from HP gland. While escaping though HP gland it does two functions such as a) balances the axial thrust, b) sealing. Steam coming out after balancing, is called balance leak off steam and connected in an intermediate section commonly in bleed section for deaerator and steam coming out from last stage of labyrinth is called gland leak off steam.

2.0 Gas Turbine (GT)

A gas turbine is a type of internal combustion engine that converts chemical energy into mechanical energy. The main elements of a typical gas turbine are:

a) Compressor
b) Combustor
c) Turbine.

The compressor, combustor, and turbine are called the core of the engine. These are usually mounted as an integral unit and operate as a complete prime mover on a so-called open cycle when air is drawn in from the atmosphere and the products of combustion are finally discharged again to the atmosphere.

2.1 Difference between steam turbine & gas turbine

Gas turbine	**Steam turbine**
Control is easier	Comparatively difficult
Quick start	Takes more time
Suitable for peak load	Suitable for base load
Space required per MW is less	More in case of steam turbine
Less weight	More weight
Boiler is not required	Boiler is must
Capital cost is less	More capital cost

2.2 Gas turbine working

The thermodynamic process used in the gas turbine is the Brayton cycle. Air and fuel are the two main inputs that are required for the working of a gas turbine. Fuel is usually natural gas, but other liquid fuels are also in use. First, air is drawn in from one end of the turbine and passed through a compressor section, Fig-6.12.

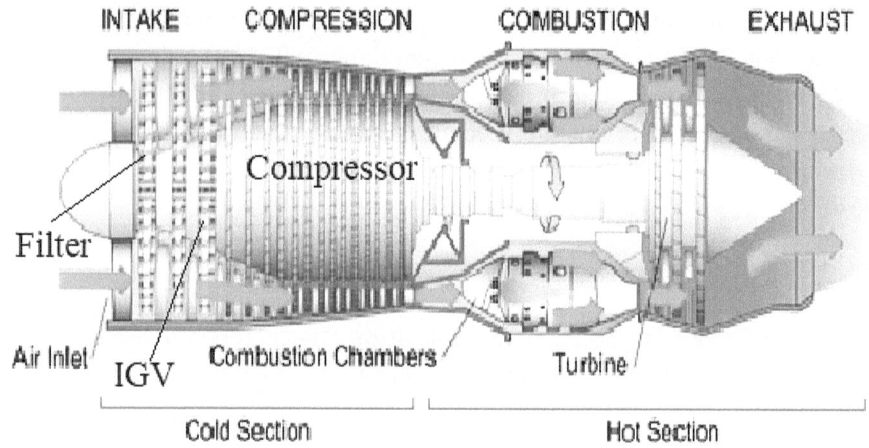

▲ **Fig-6.12: Different parts & sections of a gas turbine**

As it is compressed, the temperature of air rises and the pressure also increases. Next, fuel is injected into the turbine. Here, it mixes with the compressed air and starts to burn. The air to gas ratio (40-60) depends on various factors such as specific heating value of gas, moisture content and quality of air. The hot gas produced from the ignited mixture strikes the turbine blades, causing it to spin at very high speed. Thus, chemical energy has been converted into mechanical energy.

2.3 Gas turbine as cogeneration plant

Gas turbines are also widely used for electricity generation. They are an integral part of the combined cycle power plant, wherein, chemical energy is converted to mechanical energy and this mechanical energy is then converted into electrical energy. As explained before, the hot pressurized gas forces the blades to rotate at high speeds. This causes the drive shaft to rotate. This rotational energy is then transferred to the shaft of the generator through a gearbox.

The generator has a large magnet that is surrounded by stator coils. When magnet starts to rotate at high speed, a powerful rotating magnetic field is created which in turn causes the movement of electrons in the stator coil and thus electricity is produced.

The excess energy produced in the turbine in terms of exhaust gas can be used to produce steam which in turn can be used to run a steam turbine. Thus, the use of both the steam and gas turbines together would help in generating more power efficiently. Commonly exhaust energy is capable of generating power of about 50% of gas turbine capacity.

2.4 Gas Turbine Types

There are different types of gas turbine-driven generators. Gas turbines can be classified in several ways, common forms are:-

a) Aero-derivative gas turbines.

b) Light industrial gas turbines.

c) Heavy industrial gas turbines.

I. **Aero-derivative Gas Turbines:** Aircraft engines are used as 'gas generators', i.e. as a source of hot, high-velocity gas. This gas is then directed into a power turbine, which is placed close up to the exhaust of the gas generator. The power turbine drives the generator.

a) **Advantages of aero-derivative GT**

1. Easy maintenance since the gas generator can be removed as a single, simple module. This can be achieved very quickly when compared with other systems.
2. High power-to-weight ratio, which is very beneficial in an offshore situation.
3. Can be easily designed for single lift modular installations.
4. Easy to operate.
5. They use the minimum floor area.

b) **Disadvantages of Aero-derivative Gas Turbines**

The main disadvantages of Aero-derivative Gas Turbines are:

1. Relatively high costs of maintenance due to short running times between overhauls.
2. Fuel economy is usually lower than other types of gas turbines.
3. The gas generators are expensive to replace.

Aero-derivative generators are available in a single unit form for power outputs from about 8 MW up to about 25 MW. These outputs fall conveniently into the typical power outputs required in the oil and gas production industry, such as those on offshore platforms.

II. **Light Industrial Gas Turbines:** Some manufacturers utilize certain of the advantages of the aero-derivative machines, i.e. high power-to-weight ratio and easy maintenance.

The high power-to-weight ratios are achieved by running the machines with high combustion and exhaust temperatures and by operating the primary air compressors at reasonably high compression ratio above 7.

Less quantity of material is used and so a more frequent maintenance program is needed. Easier maintenance is achieved by designing the combustion chambers, the gas generator, and the compressor -turbine section to be easily removable as a single modular type of unit.

The ratings of machines in this category are limited to about 10 MW.

III. **Heavy Industrial Gas Turbines:** Heavy industrial gas turbines are usually to be found in refineries, chemical plants, and power utilities. They are chosen mainly because of their long and reliable running times between major maintenance overhauls. They are also capable of burning most types of liquid and gaseous fuel, even the heavier crude oils. They also tend to tolerate a higher level of impurities in fuels.

Heavy industrial machines are suitable for on-shore applications due to:

a) Lower power-to-weight ratio; means the supporting structures are larger and stronger.

b) Maintenance down time is usually much longer and maintenance is inconvenient due to the reason that the machine must be disassembled into many separate components.

c) The thermodynamic performance is usually poorer than that of the light and medium machines. This is may be due to low compression ratios in the compressor.

2.5 Fuel for Gas Turbines

The fuels usually consumed in gas turbines are either in liquid or dry gas forms and. in most cases, are hydrocarbons.

In special cases, non-hydrocarbon fuels may be used, but the machines may then need to be specially modified to handle the combustion temperatures and the chemical composition of the fuel and its combustion products.

Gas turbine internal components such as blades, vanes, combustors, seals and fuel gas valves are sensitive to corrosive components present in the fuel or its combustion products such as carbon dioxide, sulphur, sodium or alkali contaminants.

The fuels generally used are:

a) Low heating value gas.

b) Natural gas.

c) High heating value gas.

d) Refinery distillates.

e) Crude oil.

f) Residual oil.

2.6 Gas Turbine Operation – Troubleshooting

i. **Hydraulic oil** -In gas turbine operations, high pressure hydraulic oil operates the control valves on the fuel and air systems. If the hydraulic oil does not operate properly, then the control components will not operate properly, resulting in the gas turbine not operating properly. Thus, hydraulic oil system performance is crucial to proper gas turbine operation.

There are three areas of the hydraulic oil system that affect its performance:

a) **Pressure.** Electric motor-driven pumps produce high pressure oil for the hydraulic system (approximately 1200 psi pressure). Pump operation is important in achieving and maintaining pressure, making it vital for proper valve operation.

b) **Accumulator.** Pressure is maintained through transients by an accumulator, which is pre-charged and similar to a hydraulic spring. The accumulator is often overlooked by operators, but has an important role in maintaining pressure. For example, if there is a momentary lag in the motor-driven pump in producing pressure, problems can occur in the hydraulic oil system.

c) **Filters.** The hydraulic oil system filters can become plugged and create second-order problems, so filter condition is also important to proper operation.

ii. **Compressor-**The compressor function is fundamental to the performance of the gas turbine. Compressor performance dictates unit efficiency, fuel economy, combustion system performance and overall operational parameters throughout the gas turbine operation. The components/sub-assemblies of the compressor that impact its proper operation.

iii. **Impeller blades-**The rotating parts of are known as blades or vanes. Any kind of abnormality of vanes due to dirt deposition, damaged or worn result in decreased efficiency and performance. This is observed as a decrease in output pressure (compressor discharge pressure). Damaged or worn blading requires repair work to restore performance water or chemical wash systems can be used to periodically clean the compressor blades.

iv. **Turbine-**To maintain turbine performance, older gas turbines may utilize a turbine wash, similar in concept to the compressor wash. This, however, is only necessary when the unit is operating exclusively on liquid fuel or burning a lot of liquid fuel, which puts deposits on the blade units.

v. **Exhaust Pressure-**Exhaust pressure is desired to be as low as possible, with vacuum pressure being ideal. If exhaust pressure climbs too high, a safe shutdown may be initiated for necessary corrections.

vi. **Unit Vibration-**Seismic vibration probes are used on bearing housing to measure vibration. If vibration levels increase, it is a direct indication that machine performance is changing. Alarm and protection are provided for operational protection.

vii. **Degradation of blade profiles-** It can impact compressor performance over time or due to mechanical damage. Economical decision is to be taken for replacement of damaged blades.

2.7 Water wash systems

The two forms of water wash systems are Online and Offline. The online wash method is used during unit operation. If air temperatures allow, the online wash can be used by reducing unit load, opening up the Inlet Guide Vanes (IGV) and then spraying demineralized water into the air path at the compressor bell mouth. The dirt is washed off and pushed further down into the gas turbine engine and gets removed through the exhaust gas.

Relative to the online water wash, the offline method is a far more effective way to clean the compressor blades. It is used when the gas turbine is offline and cranking at slow speed. The cleaning fluid is water and chemical solvent (serving as the 'detergent') that is mixed into an emulsion and then sprayed into the compressor. When a deluge of emulsion is sprayed into the compressor, it travels all the way through the system since the unit is offline and not at high temperature, allowing the water to remain in liquid form rather than turning to steam as it eventually does in the online method. Thus, the detergent emulsion reaches all the stages of the compressor and cleans all of the compressor blades, by saturating the deposits to achieve more effective washing. The four steps of the offline water wash are:

a) Wash - Detergent spray

b) Soak - Detergent is allowed to soak and then loosen deposits

c) Rinse - Demineralized water washes the deposits down through the casing start drain valves. The rinse phase continues until the water begins to appear clean.

d) Dry - To dry the compressor, the unit is started as per normal start up procedure. Alternatively, the unit can be fired for 10-15 minutes to dry the water out of the system.

To achieve optimal levels of cleaning and more effectively restore compressor performance, a combination of periodic online and offline washing is recommended. Frequency of water wash depends on location-specific air quality.

The online water washes can be done on a more frequent basis (daily, weekly, bi-weekly, etc.) to prevent baking of debris. After an online water wash is completed, the compressor discharge pressure should increase to standard operating levels. If it does not, then the offline water wash system should be deployed to remove deposits and more comprehensively and thoroughly clean the unit. Offline water washes can also be completed routinely each time a planned shutdown is scheduled.

2.8 Efficiency assessment for gas turbine

Gas turbine works on Bryton cycle or Joule cycle. Figure-6.13&6.14 shows the components of a gas turbine in practical field. At point (1) cold air enters the compressor when it is compressed by shaft work to raise pressure at around 14 bar. At point 2 the pressurized air is fed into the combustion chamber where it is heated by combustion of fuel. At point 3 pressurized and high temperature gas is led to turbine. Gas expands in turbine and imparts shaft work in it. Part of shaft work of turbine is used for compressor and part of it is used to generate power.

At point 4 the exhaust gas is discharge to atmosphere.

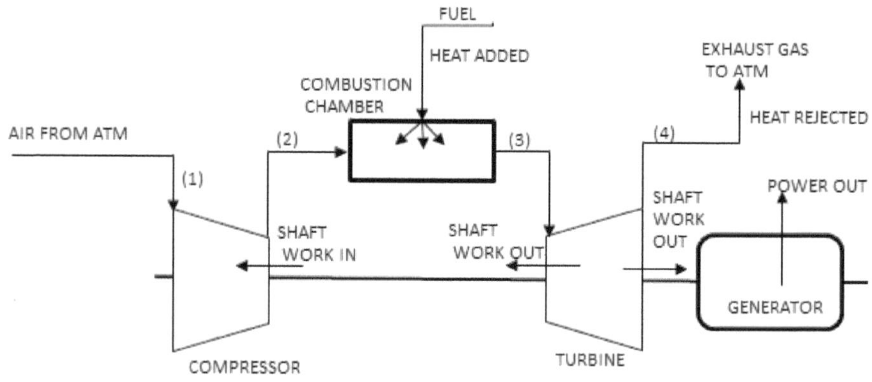

▲ **Fig-6.13: Open cycle gas turbine system**

The Bryton cycle can be explained in T-S & P-V diagram, fig-6.14.

a) Shaft work on compressor (W_C) = h_2-h_1 = C_p (T_2-T_1) kcal/ kg of air
b) Shaft work done by turbine (W_T) = h_3-h_4 = C_p (T_3-T_4) kcal/ kg of air
c) Heat supplied during the cycle ($Q_{supplied}$) = h_3-h_2= C_p (T_3-T_2) kcal/ kg of air
d) Heat rejected in atmosphere ($Q_{rejected}$) = h_4-h_1 = C_p (T_4-T_1) kcal/ kg of air
e) Efficiency = Power output/Energy input= (W_T - W_C) / $Q_{supplied}$
   = [C_p (T_3-T_4)- C_p (T_2-T_1)]/ [C_p (T_3-T_2)]
   = [(T_3-T_4)- (T_2-T_1)]/ (T_3-T_2).

It can be proved that overall gas turbine efficiency depends on pressure ratio.

a) Equation for Efficiency = $1-(1/r)^{(\Upsilon-1)/\Upsilon}$ [When, r = pressure ratio=$P_2/P_1=P_3/P_4$, $\Upsilon=C_p/C_v$, for air it is 1.25].
b) Overall gas turbine efficiency (combination of compressor and turbine) = Power output kWh x860/Fuel input x GCV of fuel (kcal/kg).

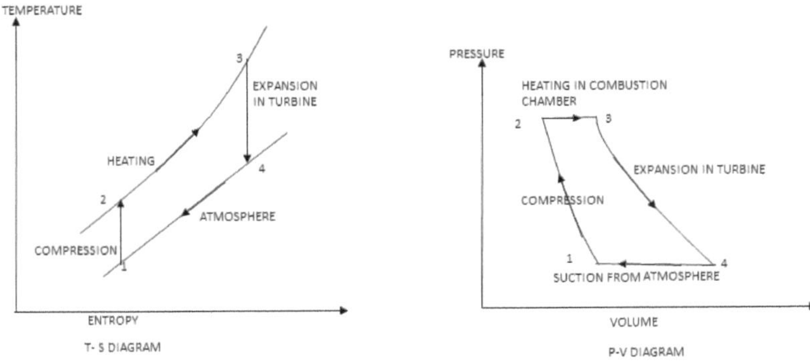

▲ **Fig-6.14: T-S & P-V curves**

2.9 Factors affecting efficiency of gas turbine

Following are the common factors that affect gas turbine efficiency.

i. **Air temperature and site Elevation**-Air is the working medium for gas turbine, hence, its performance is affected by the density and/or mass flow of the air intake to the compressor. Reference parameter for air pressure and temperature are 14.7psi/1.013bar and 15°C, Fig: 6.15 A&C.

   Inlet cooling- Lowering the compressor inlet temperature can be accomplished by installing an evaporative cooler or inlet chiller in the inlet ducting downstream of the inlet filters. Careful application of these systems is necessary, as condensation or carryover of water can exacerbate compressor fouling and degrade performance.

ii. **Humidity**- Humid air is less dense than dry air, therefore, for the same volume of air entering into the gas turbine, reduces mass flow rate. Hence, higher humidity directly affects the performance of gas turbines, reducing the power output, Fig-6.15(B).

iii. **Inlet and exhaust losses**- Provision of intake air filter, silencing, evaporative coolers or chillers at the inlet or heat recovery devices in the exhaust causes pressure losses in the system. These losses reduce GT efficiency.

   Typical effect of inlet air pressure drop & increase in exhaust pressure are as below:

   100 mm WC Inlet Drop effects:

   a) 1.42% Power Output Loss

   b) 0.45% Heat Rate Increase

   c) 1.1°C Exhaust Temperature Increase

100 mm WC Exhaust increase effects:

a) 0.42% Power Output Loss

b) 0.42% Heat Rate Increase

c) 1.1°C Exhaust Temperature Increase

iv. **Fuels-** Gas turbine is a mass flow engine and the mass flow is the sum of compressor airflow and the injected fuel flow. The heat energy of fuel is a function of the elements in the fuel and the products of combustion. Therefore, different fuels will hold different power outputs.

v. **Water & steam Injection-** Each machine and combustor is configured for water or steam injection with a limited quantity to protect the combustion system and turbine section. To achieve the desired NOx level, amount of water or steam injection is required; resulting increase in output because of the additional mass flow. It is assumed that steam is free from GT, therefore, on injection of steam heat rate improves. Since, it takes more fuel to raise water to combustor conditions than steam, water injection does not provide an improvement in heat rate.

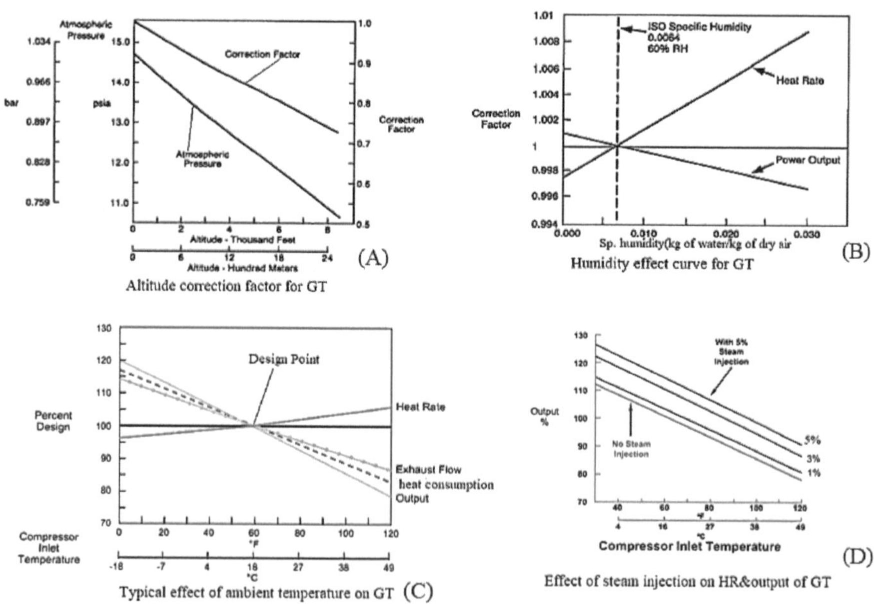

▲ **Fig-6.15(a)to(d): Different curves on Gas turbine generator**

3.0 Hydroelectric Power Plant

Generation of electricity by hydropower (potential energy in stored water) is one of the cleanest methods of producing electric power. Hydroelectricity is the most widely used form of renewable energy. It is a flexible source of electricity and also the cost of electricity generation is relatively low. The layout, basic components and working of a hydroelectric power station is discussed in this section.

3.1 Layout and Working of Hydroelectric Power Plant

The Fig- 6.16 shows the **typical layout of a hydroelectric power plant and its basic components**.

i. **Dam and Reservoir-** The dam is constructed on a large river in hilly areas to ensure sufficient water storage at height. The dam forms a large reservoir behind it. The height of water level (called as water head) in the reservoir determines how much of potential energy is stored in it.

ii. **Control Gate:** Water from the reservoir is allowed to flow through the penstock to the turbine. The amount of water, which is to be released in the penstock, can be controlled by a control gate. When the control gate is fully opened, maximum amount of water is released through the penstock.

iii. **Penstock:** A penstock is a huge steel pipe which carries water from the reservoir to the turbine. Potential energy of the water is converted into kinetic energy as it flows down through the penstock due to gravity.

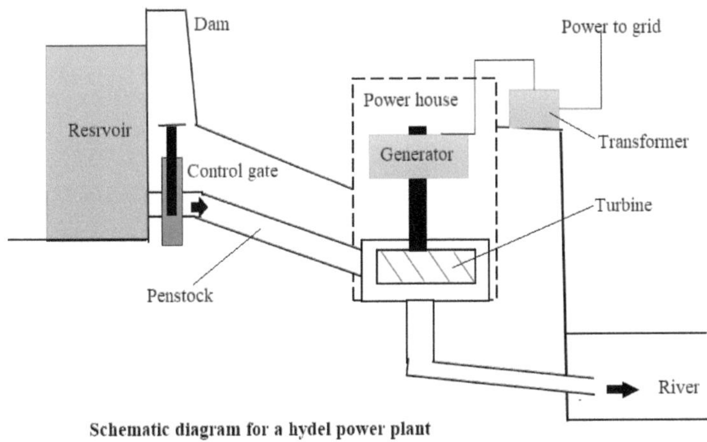

▲ **Fig-6.16: Hydel power plant**

iv. **Water Turbine:** Water from the penstock is taken into the water turbine. The turbine is mechanically coupled to an electric generator. Kinetic energy of the water drives the turbine and consequently the generator gets driven. There are two main types of water turbine;

a) Impulse turbine and

b) Reaction turbine.

Impulse turbines are used for large heads and reaction turbines are used for low and medium heads.

v. **Generator:** A generator is mounted in the power house and it is mechanically coupled to the turbine shaft. When the turbine rotates, it drives the generator and electricity is generated which is then stepped up with the help of a transformer for the transmission purpose.

vi. **Surge Tank:** Surge tanks, Fig-6.17, are usually provided in high or medium head power plants when considerably long penstock is required. A surge tank is a small reservoir or tank which is open at the top. It is fitted between the reservoir and the power house. The water level in the surge tank rises or falls to reduce the pressure swings in the penstock. When there is sudden reduction in load on the turbine, the governor closes the gates of the turbine to reduce the water flow. This causes pressure to increase abnormally in the penstock. This is prevented by using a surge tank, in which the water level rises to reduce the pressure. On the other hand, the surge tank provides excess water needed when the gates are suddenly opened to meet the increased load demand.

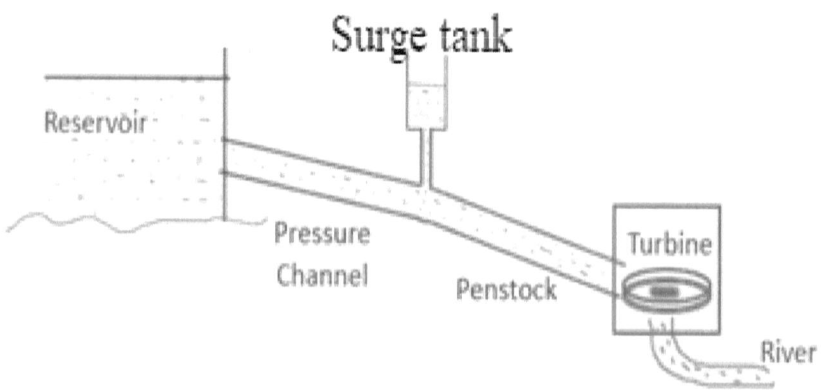

▲ **Fig-6.17: Surge tank**

3.2 Types of Hydro-Power Plants

I. **Conventional Plants:** Conventional plants use potential energy from stored water in a dam. The energy extracted depends on the volume and head of the water. The difference between height of water level in the reservoir and the water outflow level is called as water head.

II. **Pumped Storage Plant:** In pumped storage plant, Fig-6.18, a second reservoir is constructed near the water outflow from the turbine. When the demand of electricity is low, the water from lower reservoir is pumped into the upper (main) reservoir. This is to ensure sufficient amount of water available in the main reservoir to fulfil the peak loads.

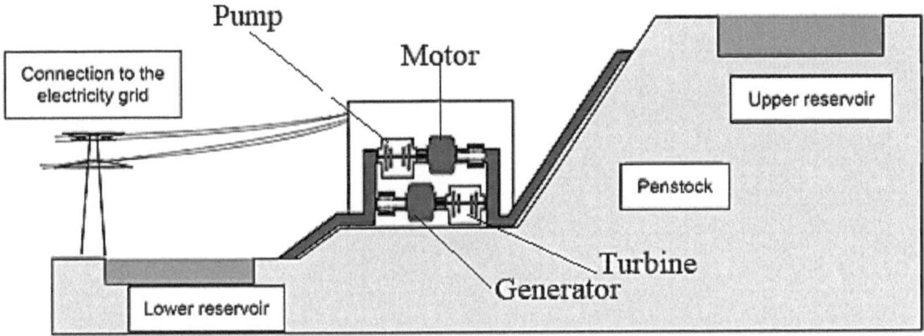

Schematic diagram of pumped storage hydel power plant

▲ **Fig-6.18: Pumped storage hydel power plant**

III. **Run-Of-River Plant:** In this type of facility, Fig-6.19, no dam is constructed and, hence, reservoir is absent. A portion of river is diverted through a penstock or canal to the turbine. Thus, only the water flowing from the river is available for the generation. Due to absence of reservoir, any oversupply of water is passed unused.

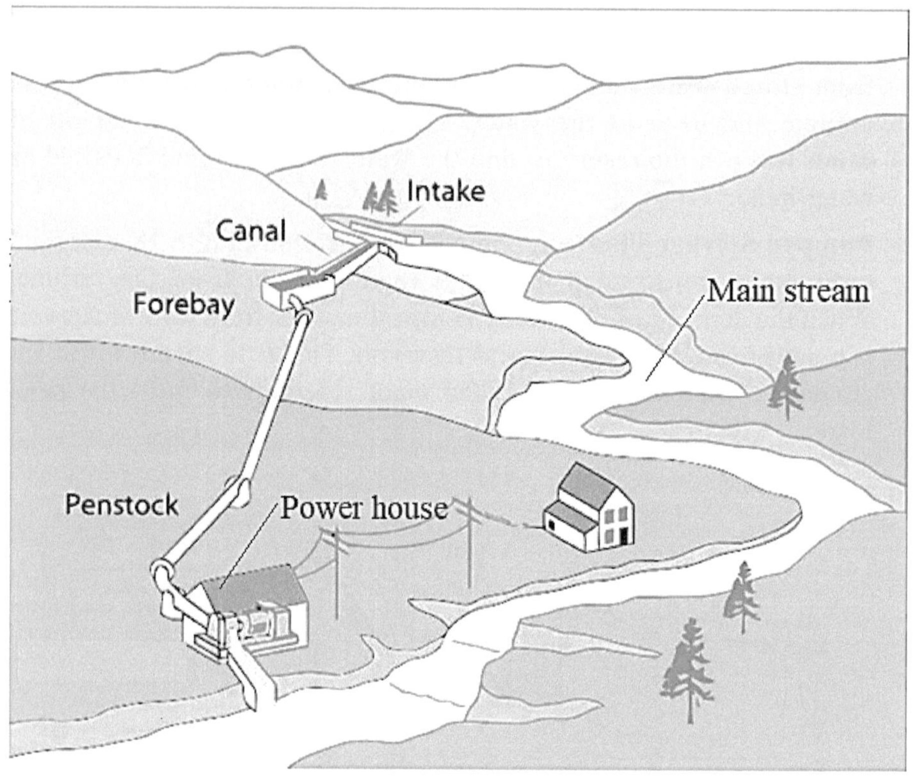

Schematic diagram for a typical Run-Of- River hydel power plant

▲ **Fig-6.19: Run-of-River hydel power plant**

3.3 Advantages and disadvantages of a Hydroelectric Power Plant

i. Advantages

a) No fuel is required as potential energy in stored water is used for electricity generation

b) Neat and clean source of energy

c) Very low power generation cost - as water is available free of cost

d) Comparatively less maintenance is required and has longer life

e) Serves other purposes too, such as irrigation

ii. **Disadvantages**

a) Very high capital cost due to construction of dam

b) High cost of transmission – as hydro plants are located in hilly areas which are quite away from the consumers

3.4 Hydro-Power Plant performance

In order to determine the hydro-power plant efficiencies, the following parameters are required.

i. **Net head**-The amount of head that is available at turbine inlet for useful work.

$$H_n = H - 0.1 \times H$$

When,

a) H_n- net head(m)

b) H-Gross head(m).

ii. **Hydraulic power**- Hydraulic Power (P_h) generated by flowing water is expressed as:

$$P_h = Q \times \rho \times g \times H_n$$

When,

a) Q=flow in m^3/s

b) ρ=density of water 1000kg/m^3

c) H_n=net head (m).

iii. Actual power generation is expressed as:

$$P = P_h \times \eta \; [\eta = \eta_t \times \eta_g]$$

When,

a) η- Plant efficiency

b) η_t- Turbine efficiency,

c) η_g- Generator efficiency.

iv. Capacity factor =Total power generation (MWh) for a given time/(Plant capacity x running hr).

v. Utilization factor = maximum power generated during a given time (P_{max} -MW)/Installed capacity (MW).

3.5 Factors affecting a hydel power plant

Following are the factors that affect hydel power plant

a) One of the most commonly occurring natural hazards impacting hydropower facilities are landslides.

b) The movement of large amounts of materials close to a site can lead to the burial of structures, weakening of foundations and dumping of large volumes of material into the water reservoir.

c) Less rain-fall results in low level in reservoir.

d) Excess rain fall- It has to be managed by releasing water through by-pass canal, to protect the dam.

Questions:

1. What are the different parts of a steam turbine?
2. Briefly discuss about support systems of steam turbine plant.
3. What are the different types of turbines based on steam action? Highlight features about any type of turbine.
4. Explain different configurations of steam turbine.
5. Briefly explain about cold start-up of a steam turbine.
6. What is reheat turbine, discuss advantages of RH turbine?
7. How to calculate turbine heat rate for non-reheat condensing turbine?
8. What is plant heat rate and plant efficiency?
9. What are the factors that affect turbine efficiency?
10. What are the losses in steam turbine- explain?
11. What are the differences between a steam turbine and a gas turbine?
12. What are the fuels used in gas turbine, explain effects of fuel in internal components?
13. Discuss in brief about washing of gas turbine.
14. Explain working of a hydro-electric power plant.
15. What are the different types of hydro-electric power plant- explaining in brief?

SECTION 07

DIESEL ENGINE, DG POWER PLANT AND COGENERATION PLANT

1.0 Diesel Engine

The diesel internal combustion engine differs from the gasoline powered Otto cycle by using a higher compression of the fuel to ignite the fuel rather than using a spark plug ("compression ignition" rather than "spark ignition"). Fig-7.1 shows the different parts of a diesel engine.

1.1 Working of a diesel cycle

In the diesel engine ideal working process, is described as below, Fig-7.1(A).

a) Air is compressed adiabatically (a-b) with a compression ratio typically between 15 and 20 that raises the temperature to the ignition temperature.

b) Fuel is injected to make the fuel-air mixture burn at a constant pressure, combustion process (b-c).

c) Then, an adiabatic expansion takes place as a power stroke (c-d).

d) Finally, a constant volumetric exhaust occurs (d-a).

e) A new air charge is taken into at the end of the exhaust, as indicated by the processes a-e-a.

Since the compression and power strokes of this idealized cycle are adiabatic, the efficiency can be calculated from the constant pressure and constant volume processes. The input and output energies and the efficiency can be calculated from the temperatures and specific heats.

Figure -7.1(C) shows the practical cycle, consisting of heat addition at constant volume and at constant pressure, two adiabatic and constant volume heat rejection.

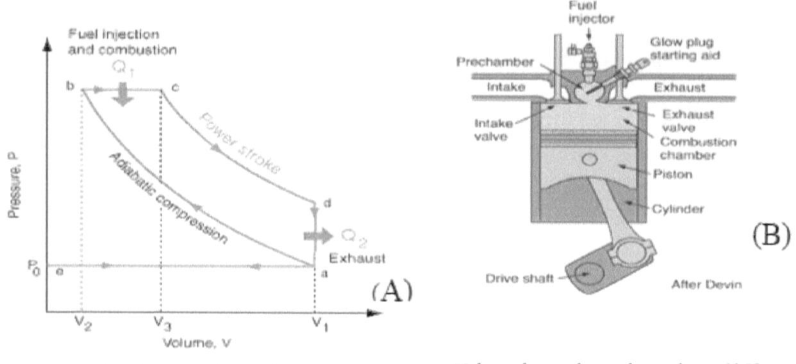

Ideal Diesel cycle

Diesel engine showing different parts

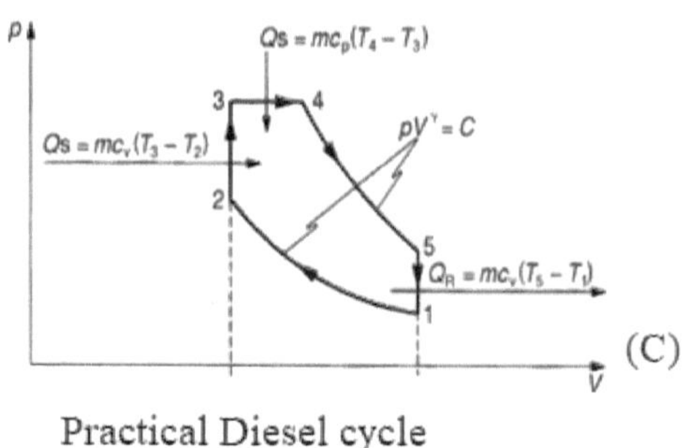

Practical Diesel cycle

▲ **Fig.-7.1: Diesel Engine & diesel cycles**

1.2 Efficiency calculation of a diesel engine

Efficiency calculation is shown through an Example:

A compression ignition engine working on the ideal dual combustion cycle has a compression ratio of 16:1. The pressure and temperature at the beginning of compression are 98 kN/m² (0.98 bar) and 30°C respectively. The pressure and temperature at the completion of heat supplied are 60 bar and 1300°C. Calculate the thermal efficiency of the cycle. The nominal values used for air at 30+273=300 K are $C_P = 1.00$ kJ/kg- K, $C_v = 0.718$ kJ/kg-K, $\gamma = 1/0.718 = 1.40$.

Solution:

a) $P_2 = P_1(V_1/V_2)^{\gamma} = 0.98x\ (16/1)^{1.4} = 47.53$ bar.

b) $T_2 = T_1(V_1/V_2)^{\gamma-1} = 303x\ (16/1)^{0.4} = 918$ K.

c) $P_2V_2/T_2 = P_3V_3/T_3$, $T_3 = 60\times918/47.53 = 1159$ K. $[V_{2=}\ V_3]$.

Now,

$P_3V_3/T_3= P_4V_4/T_4$, hence, $V_4=V_3\times1573/1159=1.358$ V3. [$P_3 = P_4$]

$P_4V_4^{\gamma}=P_5V_5^{\gamma}$, hence, $P_5=P_4(V_4/V_5)^{\gamma}=60\text{x}\ (1.358V_3/16V_3)^{1.4}=1.899$ bar.

And, $T_5=T_4(V_4/V_5)^{\gamma-1}= 1573(1.358Vc/16Vc)^{0.4}=586.5$ K [given T4=1300+273=1573 K]

a) Heat energy supplied=m x C_v (T_3-T_2) + m x C_p (T_4-T_3).
   =0.717x (1159−918) +1.004 x (1573−1159) [using a mass of 1 kg]=173+416=589 kJ/kg.

b) Heat energy rejected=m x C_v x (T5−T1) =0.717(586.5−303) =203.3 kJ/kg.

c) Air standard efficiency=1−heat rejected/heat supplied=1−203.3/589=0.6548=65.48%.

1.3 Factors affecting the Engine Performance

The factors due to which the indicated power developed by actual engines differs from that of ideal engines are as follows:

i. The working media is not air but a mixture of air and fuel in case of actual engine.

ii. The chemical composition of working media changes during combustion.

iii. The process of combustion is never at constant volume or at constant pressure.

iv. The process of compression and expansion is not adiabatic.

v. The specific heats of gases of working media vary considerably with temperature.

vi. The combustion may be incomplete.

vii. The residual gases change the composition, temperature, and actual amount of fresh charge.

viii. The amount of fresh charge is decreased due to pumping losses.

1.3.1 Heat Transfer & heat loss in diesel engine

i. **Heat transfer:** Heat transfer takes place between the gases &engine cylinder walls and the other parts of the engine coming in contact with the gases as described below.

   a) During combustion, expansion, exhaust, and the later part of the compression.

b) Heat transfer takes place from the gases to the engine walls and from the engine wall to the cooling water or ambient air.

c) During suction and the earlier part of the compression, heat transfer takes place from the walls to the gases.

ii. **Heat loss**

a) The heat lost to the walls during latter part of compression is almost equal to the heat received by the gases from the walls during early part of compression.

b) Amount of heat lost during exhaust stroke is unavailable and it is unavoidable.

c) The heat lost during combustion and expansion lowers the thermal efficiency of the engine.

iii. **Factors responsible for heat loss:** The factors that affect the heat losses to the walls are as follows:

a) **Duration of combustion of the charge:** This increases the heat loss.

b) **Temperature of combustion:** Temperature of combustion depends upon the type of fuel, compression ratio and the load on the engine. It increases with load and compression ratio, hence, increases the thermal loss.

c) **Speed of the engine:** Increase of the engine speed decreases the duration of combustion hence, decreases the heat loss.

d) **Shape of the combustion space:** Increase in ratio of combustion chamber surface area to volume decreases the heat loss. However, turbulence and flame propagation also effect the heat transfer to combustion chamber wall.

e) **Size of the cylinder:** An increase in the cylinder size decreases the ratio of surface to volume but increases the frame travel. This increases the combustion duration and hence, engine speed is decreased.

f) **Ignition timing and fuel injection timing in C.I. engines:** Proper ignition and injection timings give rise in quicker combustion with less after burning heat loss. The heat flow from the walls to fresh charge during suction stroke increases the temperature of the charge hence, decreases **the quantity of charge**. This decreases the power that the engine can develop.

1.3.2 Residual Gas

a) Residual gases, left in the compression space from the previous cycle, dilute the fresh charge by increasing the amount of inert gases in it. This affects the ignition and combustion.

b) Residual gases also lower the volumetric efficiency of the suction stroke and raise the temperature of the charge results lowering the amount of fresh charge induction.

1.3.3 Valve Resistance

In theoretical cycle of four-stroke engines, it is assumed that the exhaust and intake pressure are equal to atmospheric. But practically, the exhaust pressure is higher and the suction pressure is lower than atmospheric pressure due to the resistances in exhaust and intake manifolds and valves. The valve resistance affects the volumetric efficiency and causes pumping losses. The pumping losses increase with an increase in speed.

1.3.4 Valve Timing

In ideal cycle it is assumed that opening and closing of intake and exhaust valves take place on dead centres. In actual case, the exhaust valve closes and intake valve opens approximately on TDC, but the opening of the exhaust valve and the closing of the intake valve vary considerably from the BDC, depending principally on the desired speed. The net result due to deviations of valve opening and closing other than at dead centres is that the indicator diagram is rounded at the exhaust corner. This reduces the work output by 1 to 2%.

1.3.5 Combustion Time

In ideal cycle it is assumed that, the time of combustion is zero for constant-volume process and combustion occurs at a rate necessary to maintain constant pressure during the constant pressure process. Actually, combustion process requires an appreciable amount of time, which depends upon various factors. The increase in the combustion time decreases the ideal efficiency by 2 to 3%.

1.3.6 Incomplete Combustion

A volumetric analysis of the constituents of the products of combustion indicates an incomplete combustion that amounts to about 2% of the heating value of the fuel. Mixture with excess air tends to reduce this loss to zero; on the other hand, rich mixtures result in considerable unburnt fuel due to oxygen deficiency.

1.3.7 Atmospheric Conditions

The temperature of air, humidity of air and barometric pressure affect the air charge. The weight of the air charge found to be inversely proportional to the square root of the temperature, especially in high-speed automobile engines. For obtaining the performance at the standard conditions, the following corrections on pressure, temperature and humidity are to be adopted.

Barometric correction (bp) correction = BP x $\sqrt{[(273+t)/(273+25)]}$x 760/(P-Pv)

When,

a) t- test temperature, 25°C- design temperature

b) P- measured atmospheric pressure in mmHg.

c) Pv- Vapour pressure at t°C.

1.4 Performance Curves for Diesel engine

The performance for a combustion engine is generally used for designating the relationship between:

a) power

b) speed

c) fuel consumption

In variable speed engines (like automobile engines), the rated power at a particular speed does not provide enough information. Under such situations, the performance curves help to obtain necessary information. Typical performance curves for internal combustion engine for diesel engines used in automobiles is shown in Fig-7.2. It is a common practice to give the automotive diesel engine three ratings:

a) Maximum rating for short intervals of operation

b) Rated output for larger period of operation than the former

c) Continuous output rating for operation with no time limit.

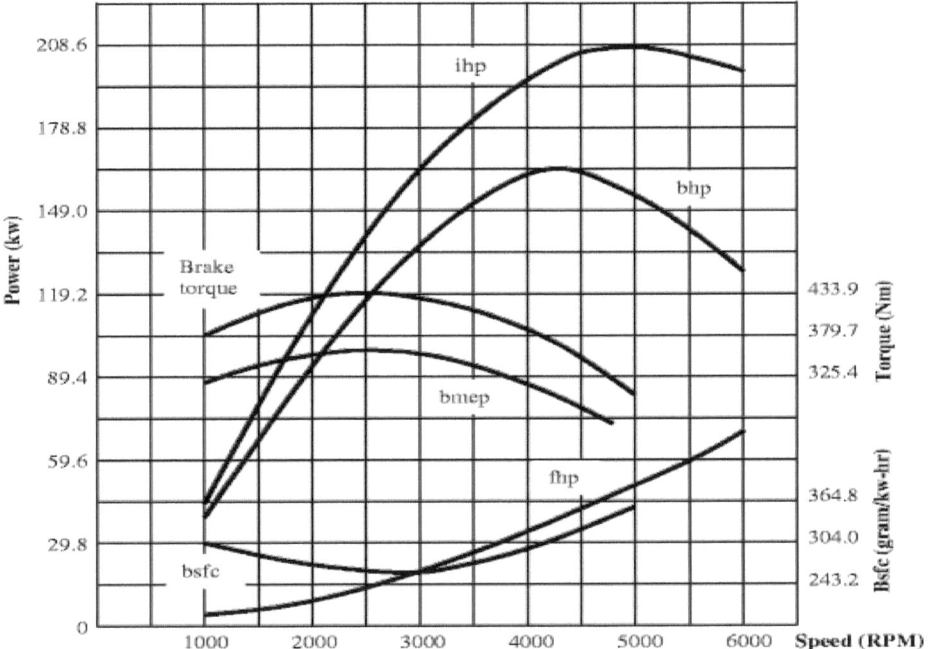

Typical performance curve for CI engine (diesel)

▲ **Fig-7.2: Performance curve for CI engine**

1.5 Energy Losses (Heat Balance) in diesel engine

Only a part of the energy supplied to the engine is transformed into useful work where as the rest is either wasted or utilized for heating purposes. The main part of the unutilized heat goes to exhaust gases and to the cooling system. In order to draw a heat balance chart for an engine, tests should be conducted to give the following information.

i. Energy supplied to an engine which is known from the heating value of the fuel consumed.

ii. Heat converted to useful work.

iii. Heat carried away by cooling water.

iv. Heat carried away by exhaust gases.

v. Heat unaccounted for (radiation etc.)

Description	(%) for Diesel Engine
Heat converted to useful work (IHP)	36 - 45
Heat carried away by cooling water	28- 35
Heat carried away by exhaust gases	20-29
Heat unaccounted for	5-7
Total (= Energy supplied)	100

1.6 Performance derivation of DG power Plant

i. Notations

a) Pm -Mean Effective pressure – (Pascal-Pa, 1 bar = 10^5 Pa)

b) L-Stroke length (m)

c) A -Cylinder Area- (m^2)

d) N- Rotation per Minute (RPM)

e) n- Rotation/ second (RPS, RPS=RPM/60)

f) Q- fuel flow (kg/s)

g) CV-Calorific value of fuel (Kcal/kg or KJ/kg, 1 kcal=4.18 KJ)

h) Va- Actual cylinder volume (m^3)

i) Vs- Swept volume (m^3)

j) r- Pressure ratio

k) X- number of Cylinder

ii. Definition

a) **Indicated Horse power (IHP)**: It is computed as work done which is equal to force multiplied by swept volume.

IHP= Pm x Lx A x n, for 2 stroke engine;

IHP=Pm x Lx A x n/2, for four stroke engine.

For multi Cylinder, the equations are to be multiplied by X.

b) **Brake Horse Power (BHP):** BHP is derived after deduction of Friction Horse Power (FHP).

FHP is the losses due to friction, resistance in fly-wheel rotation, power required to drive fuel pump, LO pump etc. FHP is 15-30%.

Hence, BHP = IHP- FHP.

Also, BHP =Torque (T) x2π x n. [T = Force x r = F x r [F-force, r-Torque arm radius]

c) **Brake Mean effective pressure (BMEP):** It is computed by using basic engine work (BHP) equation and expressed as:

Pm =BHP/ (L A n), for 2 stroke engine

Pm = 2xBHP/ (L A n), for 4 stoke engine.

d) **Mechanical Efficiency:** it is derived as: BHP/IHP.

e) **Thermal Efficiency:** It is the ratio of IHP or BHP to heat energy supplied through fuel.

1. Thermal Efficiency (Indicative)= $\eta_{I.thermal}$ =IHP/(Q x CV)
2. Thermal Efficiency (brake) = $\eta_{B.thermal}$ =BHP/(Q x CV).

f) **Volumetric efficiency:** the ratio of air volume drawn into the cylinder to the cylinder's swept volume (Vair/Vs), or, it is the ratio of Swept volume to Actual cylinder volume; $v_{olumetric}$ = Vs/Va

g) **Relative efficiency:** it is the ratio of actual thermal efficiency to Ideal thermal efficiency.

Efficiency (relative)= = $\eta_{relative}$ == $\eta_{thermal}$ /Ideal thermal efficiency= $\eta_{thermal}$ /(1-1/r)$^{\gamma}$.

h) **Specific fuel consumption**: Quantity of fuel required to generate one-unit power. SFC=kg fuel/kWh

Example: An eight-cylinder automobile engine of 85.7 mm bore and 82.5 mm stroke with a compression ratio of 7 is tested at 4000 r.p.m. on a dynamometer which has a 0.5335 m arm. During a 10 minutes test at a dynamometer scale beam reading of 400 N, 4.55 kg of gasoline for which the heating value is 46,000 kJ/ kg are burnt, and air at a temp of 294K and pressure 10 x 10^4N/m^2 is supplied to the carburettor at the rate of 5.44 kg per min. Find (a) the b.h.p. delivered, (b) the b.m.e.p., (c) the b.s.f.c, (d) the specific air consumption, (e) the brake thermal efficiency, (f) the volumetric efficiency, (g) the air-fuel ratio.

Given:

a) Four stoke Engine

b) Number of cylinder (X) =8

c) Stoke length (L) = 82.5 mm=0.0825 m

d) Area (A) = 0.785 x $(0.0857)^2$= 0.006 m^2

e) RPM(N) = 4000, RPS(n) = 4000/60 = 66.6

f) Pm = Force/area = 400/0.006 = 66666.6 N/m^2(Pa)

g) Q = 4.55kg for 10 min = 4.55/(10x60) =0.0076 kg/s=27.36 kg/h

h) CV=46000KJ/kg

Total swept volume (for all 8 engines) = A x L x n x X/2= 0.006 x0.0825x66.6x8/2 m^3/s = 0.131m^3/s

Torque = Force x Length= 400 x 0.535 =214N-m

a) Air temp. =294K
b) Supplied air pressure =10^5 Pa
c) Air flow=5.44kg/min=0.09 kg/s.

Air flow in volume(m^3/s) = V= w x RT/P = 0.09 x287x294/10^5=0.075 m^3/s

a) BHP=2xπ x n x T/1000=2x3.14x66.6x214=89500J/s=89500/1000kW=89.5kW.
b) BMEP=2x BHPx1000/ (X L A n) (4-stoke,8 cylinder)

   =2x89.5/(8x0.0825x0.006x66.6) =67871 N/m^2.
c) Brake Sp. Fuel Consumption=Kg /BHP-h=27.36/89.50=0.306 kg/BHP-h
d) Break Thermal efficiency=89.50/(0.0076x 46000)x100=28%
e) Sp. Air consumption=0.09/0.0076=11.842 kg air/kg fuel
f) Air consumption/BHP-h= 0.09 x3600/89.5=3.6 kg/BHP-h
g) Volumetric efficiency=air flow/swept volume=0.075/0.131=57%.

2.0 Diesel Power Plant

Diesel power plant uses a diesel engine as a prime mover to rotate alternators and produce electrical energy and this power plant is known as a diesel power plant. Due to the combustion of diesel, rotational energy is generated using crank shaft. The alternator is connected with the same shaft of the diesel engine, and the alternator is used to convert the rotational energy of the diesel engine into electrical energy.

In most cases, the diesel power plant is used to generate electrical energy for small-scale power generation and at the load end. When the grid power is not available, the diesel engine is used to supply load in emergency conditions.

2.1 Components, systems of Diesel Power Plant

Figure 7.3 shows a schematic diagram of a diesel power plant.

Different components or systems used in a diesel power plant is as listed below.

a) Diesel engine
b) Air intake system
c) Exhaust system
d) Cooling water system
e) Fuel supply system
f) Lubrication system
g) Diesel engine starting system

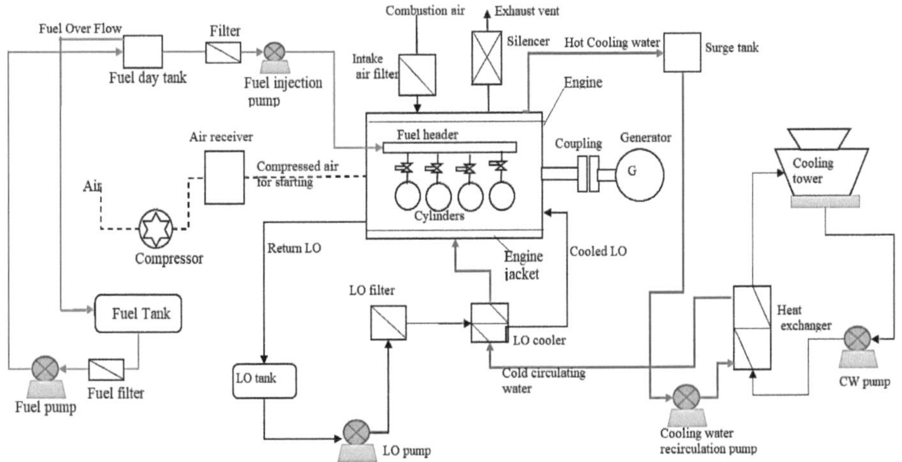

▲ **Fig-7.3: DG power plant with different parts and auxiliaries**

i. Diesel Engine

A diesel engine is the main component of a diesel power plant. It is used to generate mechanical power in the form of rotation energy with the help of the combustion of diesel. An alternator is connected to the same shaft as the diesel engine. The required capacity of a diesel power plant can be calculated by the equation:

Capacity of Plant = (Connected Load × Demand Factor) / (Diversity Factor)

There are two types of diesel engines;

a) Two-stroke engines

b) Four-stroke engines

ii. **Two stroke vs four stroke engine**

In two-stroke engines, every revolution of the crankshaft, one power stroke is developed. In four-stroke engines, one power stroke is developed every two revolutions of the crankshaft.

Compared to four-stroke engines, two-stroke engines have a low weight-to-power ratio, are more compact, easy to start, and have low capital cost. But the thermodynamic efficiency of a two-stroke engine is less compared to four-stroke engines. Two-stroke engines require more cooling water and consume more lubricants.

Four-stroke engines are more preferred over two-stroke engines for the application of small-scale generation and DG sets. And for large-scale production, two-stroke engines are preferred.

The diesel engine power plant below 3 MW capacity is used as standby plants and 3 to 25 MW plants are used as baseload plants. Generally, in this type of plant, four-stroke engines are used. The plants used for baseload plants having a capacity of above 10 MW capacity use two strokes engines.

iii. **Air Intake System**

Large diesel engine power plant requires air in the range of 4-8 m^3/kWh. Air filters are used in the air intake systems to remove dust from natural air.

The air filters are made of cloth, wood, or felt. In some cases, oil bath filters are used. In oil bath filters, the dust particles are oil-coated. Air intake system is designed to have minimum pressure loss during airflow. High pressure drop results in increased fuel consumption and reduction of engine capacity. A periodic cleaning reduces pressure due to clogged filter.

iv. **Exhaust System**

Exhaust system is provided to remove the gases generated during combustion and to discharge into the atmosphere. It is designed to have minimum pressure drop during exhaust. Excess back pressure in exhaust system results in increased fuel consumption and reduction of power output of diesel engines.

To reduce the noise level, the exhaust system must be provided with mufflers and silencers. Insulation is used on the exhaust system to avoid any high temperature accident.

v. **Cooling Water System**

The IC engine works by burning fuel with air and the percentage utilization of energy is as below:

a) 30-37% – useful work

b) 30-35% – carried by exhaust gases

c) 0-12% – lost by radiation, convection, and conduction

Balance 22-30% – heat energy flows from gases to cylinder walls. Hence, to avoid overheating of the engine, it requires a cooling system. There are two types of cooling systems:

a) Direct cooling- Such cooling is also known as air cooling used for small-capacity engines. Cooling fins and baffles are used to remove heat from the engine.

b) Indirect cooling- Indirect cooling is also known as water cooling. A water-cooling system is used for large and medium capacity engines. The water-cooling system is used a water jacket, and piping connections. Fig-7.3 shows water cooling system.

vi. **Fuel Supply System**

In a diesel power plant, diesel is used as a fuel. Functions of the fuel supply system are as follows.

a) Storage tank is required to store diesel.

b) Fuel must be filtered before supplying fuel to the engine

c) Metering of fuel is necessary.

d) According to the load in each cycle, it must inject the exact quantity of fuel.

e) A return path is provided for unused fuel.

f) In a multi-cylinder engine, it is required atomization of fuel and even distribution of fuel to each cylinder.

There are three types of mechanical fuel injection systems:

a) Common rail system

b) Individual pump system

c) Distributor system

Fig-7.3 shows common rail fuel injection system.

vii. **Lubrication System**

In the IC engine, the piston-cylinder arrangement faces a very large variation of temperature. It works at a maximum temperature of around 2000°C and lubricating material in piston ring cannot work at such elevated temperature.

Therefore, lubrication system is necessary for the IC engine to reduce friction loss. The lubrication system prevents direct contact between two metals and reduces the wear and tear in moving parts. The below-listed components of the IC engine are lubricated:

a) Piston and cylinder

b) Main crankshaft bearings

c) Cam, camshaft, and its bearings

d) Ends of bearings at connecting rod

There are three types of lubricating systems such as:

a) Mist or charge lubricating system

b) Wet sump injection system

c) Dry sump injection system

viii. **Starting system**

During start of a DG, the temperature and pressure of the cylinder are not sufficient to initiate the combustion. To make the condition suitable, a starting system is provided with the engine shaft. There are several methods introduced to start a diesel engine. Some of these methods are listed below.

a) Hand or kick-starting

b) Electrical starting using motor

c) Compressed air (fig-7.3 shows compressed air starting)

d) Auxiliary petrol engine

e) Hot bulb ignition

f) Special cartridge starting

Commonly the electrical starting method is used to start a diesel engine. In this method, a battery is used with a series-wound DC motor (starting motor). This arrangement is designed to operate on a large current at low voltage. The starting motor is connected with the engine flywheel through gears and supplies torque till the engine starts.

2.2 Diesel Generator Working Principle

To produce AC (alternating current) power, DG under goes through a series of stages. When the diesel generator starts, it goes through its four processes: suction, compression, power, and exhaust. As a result, chemical energy of fuel is transformed into rotational mechanical energy. This mechanical energy is used to turn a crankshaft which is coupled with alternator's rotor. The rotor and stator are the two crucial parts of an alternator/ generator and electricity is generated by mutual electromagnetic action of rotor and stator.

In an alternator, electricity is generated when a magnetic field moves inside the coil's winding. There are different types of alternators depending on the application and design.

Electric generators are machines that provide electricity when there is no power supply from the power grid. The armature or winding of the wire in a generator rotates within a fixed magnetic field and hence, generates electricity.

2.3 Site Selection of Diesel Power Plant

The factors affecting a selection of a location for diesel power plant are listed below.

1. **Bearing capacity:** The diesel engine is placed on a foundation. If the bearing capacity of selected land is high then it does not require high depth for a foundation. And it will save the initial cost of a power plant.
2. **Transportation facility:** The plant requires heavy pieces of machinery. Hence, the selected site must have an adequate transportation facility.
3. **Labour:** Large capacity diesel power plant requires several labours.
4. **Availability of water:** The diesel power plant requires water for cooling purposes.
5. **Future expansion:** There is some extra land available for future expansion.
6. **Availability of fuel:** This plant requires a high volume of fuel (diesel). So, a site should be selected where fuel is available easily.
7. **Distance from the populated area:** The operation of a diesel engine pollutes nearby areas. Hence, the plant must be located at a far distance from the human being.
8. **Distance from load centre:** To avoid transmission loss, the site should be selected near the load centre.

2.4 Advantages & Disadvantages of Diesel Power Plants

I. Advantages

The advantages of diesel power plants are listed below.

a) It can start and stop quickly when required.

b) This plant can be located at any place and it is easy to install for a small capacity power plant.

c) It does not require more space.

d) For varying loads, this plant responds quickly.

e) The water is required only for cooling purposes. So, a very little quantity of water is required.

f) The thermal efficiency of this plant is higher than a steam power plant.

g) The diesel power plant can be efficiently used up to 100 MW.

h) Less manpower is required.

i) It can burn a wide range of fuel.

j) Fewer fire chances.

k) Relatively low maintenance

l) Strength: More strong and reliable.

m) Fuel easily accessible: Portable diesel generators are frequently carried behind diesel vehicles.

II. Disadvantages

The disadvantages of diesel power plants are listed below.

a) The generation cost per unit is very high. As the operation of this plant depends on the price of diesel. And diesel prices are high.

b) The capacity of a diesel power plant is less compared to a steam power plant and hydroelectric power plant.

c) It creates noise pollution and carbon pollution by the combustion of diesel.

d) It requires high maintenance and lubrication costs.

e) This plant is not capable to meet continuous overload demand.

f) The life of this plant is less compared to other power plants.

g) Since diesel engines may be somewhat noisy, they're frequently located away from work areas.

h) Diesel generators are considered too heavy and massive to be termed mobile and compact.

2.5 Application of Diesel Power Plants

The applications of diesel power plants are as follows:

i. **Peak load plant**

   The diesel power plant is used with thermal power plant and hydroelectric power plant to meet peak demand. It reduces the per-unit cost of power generation. It can easily start and stop with demand and varying with load variation.

ii. **Emergency plant**

   The DG power plant can be used as an emergency plant. When the power of the grid is not available, the diesel engine is used as a backup plant for emergency conditions.

iii. **Mobile plant**

   The small and medium capacity of the diesel power plant can be fixed on a truck or trailer. This plant can be used as a mobile power plant to supply power when grid power is not available. This plant is also used as an emergency plant while power failure.

iv. **Stand by unit**

   The DG power plant can be used as a stand-by unit with a hydroelectric power plant. When the water is not available sufficiently in the hydro power plant, to meet power demand the diesel power plant operates parallel with the hydropower plant.

v. **Power plant for small industries**

   This plant can be used to run a small industry for short periods where the reliability of power is essential throughout the day.

vi. **Nursery station**

   In some areas where a grid is not available, the diesel power plant is used as a temporary solution to supply power and it is removed when the grid is connected.

2.6 Energy Saving Measures for DG Sets

a. Ensure steady load conditions on the DG set, and provide cold, dust free air at intake, improve air filtration.

b. Ensure fuel oil storage, handling and preparation as per manufacturers' guidelines/oil company data.

c. Consider fuel oil additives in case they benefit fuel oil properties for DG set usage.

d. Calibrate fuel injection pumps frequently.

e. Ensure compliance with maintenance checklist.

f. Ensure steady load conditions, avoiding fluctuations, imbalance in phases, rectify harmonic loads.

g. In case of a base load operation, waste heat recovery system can be provided for steam generation or a refrigeration chiller unit can be incorporated.

h. Jacket Cooling Water heat can be used in vapour absorption system.

i. Partial use of Bio-mas gas may reduce fuel costs.

j. Parallel operation among the DG sets improves loading and fuel economy.

k. Carry out regular field trials to monitor DG set performance, and maintenance planning as per requirements.

3.0 Cogeneration plant

3.1 Introduction

Cogeneration is an energy system that produces heat and electricity simultaneously in a single plant, powered by just one primary energy source. Such a system is more energy efficient than achieving from two separate production sources. In this way, the dissipation of energy into the environment, produced by combustion processes, is reduced but is recovered and reused. Cogeneration systems involve the combustion of fuels such as natural gas, diesel, biogas, bio-methane, vegetable oil, or biomass.

3.2 Working of a cogeneration system

Conventional power plants generate electricity by heating water to the boiling point, thereby producing steam to drive a turbine that creates the kinetic energy needed for electricity. The water is usually heated by using a fossil fuel like coal, oil, or natural gas. Energy is wasted in every step of this process, particularly, when a huge quantity of heat is dumped in the condenser and ultimately the heat is released into the atmosphere.

Some 60% of energy is wasted during traditional electricity generation, means that energy efficiency runs at about 30% (some of the energy dissipates during transmission).

Whereas, a cogeneration plant captures and uses this dumped heat to produce hot water and is supplied to consumers for heating purpose. The waste of energy in cogeneration is only 10%-30% of total input energy, means, energy efficiency improves to 70%-90%.

3.3 Classification of Cogeneration Systems

Cogeneration systems are normally classified according to the sequence of energy use and the operating schemes adopted. A cogeneration system can be classified based on the sequence of energy use as:

a) Topping Cycle (Power is at the top of the sequence)- Energy is used for Power generation first, then the waste heat or heat extracted from power generating system is used for process heat supply. STG with back pressure, controlled extraction, GTG with WHRB, DG with WHRB are such example.

b) Bottoming Cycle (Power is at the bottom of the sequence)- Energy is used for process first and the waste heat is used to generate power. Cement, steel, petro-chemical process etc. are such examples.

3.4 Factors Influencing Cogeneration Choice

The selection and operating scheme of a cogeneration system is very much site-specific and depends on several factors, as described below. STG based different cogeneration schemes (topping cycle) are shown in Fig-7.4, 7.5 & 7.6. Fig-7.7 shows a bottoming cycle cogeneration plant.

i. Base electrical load matching -In this configuration, the cogeneration plant is sized to *meet the minimum electricity demand* of the site. The rest of the needed power is purchased from the utility grid. The thermal energy requirement of the site could be met by the cogeneration system alone or by additional boilers. If the thermal energy generated with the base electrical load exceeds the plant's demand, excess thermal energy can be exported to neighbouring customers.

ii. Base Thermal Load Matching- In such configuration, the cogeneration system is sized to supply the *minimum thermal energy requirement* of the site. Stand-by boilers or burners are operated during periods when the demand for heat is higher. The prime mover installed generates electricity at full load at all times. If the electricity demand of the site exceeds, then the remaining amount can be purchased from the grid. The excess electricity can also be sold to the power utility.

iii. Electrical Load Matching -In this scheme, the facility is totally independent and called as "stand alone". *Entire power requirements* of the site, including the reserves needed during scheduled and unscheduled maintenance, are to be taken into account while sizing the system. If the thermal energy demand of the site is higher than that generated by the cogeneration system, auxiliary boilers are used.

iv. Thermal Load Matching -The cogeneration system is designed to meet *the thermal energy requirement* of the site at any time as a first priority and the prime movers are operated following the thermal demand. During the period when the electricity demand exceeds than the generation capacity, the deficit can be compensated by power purchased from the grid. Similarly, electricity produced in excess at any time may be sold to the utility.

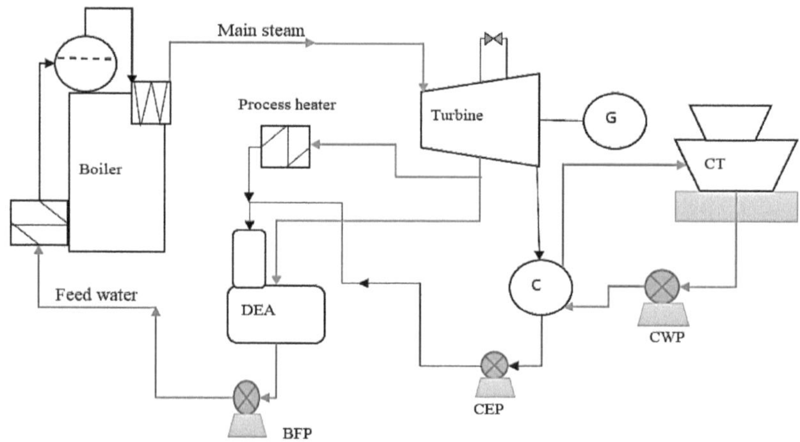

Schematic diagram of a STG base Co-generation plant having condensing cum controlled extraction for process

▲ **Fig-7.4: STG base Co-generation plant having condensing cum controlled extraction**

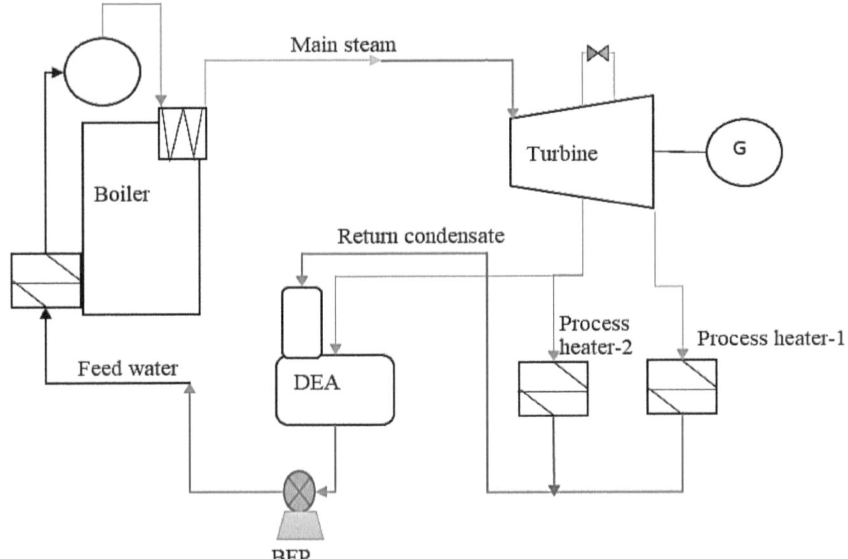

Schematic diagram for STG based Cogeneration plant with back pressure and controlled extraction

▲ **Fig-7.5: STG-based Cogeneration plant with back pressure and controlled extraction**

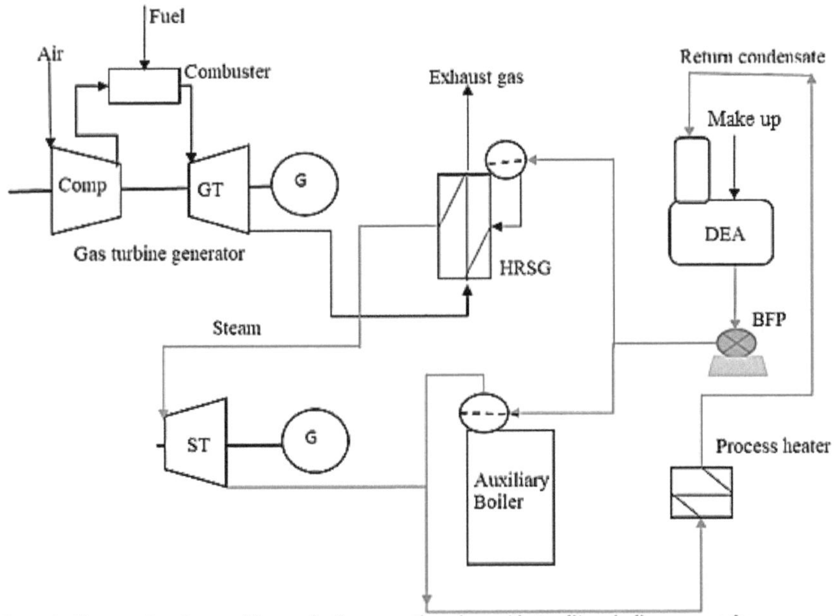

Schematic diagram showing combine cycle Co-generation plant with Auxiliary boiler support for process steam

▲ **Fig-7.6: Combine cycle Cogeneration plant with Auxiliary boiler support**

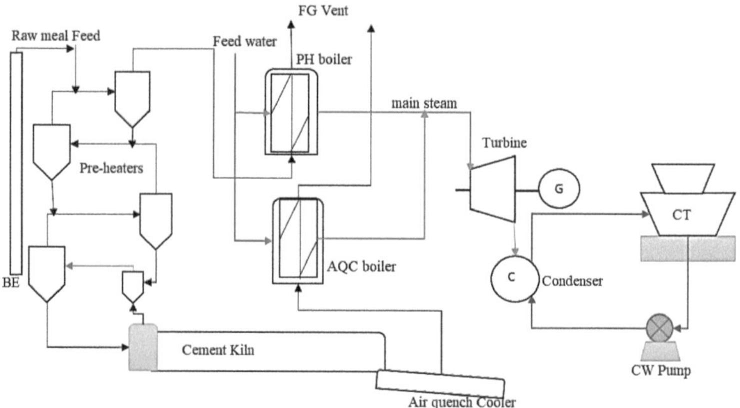

Schematic diagram of Cement base Cogeneration plant (bottoming cycle) showing WHRS

▲ **Fig-7.7: Cement base Cogeneration plant**

V. Heat-to-Power Ratio -Heat-to-power ratio is one of the most important technical parameters influencing the selection of the type of cogeneration system. The heat-to-power ratio of a facility should match with the characteristics of the cogeneration system to be installed. It is defined as the ratio of thermal energy to electricity required by the facility. Though, it can be expressed in different units such as Btu/kWh, k Cal /kWh, lb per hr/kW, etc., common unit as kW is used in the table-7.1 to represent the basis of heat & power. Steam turbine cogeneration system can offer a large range of heat-to-power ratios.

▼ **Table 7.1: Different types of Cogen system**

Table showing different types of cogen system

Sl No	Heat to power ratio (kW_{therm}/ kW_{elec})	Cogeneration system	Power output (%) (w.r.t to fuel input)	Overall efficiency (%)
1	4-14.3	Back pressure STG	14-28	84-92
2	2-10	Extraction condensing STG	22-40	60-80
3	1.3-2.0	Gas turbine Generator (GTG)	24-35	70-85
3	1-1.7	Combined cycle	34-40	69-83
4	1.1-2.5	IC engine (Reciprocating)	33-53	75-85

3.5 Efficiency Derivation of Cogeneration Plant

The following equations are used for topping cycles for assessment performance of the Topping cycle cogeneration plant.

Plant Heat rate (PHR) = [Quantity of fuel/hr x CV+ Quantity of return condensate x Enthalpy+ Make up quantity x Enthalpy- (Extraction Process steam x enthalpy + Exhaust process steam x enthalpy)]/Power Generation.

Unit: Kcal/kWh

Efficiency = 860/PHR.

Bottoming cycle performance assessment as:

a) Specific heat consumption for production = Kcal/ Kg or, KJ/kg.

b) Heat rate for WHRS system= Waste Heat Available/Power generation (KCal/hr/kWh)

3.6 Benefits of cogeneration

A cogeneration system (also called a **Combined Heat and Power system**, or CHP), can deliver significant benefits for commercial and industrial customers, because it produces heat and electricity at the same time. Using the same fuel to generate both heat and electricity therefore improves energy efficiency, delivers **environmental benefits** and ensures savings. Cogeneration power plants generally operate at between 50 to 70% higher efficiency rates than traditional power plants.

Cogeneration systems can:

a) Improve the overall efficiency of energy usage by combining the production of heat and electric energy from a single plant

b) Reduce energy costs

c) Reduces emissions

d) Reduce risks of power cuts due to grid problems

e) Use renewable energy sources like biomass

f) Be adapted to fit the needs of all sorts of users, including residential

g) Promote energy self-sufficiency and reduce energy imports.

Questions:

1. Explain in brief about working of a DG engine.
2. What are the factors affecting the performance of a DG engine?
3. Explain BHP, IHP, BMEP of DG engine.
4. What are the different types of efficiencies encountered in DG engine?
5. Briefly explain a DG power plant.
6. What are the advantages of DG power plant?
7. Explain disadvantages of a DG power plant.
8. Discuss about application of DG power plant.
9. What is a cogeneration plant- explain.
10. What is topping cycle & bottoming cycle cogeneration?
11. What is heat to power ratio, discuss how this help configuring co-generation plant?
12. Explain the benefits of co-generation plant.

SECTION 08

RENEWABLE ENERGY SOURCES-SOLAR ENERGY, WIND ENERGY, GEOTHERMAL ENERGY & BIOMASS ENERGY PLANT

1.0 Renewable Energy Sources

1.1 Introduction

Several renewable energy sources are at present in use to reduce consumption of fossil fuel for combating global warming. Some of the renewable energy sources are used in conventional ways by burning for heating purpose, some are being used as co-fuel for power generation and energy source for process plant. Out of all the sources, this section covers the renewable energy sources that has the potential to replace consumption of fossil fuel in future. Hence, Solar energy, wind mill & geothermal energy have been identified for this section for discussion.

2.0 Solar Energy

Solar energy is the radiation from the Sun capable of producing heat, causing chemical reactions, or generating electricity and it is one of the vast renewable energies. Once the sunlight passes through the earth's atmosphere, most of it is in the form of visible light and infrared radiation. Plants use it to convert into sugar and starches; this conversion process is known as photosynthesis. Solar cell panels are used to convert this energy into electricity. The total amount of solar energy received on Earth is vastly more than the world's current and anticipated energy requirements.

2.1 Types of Solar Energy

Solar energy can be classified into two categories depending upon the mode of conversion and type of energy it is converted into, they are- Passive solar energy and Active solar energy.

i. Passive solar energy refers to trapping the sun's energy without using mechanical devices such as drying in the sun.

ii. Active solar energy uses mechanical devices to collect, store, and distribute it. Active solar energy can be sub-classified as:

   a) Solar thermal energy: This energy is obtained by converting solar energy into heat.

   b) Photovoltaic solar power: It is the energy obtained by converting solar energy into electricity.

   c) Concentrating solar power: This is a type of thermal energy obtained from solar energy to generate solar-thermal electricity.

2.2 Use of solar energy in several ways

Solar energy is used in a number of ways:

i. As heat for making hot water, heating buildings and cooking.

ii. To generate electricity with solar cells or heat engines.

iii. To take the salt away from sea water.

2.3 Photovoltaic effect

The photovoltaic effect is a process that generates voltage or electric current in a photovoltaic cell when it is exposed to sunlight. This effect is used in solar panels to generated electrical energy.

2.4 Photo electric Process

Solar cells are composed of two different types of semiconductors: a p-type and an n-type and are joined together to create a p-n junction. By joining these two types of semiconductors, an electric field is formed in the region of the junction as electrons move to the positive p-side and holes move to the negative n-side. This field causes negatively charged particles to move in one direction and positively charged particles in the other direction.

Light is composed of photons, which are small bundles of electromagnetic radiation or energy. When light of a suitable wavelength is incident on these cells, energy from the photon is transferred to an atom of the semiconducting material in the p-n junction. Specifically, the energy is transferred to the electrons in the material. This causes the electrons to jump to a higher energy state known as the conduction band. This leaves behind a "hole" in the valence band from where the electron jumped up. This movement of the electron as a result of added energy creates two charge carriers, an electron-hole pair.

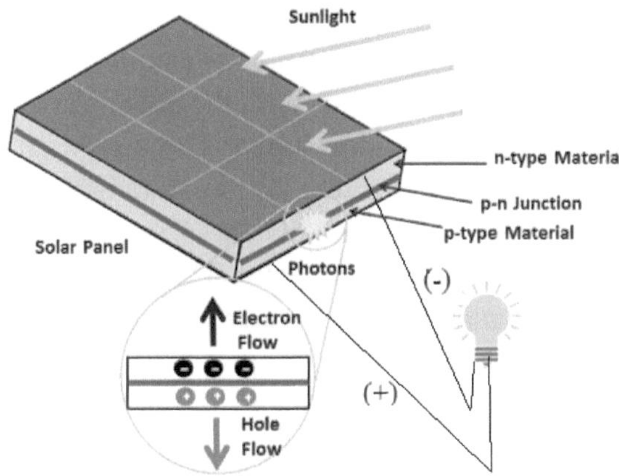

Schematic diagram showing working of a solar cell

▲ **Fig-8.1: Working of a solar cell**

In their excited state in the conduction band, these electrons are free to move through the material. Because of the electric field that exists as a result of the p-n junction, electrons and holes move in the opposite direction. This motion of the electron creates an electric current in the cell. Once the electron moves, there's a "hole" that is left and this hole can also move, but in the opposite direction to the p-side. It is this process which creates a current in the cell Fig: 8.1.

2.5 Theoretical explanation of photo electric effect

Experiments showed that increasing the light frequency increased the kinetic energy of the photoelectrons, and increasing the light amplitude increased the current.

Based on these findings, Einstein proposed that light behaves like a stream of photons with an energy of $E = h\nu$ [h- Planck's constant, ν- frequency].

The work function, Φ, is the minimum amount of energy required to induce photo-emission of electrons from a specific metal surface.

The energy of the incident photon must be equal to the sum of the work function and the kinetic energy of a photo-electron: $E_{photon} = \frac{1}{2} mv^2 = KE_{electron} + \Phi$ [m &v are mass and velocity respectively of photo-electron].

2.6 Solar Water heater

Solar heaters are the devices that heat up water using solar radiation or sunshine as source of energy to heat water. These are cheap and cost-effective way to generate and supply hot water for domestic use.

2.6.1 Working of a Solar Water Heater

The solar water heater absorbs light by means of a collector placed on the roof under sun-rays and converts it into heat. With the advancement of technology, new solar water heaters have been developed. These new solar water heaters work on the same principle of heating by solar radiation but have much sophisticated system including pumps, insulated storage tanks, temperature gauges, anti-freeze valves, and solar collector. This new system absorbs more solar energy from the sun and thus heats water quicker.

2.6.2 Parts of a solar heater

Solar water heaters are manufactured in different designs but all of them have following common components, Fig-8.2.

a) **A Collector** – It collects solar energy and embedded with tubes for water to flow inside it. Cold water enters at the bottom of the collector and gradually rises up as it gets heated up by sun rays and finally enters in the storage tank.

b) **Insulated Storage Tank** – It has inlets and outlets connected to and from the heater.

c) **Storage cum backup heater-** This is an additional attachment in the solar heater system to ensure continuous hot water supply at a fixed temperature. One electric heater & cold water mixing valve are a fitted with it to maintain fixed water temperature.

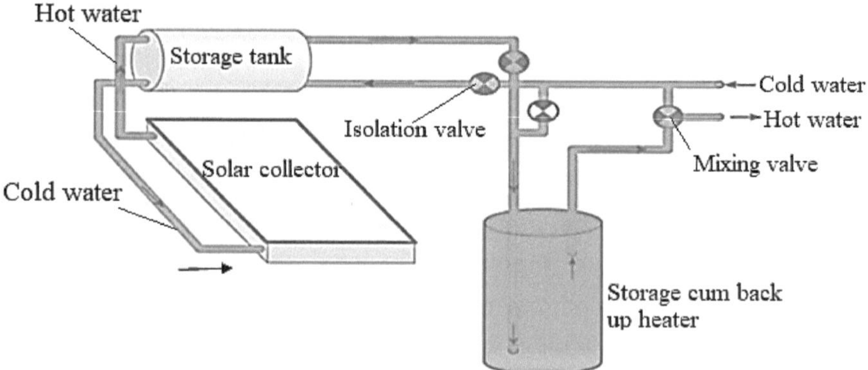

▲ **Fig-8.2: Natural circulation solar heater**

2.6.3 Circulation Systems in Heat Water

Different types of circulation systems are used in solar water heaters to heat water:

a) Forced-circulation- In this system, controllers, electric pumps and valves are used to force water from the collector to the storage tank.

b) Passive/ Natural Circulation System-In this system, water automatically moves from the collectors to the storage tank as it heats up. This process happens because of convection and density difference. There is no need of any electric pump.

c) Direct Circulation System-In this system, water is circulated through solar collectors when it is heated by heat of the sun. This hot water can be piped to a storage tank, or it can be directly used.

d) Closed-loop Circulation System (indirect system)-In this system, non-freezing liquid is stored in collectors. Heat from the Sun heats this liquid that passes through a heat exchanger in the storage tank. This process transfers heat from the non-freezing liquid to water in the tank. The non-freezing liquid then cycles back to the collectors.

e) This system is best suitable for places with very cold climate or where there is snowfall.

2.6.4 Performance assessment

Performance assessment of water heater can be explained using an example as below.

Example: Daily hot water requirement of a hostel resort is 5000 litres. Solar radiation available on the location is 600 W/m^2. Temperature of feed water is 30°C and the final hot water temperature is designed to be 60°C. The area of the solar collector is 75 m^2. Assess the performance of the collector. Assume the required data for hot water generation.

Given data:

Solar insolation during the study period, S: 600 W/ m^2

Capacity of hot water system, M: 5000 Litre/day

Feed water temperature, T_a: 30°C

Temperature of hot water, T_h: 60°C

Area of the collector area, A_c, m^2: 75 m^2

Solution:

Useful solar heat energy available:

Daily available solar radiation, S_D, (kJ /day)

= S x A_c x 8 x 3600/1000 (Assuming 8 hours of daily operation).

=(600x3600) x 75x8/1000

= 1,296,000/4.18 =310,048 kcal/day

Heat energy required to heat the water = M x Cp x ΔT =5000x1x (60-30)=150,000 kcal/day

System performance

Thermal efficiency of solar water heater,% = Output/Inputx100

=(150,000/310,048)x100=48.38%

2.6.5 Benefits of solar heater

Solar water heating systems are eco-friendly and can significantly reduce energy bills. They use free and renewable solar energy, reducing consumption on fossil fuels. Additionally, they require minimal maintenance and have a long lifespan, contributing to both environmental and financial savings.

2.7 Solar energy in power generation using PV cell

The process of creating power from sunlight starts with the larger part of a solar installation, the solar panels Fig-8.3. A typical solar panel is made from semiconductors either monocrystalline or polycrystalline silicon housed in a metal panel frame with a glass casing. They are attached by wire to a circuit, Fig-8.4. As light strikes the semiconductor, light energy is converted into electricity that flows through the circuit.

When sunlight strikes the thin layer of silicon on the top of a solar panel, it knocks electrons off the silicon atoms, known as photovoltaic effect. Electrons are negatively charged, which means they are attracted to one side of the silicon cell. This creates an electric current that is captured by the wiring in a solar panel.

Loose electrons that have been collected at individual panels, the resulting current is known as direct current, or DC. DC can be used to supply power to the devices such as: electric vehicles or charging battery.

Largely power used for household and industries is Alternating Current (AC). Hence, DC needs to be converted to AC before use. This conversion is done by an important device called inverter. Once the electricity passes through inverter, DC is changed to AC electricity and flows to electric panel and meter. Finally, it can be distributed throughout home or into the electric grid, Fig-8.3.

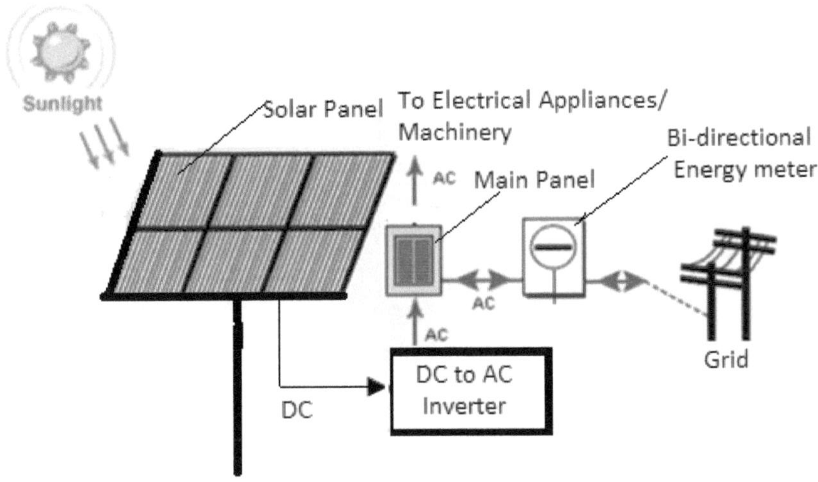

▲ **Fig-8.3: Solar power generation system**

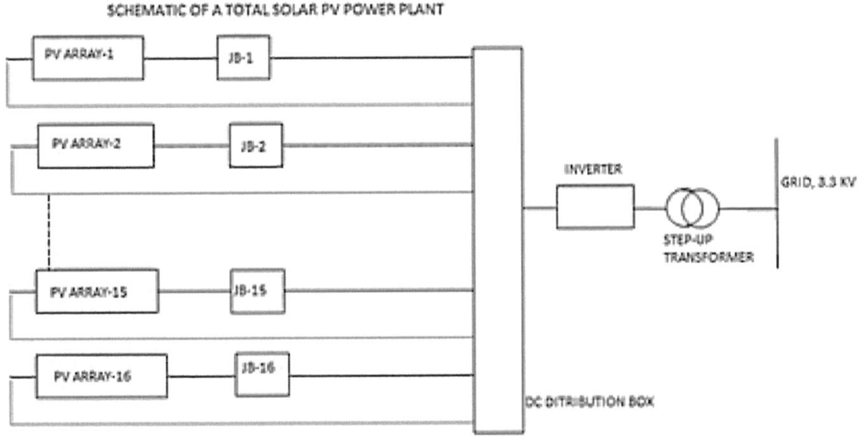

▲ **Fig-8.4: Set up of a Solar power plant**

2.8 Working of a Solar Thermal power plant

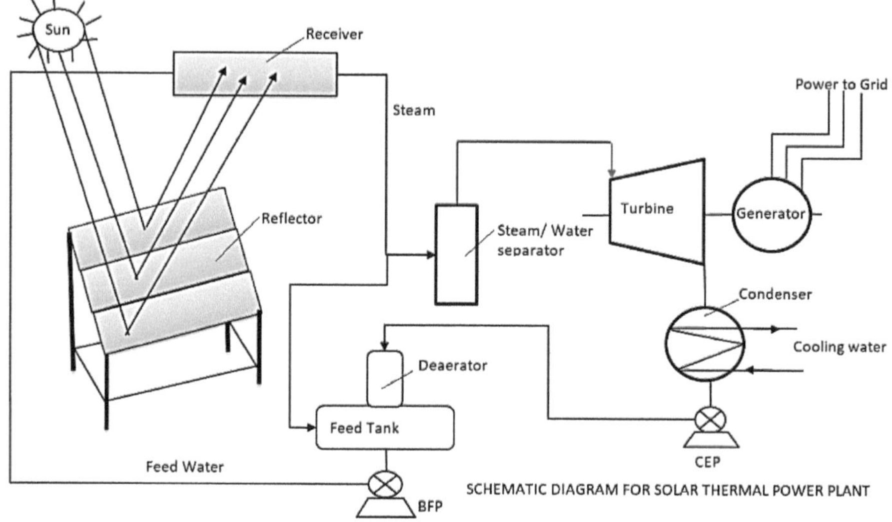

▲ **Fig-8.5: Solar Thermal Power plant**

Fig-8.5 shows a schematic diagram of a solar thermal power plant. Sun rays get reflected by reflectors. The reflectors are provided with tracking of sun rays. Reflected rays heat up the water line passing through the receptor. Water, while passing through the receptor converts into, steam.

Steam is supplied to steam turbine to generate electrical power. Steam separator, Condenser, deaerator, CEP & BFP are provided as required in Rankine cycle and to reuse water as feed for receptor.

2.9 Solar Energy Advantages and Disadvantages

i. Advantages of solar energy

a) Clean: It is considered to be the cleanest form of energy as there is no combustion of fossil fuels which is one of the causes of global warming.

b) Renewable: There is ample energy available on earth as long as the sun exists.

c) Reliable: The energy can be stored in the batteries, so it is reliable.

d) Reduction in energy costs.

e) Solar power imparts no fuel costs.

f) Free energy because it can be trapped easily.

g) Solar power generation releases no water or air pollution, because there is no combustion of fuels.

h) Solar energy can be used very efficiently for heating (solar ovens, solar water and home heaters) and day lighting.

i) Distributed point-of-use photovoltaic systems eliminate expensive long-distance electric power transmission losses.

j) PV system is efficient in their conversion of solar energy to usable energy than biofuel from plant materials.

ii. **Disadvantages of solar energy:**

a) The production is low during morning, evening, winters and on cloudy days.

b) Installation and the initial cost of the materials are expensive.

c) Space consumption for installation is more.

d) Efficiency is lesser

2.10 Efficiency of Solar power

The global formula to estimate the electricity generated in output of a photovoltaic system used in solar power plant for generating solar energy for home/industrial/commercial roof top or ground mounted is: E = A x r x H x PR

When,

a) E = Energy generation (kWh)

b) A = Total solar panel Area (m^2)

c) r = solar panel yield (%), given by the ratio: electrical power (in kWp) of one solar panel divided by the area of one panel.

d) H = Annual average solar radiation data on tilted panels (shadings not included).

e) PR = Performance ratio, coefficient for losses (range between 0.5 and 0.9, default value = 0.75).

2.10.1 Explanation on Solar Panel Yield

a) Yield is given for Standard Test Conditions (STC): radiation=1000 W/m^2, cell temperature=25°C, Wind speed=1 m/s, Air Mass (AM)=1.5.

b) The unit of the nominal power of the photovoltaic panel in these conditions is called "Watt-peak" (Wp or kWp=1000 Wp or MWp=10^6 Wp).

c) Air Mass (AM) is a measure of how much atmosphere the sun's rays have to pass through on their way to the surface of the earth. Since particles in the atmosphere absorb and scatter light rays, the more atmosphere solar radiation passes through on its way to us, the less solar energy we can expect to get. It is 1 when sun is at top of head (Zenith) and 2 in the morning or evening when zenith angle is 60°.

Example to derive yield

Given: Solar panel yield of a PV module of 250 Wp with an area of 1.6 m^2 is 15.6%, Justify.

a) Peak Solar energy= 1000 W/m^2.

b) Actual Power generation= 250/1.6=156 Watt/m^2

c) Yield= (156/1000) x100= 15.5%.

3.0 Wind Power- working and performance assessment

3.1 Introduction

Wind turbines work on the principle- as wind blows over the blades, it rotates wind turbines. The generator which is coupled with the wind turbine also spins to generate electricity. Wind is a form of solar energy caused by a combination of the events:

a) Unevenly heating of atmosphere

b) Irregularities of the earth's surface

c) The rotation of the earth.

Wind flow pattern and speed vary greatly across the earth surface and the variations due to bodies of water, vegetation, and differences in terrain. The terms "wind energy" and "wind power" both describe the process by which the wind is used to generate mechanical power or electricity.

3.2 Working of a Wind mill

A wind turbine converts wind energy into electricity using the aerodynamic force from the rotor blades. When wind flows across the blade, the air pressure on one side of the blade decreases. The difference in air pressure across the two sides of the blade creates both lift and drag. The force of the lift is stronger than the drag and this causes the rotor to spin.

The rotor is connected to the generator, either directly or through a shaft and a series of gears (a gearbox) that speeds up the rotation of generator resulting generation of electricity.

3.3 Types of Wind Turbines

The majority of wind turbines fall into two basic types:

a) Horizontal-axis Wind turbines (HAWT)

b) Vertical-axis Wind turbines (VAWT)

3.4 Applications of Wind Turbines

Modern wind turbines can be categorized by when they are installed and how they are connected to the grid:

a) Land-based wind

b) Offshore wind

c) Distributed wind

3.5 Wind mill & its capacity

Wind strength largely influences electricity generation. According to a study, a typical wind turbine has a 2-3 MW (megaWatts) capacity and can produce over 6 million kWh (kiloWatt-hours) electricity per year. The wind turbine has an average 35% to 45% efficiency reaching 50% in the best possible scenarios. It is known that maximum efficiency is 59.6% which is Betz limit and actual efficiency is generally 50% to 70% of maximum efficiency.

The faster the wind, the more power the wind turbine generates. Eight times more electricity is generated at double speed of the wind. However, there is a limitation of wind speed and wind turbines need to be stopped if the wind is too strong to damage.

Total power $P_{total} = \frac{1}{2} \rho A V^3$ Watts when ρ is density of air, A is the area covered by turbine blade of diameter, D and V is the velocity of wind stream.

3.6 Relative Velocity & maximum speed of Wind Turbines

Relative velocity of the wind is given by the formula: $V_{RELATIVE} = V_{WIND} - V_{BLADE}$. So, the relative velocity is the difference between the velocity of the wind and that of a blade. That's why blades are tilted in such a way that they align to the relative wind speed.

As the blade velocity increases to the tip, the relative velocity becomes more inclined towards the tip, which means a continuous twist is given to the blade from the root to tip. A gearbox is fitted before the generator, which helps achieve a high-speed ratio of 1: 90 (cut-off speed of 80 km/h). A brake is also fitted in a nacelle (A **nacelle** is a cover housing that houses all of the generating components in a wind turbine, including the generator, gearbox, drive train, and brake) that helps tackle the wind arrest during windy conditions.

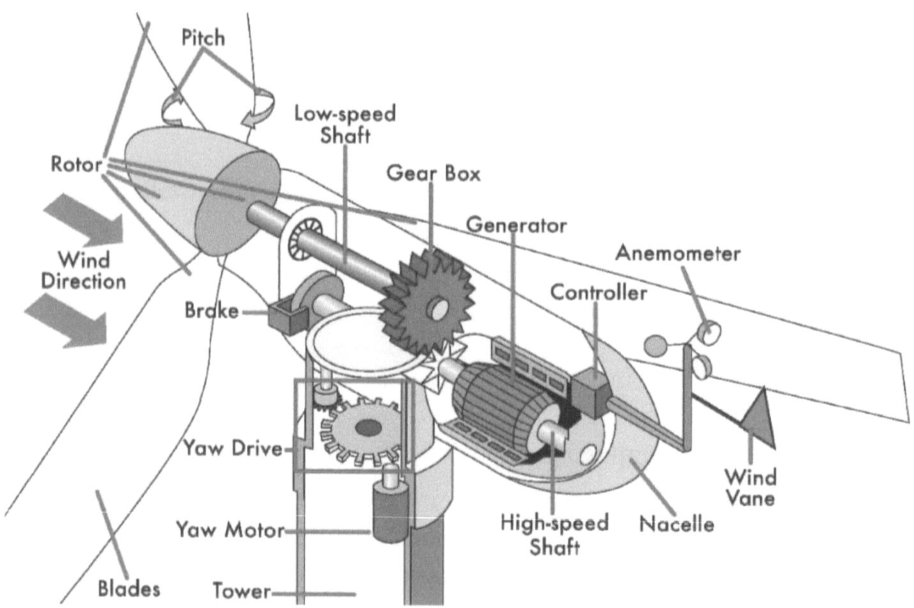

▲ **Fig- 8.6: Showing different components of a horizontal Axis Wind turbine**

3.7 Components of a horizontal axis Windmill

Following are the components of windmill, Fig-8.6:

a) **Rotor:** Rotor is also known as a propeller.

b) **Blades:** Blades are one of the essential components of the windmill as they control the functioning of rotor speed.

c) **Anemometer:** This component is used for measuring the wind speed.

d) **Tower:** This is the support system holding the blades and propeller together.

e) **Yaw drive**: The yaw drive is an important component of a horizontal axis wind turbine. To ensure the wind turbine is producing the maximal amount of electric energy at all times, the yaw drive is used to keep the rotor facing into the wind as the wind direction changes.

f) **Gear box**- It includes train of gears and increases the speed to match generator rpm.

g) **Generator**- It converts the mechanical energy of turbine to electrical energy. Permanent or electromagnet is used in the rotor to generate magnetic field so as to collect power from stator to be delivered to the grid through step up transformer.

h) **Nacelle**- It encases all the assemblies & sub-assemblies of wind turbine.

i) **Controller**- It is meant for controlling the entire operation of wind turbine and its protection.

3.8 Uses of horizontal axis Windmill

They are used in the following places:

a) They are used for pumping water,

b) For grinding grains, and

c) Generating electricity.

3.9 Vertical Axis Wind Turbine

The Vertical Axis Wind Turbine is a type of wind turbine and it is most frequently used for residential purposes to provide a renewable energy source to the home. This turbine includes the rotor shaft and two or three blades when the rotor shaft moves vertically. In this turbine, the generator is placed at the bottom of the tower whereas the blades are covered around the shaft.

3.9.1 Components of a Vertical axis wind turbine

Vertical axis wind turbine components are a) blade, b) shaft, c) bearing, d) frame & e) blade support. Except the orientation, the functions of the stated components are same as horizontal axis wind turbine. Different types of vertical axis wind turbines are shown in fig-8.7.

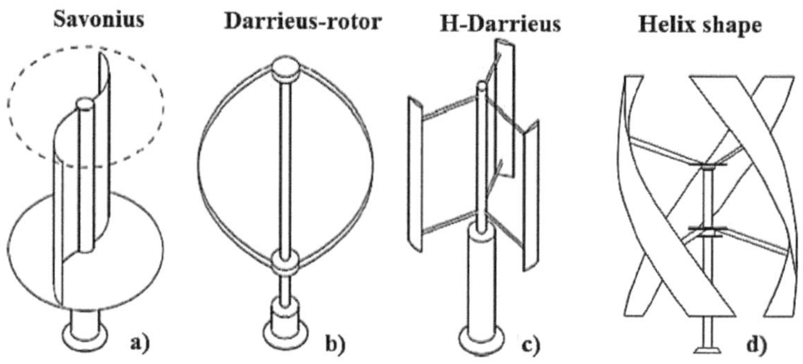

▲ **Fig- 8.7: Vertical axis wind turbine**

3.9.2 Advantage & disadvantages of Vertical axis wind turbine

i. **Advantages**

The advantages of a vertical axis wind turbine are as follows.

a) It can generate power round the clock as long wind blows.

b) It can generate electricity in any direction of the wind.

c) It doesn't require a strong supporting tower because the gearbox, generator & other components are arranged on the ground.

d) As compared to horizontal axis turbines, these are cheaper to design.

e) Installation is easy as compared to other types.

f) It is portable and is designed with fewer speed blades to reduce the risk to birds & people.

g) It works in all weather conditions like variable winds & mountain conditions.

h) Its operation is simple; can be installed close to the earth so that maintenance and the cost for construction can be reduced.

i) These are economical, quiet, efficient & ideal for residential energy sources, particularly in urban areas.

ii. **Disadvantages**

The disadvantages of a vertical axis wind turbine are as follows.

a) As compared to horizontal axis turbine, the efficiency level is lower because of the drag that happens in the blades when they rotate.

b) The efficiency of rotation is low.

c) Lower accessible wind speed.

d) Some animals or birds may interrupt its rotation because it is arranged in an open area.

e) It has high vibration due to the flow of air close to the ground makes the turbulent flow.

f) Noise pollution is more compared to HAWT.

3.9.3 Applications

The applications of a vertical axis wind turbine are:

a) Used in small wind projects.

b) Used in residential applications.

c) These turbines are used to generate power even in unstable weather conditions like gusty wind & turbulence.

3.10 Wind Efficiency and Wind Capacity Factor

i. **Wind Efficiency and Its Limit**

Wind efficiency is the amount of kinetic energy in the wind that is converted to mechanical energy and electricity. The maximum theoretical limit is 59.6%. The wind requires the rest of the energy to blow past the blades.

It is, however, not possible for any machine to convert all of the trapped 59.6% of kinetic energy from wind to electricity. Due to efficiency factors of gears, bearings, generator etc. the amount of energy that is finally converted to power is 35-45%.

ii. **Wind Capacity Factor**

The wind capacity factor is the amount of energy produced by a generator as against what it could produce if it functioned all the time at peak capacity.

Wind capacity factor varies from place to place and at different times of the year, even with the same turbines, since it depends on the speed of wind, its density and swept area. Wind capacity factor can be optimised by choosing places when ideal wind conditions prevail the whole or greater part of the year. So, it is important to consider wind capacity factor and the conditions that causes it to maximise power output.

iii. **Conditions affecting wind capacity factor**

a) Wind speed below 45 KMPH produces little energy, even small increase in speed can generate into substantial increase in power generation as power generated(P) is the cube of the wind speed i.e. turbine RPM (N) ($P \alpha N^3$).

b) Air density is more in cooler regions and at sea level than in mountains. So, the ideal places with high wind density are seas with colder temperatures. This is one reason for the large-scale expansion in off-shore wind generation.

c) Larger and taller turbines can take advantage of more wind higher above the ground and by the increased span of their blades.

The capacity factor is constantly being increased with improved technology. Wind turbines built in 2014 reached a capacity factor of 41.2% as compared to 31.2% for turbines built between 2004-2011,

4.0 Geothermal energy- Working, types and performance assessment

4.1 Introduction

Geothermal energy is the thermal energy already stored inside sub-surface of the earth. It is a kind of renewable energy extracted with water and/or steam to the earth's surface. The geothermal energy of the earth's crust originates from the original formation of the planet and from radioactive decay of materials and continual heat loss from the earth's formation. Geo-thermal is a form of energy conversion in which heat energy from within the earth is obtained and harnessed for cooking, bathing, space heating, electrical power generation, and many more uses.

4.2 Applications of geothermal energy

I. **Direct Uses-**Geothermal resources, which is used directly, is at low-temperature, at about 50 to 150°C. Low-temperature geothermal water and steam have been used to warm domestic buildings from a central supply source. Also, most swimming pools, greenhouses, therapeutic facilities at spas and aquaculture ponds have been heated with geothermal resources. Other common applications of direct uses of geothermal energy include cooking, industrial uses such as drying fruits, vegetables, and timber, milk pasteurization, and large-scale snow melting.

II. **Geothermal heat pumps-** Geothermal heat pumps (GHPs) are used to heat buildings in the winter and cool them in the summer. Most GHPs are found at shallower depths such as 6 meters of the earth's surface. Consequently, the heat can be used to warm buildings during winter, when the air temperature falls below that of the ground. Similarly, during summer, warm air is drawn from a building and circulated underground, where it loses much of its heat and is returned with colder temperature.

4.3 Geothermal energy and Electrical power generation

Geothermal energy is used to generate electricity based on the temperature and the fluid (steam) flow. Electricity is produced in three ways in geothermal power plants. These three different designs based on the type of availability of geothermal energy which is used to drive electrical generators. The excess water vapour produces at the end of each process is condensed and returned to the ground, when it is reheated for later use. This is why geothermal power is considered a form of renewable energy.

4.4 Working of geothermal power plant

Wells up to a 1KM deep or more are dig into underground reservoirs so that geothermal resources can be obtained. These resources can be exploited from naturally occurring heat, rock, and water permeability. These geothermal resources are further used to drive turbines linked to electricity generators. Geothermal power plants are available in three different designs; dry steam, flash, and binary.

4.4.1 Dry steam geothermal power plant

In such design, dry steam received from the ground, is directly used to drive a turbine. The heated water vapour is piped directly into a turbine that drives an electrical generator, Fig-8.8.

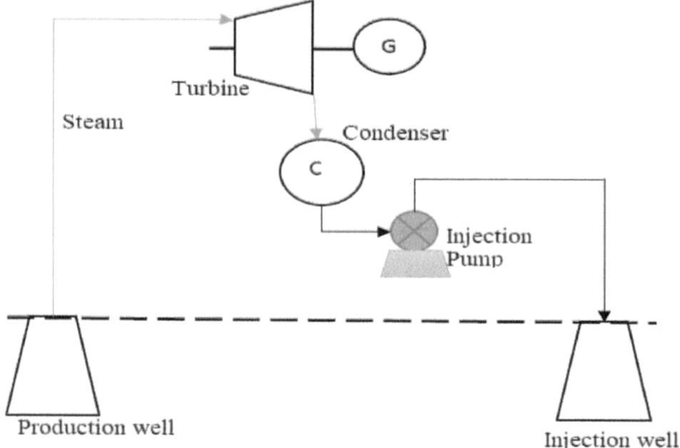

▲ **Fig-8.8: Dry steam geothermal power plant**

4.4.2 Flash steam geothermal power plant

In Flash steam geothermal plant, pressurized high-temperature water is drawn from beneath the surface into the vessel at the surface known as flash tanks. This is where the sudden decrease in pressure causes the liquid water to flash or vaporize into steam. The flash steam thus generated is then used to drive the turbine-generator set to generate power, Fig-8.9.

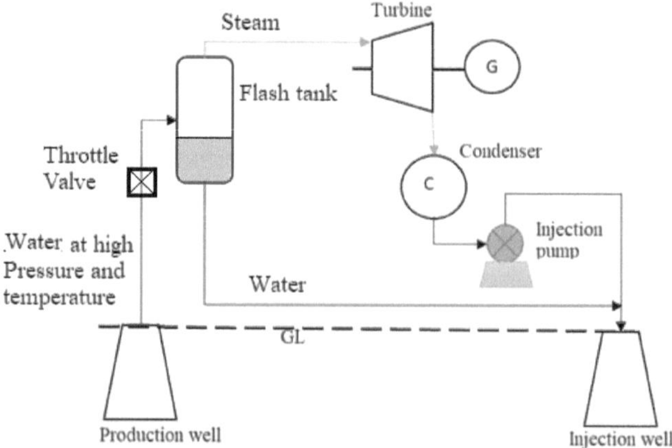

▲ **Fig- 8.9: Flash steam geothermal power plant**

4.4.3 Binary-cycle geothermal power plant

The binary-cycle power plants employ a secondary working fluid such as ammonia and hydrocarbon within a closed loop of pipes to power the turbine-generator set. In this process, geothermally heated water is drawn up through a different set of pipes, and much of the energy stored in the heated water is transferred to the working fluid through a heat exchanger. The working fluid then vaporizes and passes through the turbine in order to rotate it. It then re-condensed and piped back to the heat exchanger, Fig-8.10.

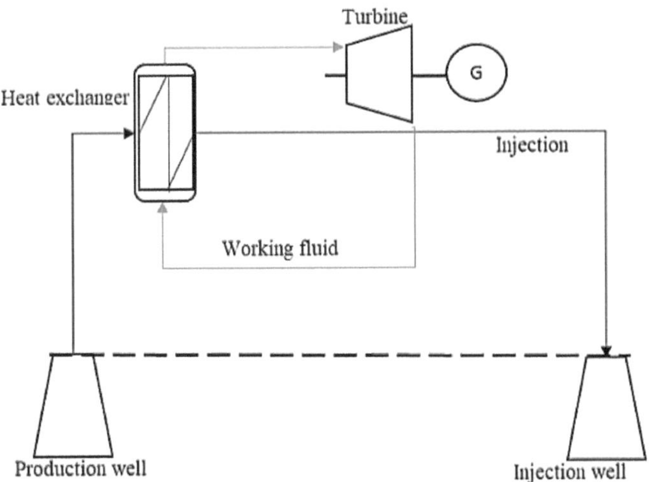

▲ **Fig-8.10: Binary cycle in Geothermal power plant**

4.5 Advantages and disadvantages of geothermal energy

i. **Advantages**

Below are the advantages of geothermal energy in its various applications.

a) Environmentally friendly than conventional fuel source such as fossil fuels.

b) Geothermal energy is renewable; they can last decades.

c) It is sustainable, stable and a reliable source of energy.

d) The energy can be used for heating and cooling in direct usage.

e) No fuel is required.

f) The energy generation process has been improved by new technologies, making it a rapid evolution.

g) Its power output is easy to predict.

ii. **Disadvantages**

Despite the advantages of geothermal energy, some limitations still occur. Below are the disadvantages of geothermal energy in its various applications.

a) Location is restricted, that is, geothermal plants must be built in places where energy is accessible.

b) Environmental side effects due to the digging as some gases stored under the earth are released into the atmosphere.

c) Geothermal energy can trigger earthquakes due to the alteration in the earth's structure as a result of digging.

d) High installation cost of geothermal plants.

e) High sustainability of geothermal energy fluid is required to be pumped back into the underground reservoirs faster.

5.0 Biomass energy

5.1 Introduction

Biomass energy, when solar energy is utilized indirectly, has been the major source of energy for human beings throughout the history of civilization. Biomass energy is the outcome of the photosynthesis.

5.2 Biomass as fuel

Biomass is produced by means of photosynthesis which is light-dependent process and chemosynthesis which is light-independent process. The majority of the biomass consists of green plants that harness solar energy and convert it into chemical energy by means of photosynthesis and is stored as biomass energy.

The biomass serves as a primary fuel as well as a secondary fuel. The processing of biomass and its further utilization makes it a secondary fuel. Many countries, including India, are exploring biomass as an alternative energy source (cow dung, poultry doweling, food waste, and biodegradable sewage waste, etc.).

Since the discovery of fire, bioenergy has been one of the most widely used forms of renewable sources of energy worldwide. Biomass provides fuel flexibility to match a broad range of energy demands. It can be stored and has a benefit over other sets of renewable energies.

Presently, all types of biomasses collectively provide nearly 14% of the global primary energy supplies and represent almost 80% of the world's renewable energy share. In some developing countries, the major share is from bioenergy (around 90% of energy supply), with the use of traditional biomass for cooking and heating (UNEP, 2016).

5.3 Biomass resources

Entire Biomass and biomass wastes are categorized as below:

i. Forestry waste: Logs, wood, chips, bark, and leaves are the major forestry waste. Sawdust produced during the processing of timber also adds to forest waste.

ii. Agricultural residues: The biomass produced as a by-product of processing, and harvesting of agricultural crops is known as agricultural residue.

iii. Agro-industrial waste: Paper mills, molasses, pulp waste from food processing industries, and textile fiber waste adds to agro-industrial waste

iv. Municipal waste: The waste collected from household consists mainly the organic portion that can be utilized for energy recovery. Food and kitchen waste, green waste, and other biodegradable portion of waste constitutes municipal solid waste. Sewage and animal manure from households also has energy potential.

v. Industrial wastes: The waste and wastewater from the industries like paper and pulp industry, dairy industry, breweries, vegetable packaging industry, and confectionary industry can also be used as energy resources. The food industry wastes from hotel, restaurants and community kitchens are also a potential bioenergy source. The bagasse, a byproduct residue from sugar mills after the juice extraction is used largely for cogeneration to produce electricity. The calorific value, moisture content, fixed carbon content, ash content are important properties in energy generation from the biomass.

5.4 Biogas

It is a clean fuel produced in anaerobic digestion of organic origin whose combustion produces fewer pollutants than other combustible energy sources. Although the production of energy requires a sufficient area of land and water, in its various forms, biomass energy appears to have a bright future as a source of energy.

5.4.1 Biogas potential of India

The biogas potential in India is estimated to be 12 million biogas plants of 2 cubic meters capable of producing 17,000 MW energy, with 4.75 million installed plants (42.7% of the target). State nodal departments and agencies are implementing the National Biogas and Manure Management Program. Khadi and Village Industries Commission, Mumbai is also implementing the program country-wise through its states' offices.

5.4.2 Feedstock for biogas production

The biogas is produced from organic wastes irrespective of the composition and with limited feedstock preparation. The substrate with a range of moisture content can be used for biogas generation. The solid or slurries, concentrated or dilute liquids of organic wastes can be utilized.

The manures are widely used feedstock for biogas production and its biogas production potential is given in Table 1.

▼ **Table-1. Potential biogas yield from different feedstocks**

Feedstock	Biogas yield per ton of fresh matter(m^3)
Cattle waste	55-68
Buffalo waste	0.54
Piggery waste	0.18
Chicken waste	126
Horse manure	56
Pig slurry	11-25
Sewage sludge	47

Source: Sustainable Energy Authority of Ireland (2002).

5.4.3 Composition of Biogas:

Biogas is a mixture of gases mainly composed of methane and carbon dioxide (Table 2). Biogas is characterized based on its chemical composition and the physical characteristics which result from it.

▼ **Table-2. Composition of Biogas from cattle dung**

Component	**(% by volume)**
Methane	50-70
Carbon di oxide	20-40
Nitrogen	0-5
Hydrogen	0-1
Hydrogen Sulfide	0.1-0.5

Methane is the chief combustible component and mainly determines the properties of biogas. The relative percentages, quantity, and quality of gas produced during anaerobic digestion depend on the quality of feed material and the process conditions. Biogas is normally rich in methane (about 65%) with the major impurities of hydrogen sulfide, carbon dioxide, and water. The carbon dioxide present in the digester is not necessarily a contaminant, but it dilutes the energy content of biogas, lowering its calorific value.

5.4.4 Components of Biogas Plants

a) **Mixing tank:** The feed material (dung) is collected in the mixing tank. Sufficient water is added and the material is thoroughly mixed till homogeneous slurry is formed.

b) **Inlet pipe:** The substrate is discharged into the digester through the inlet pipe/tank.

c) **Digester:** The slurry is fermented inside the digester and biogas is produced through bacterial action.

d) **Gas holder or gas storage dome:** The biogas gets collected in the gas holder, which holds the gas until the time of consumption.

e) **Outlet pipe:** The digested slurry is discharged into the outlet tank either through the outlet pipe or the opening provided in the digester.

f) **Gas pipeline:** The gas pipeline carries the gas to the point of utilization, such as a stove or lamp.

5.4.5 Different types of biogas plants in India

The main approved models of biogas plants in India are grouped into the following two categories, Fig-8.11 (a) & (b). These are i) Floating drum ii) Fixed dome

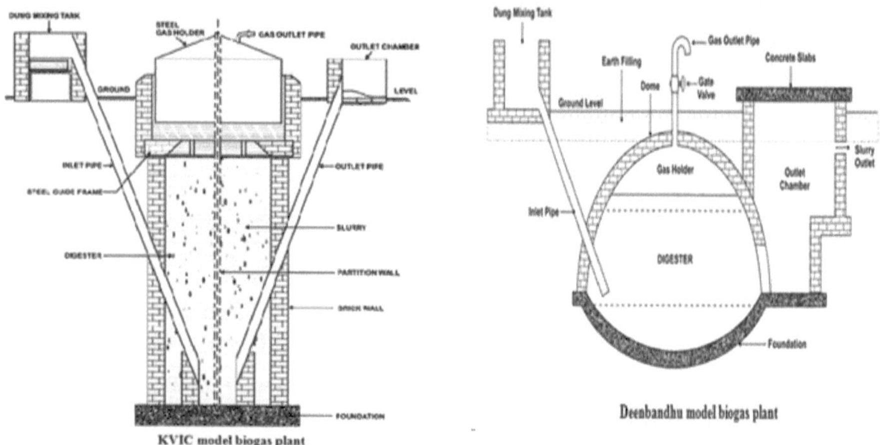

▲ Fig-8.11 (a): Floating drum biogas plant (KVIC Model) Fig-8.11(b): Fixed dome biogas plant (Deenbandhu Model)

Initially the digester is filled with a uniformly premixed mixture of dung and water (1:1 ratio) and the digester may be filled in three or four days or more time depending upon the availability of the dung. In order to facilitate gas production, addition of 5 to 10% inoculums, taken from a running biogas plant, will hasten the process by three to four days. In case no inoculums are available, sewage sludge can also be added. The first two or three installments of gas will not burn because of excessive CO_2.

When the cattle dung is used as feed stock, the biogas plant is to be filled with a homogenous slurry made from a fresh dung and water in a ratio of 1:1 up to the level of the second step in the outlet chamber.

As the gas generates and accumulates in the empty portion of dome of the biogas plant, it presses down the slurry of the digester and displaces it into the outlet chamber. The slurry level in the digester falls, whereas in the outlet chamber, it starts rising with the formation of gas. This fall and rise continues till the level in the digester reach the upper end of the outlet opening, and at this stage, the slurry level in the outlet chamber will be at the slurry outlet. Any gas produced after this stage will escape through the outlet chamber till the gas is not used. When the gas is used, the slurry which was earlier displaced out of digester and stored in the outlet chamber begins to return into the digester. The difference in levels of slurry in digester and the outlet chamber exerts pressure on the gas which makes it flow through the gas outlet pipe to the points of utilization of biogas.

The requirement of biogas plant dependent on:

a) average daily feedstock

b) hydraulic retention time

c) digester volume

Example-1: Cattle dung requirement for a biogas plant

One kg of cow dung along with an equal quantity of water under anaerobic conditions in a day produces 0.04 m^3 or 40 liters of biogas.

Example-2: Find out the quantity of cow dung required for a biogas plant that generates 1m3 of biogas with a production of 40 L biogas per kg of cow dung per day.

Solution:

Capacity of biogas plant = 1m^3

Biogas production from one kg of cow dung per day = 40 L= 0.04 m^3/kg of cow dung Therefore the quantity of cow dung required for 1 m3 = 1m^3/ (0.04 m^3/kg of cow dung) = 25 kg of dung.

Example-3: Calculate digester volume for a daily feed of 25 kg of cow dung for an HRT of 40 days.

Solution:

Digester volume (L)= daily feed (L/day) x retention time (days).

Fermentation slurry = 25 kg cow dung +25 L water = 50 L Digester volume (L) = 50 (L/day) x 40 days = 2000 L= 2 m^3

5.4.6 Benefits of Biogas Technology

Production of biogas provides a versatile carrier of renewable energy, as methane can be used for the replacement of fossil fuels in both heat and power generation and as a vehicle fuel. Biogas can be used for Cooking, Lighting, Power Generation, Transport Fuel and as a fertilizer.

5.4.7 Constraints

Highly skilled personnel Improper preparation of effluent solids leading to blockage and scum formation, temperature fluctuations, maintenance of pH for optimal growth of methanogenic bacteria, C/N ratio, dilution ratio of influent solids content, corrosion of gas holder, pin-hole leakages (digester tank, holder, inlet, outlet), etc are some of the technical problems in biogas production and usage.

a. **Financial Constraints**-Digester design; high transportation costs of digester materials; installation and maintenance costs; increasing labour costs in distribution of biogas products for domestic purposes.

b. **Social Constraints**-Social constraints and psychological prejudice against the use of raw materials like night soil also prevents biogas from its wide use.

c. **Environmental constraints**-Most of the digesters in our country operate normally at ambient conditions. Northern India records a shortfall in biogas output during winters and in some other parts of the country, especially in dry tracts also affect the digester performance due to higher temperature.

There are different ways of enhancing biogas production involve the use of additives, recycling of digested slurry and slurry filtrate, Variation in operational parameters like temperature, hydraulic retention time (HRT) and particle size of the substrate and use of fixed film/bio-filters etc.

Questions:

1. What are different types and uses of solar energy?
2. How PV cell works?
3. Describe different components of solar water heater.
4. How power is generated in PV cell?
5. Describe solar thermal power plant.
6. What are the advantages and disadvantages of solar energy?
7. How to derive solar panel efficiency?
8. Explain working of horizontal axis windmill.
9. What are the advantages and disadvantages of vertical axis wind mill?
10. What are the different kind of geo-thermal power plant?
11. Explain binary cycle geo thermal power plant.
12. Why Biomass is considered alternate source of renewable energy?
13. What are the constraints in efficient use of a Biogas plant?
14. How Biogas plants can change the economic development of rural communities?

SECTION 09

COOLING TOWER, HEAT EXCHANGER, CONDENSER & EVAPORATOR

1.0 Introduction

For any industrial production process, a heat source and a sink is required to have a stability in it. Heat transfer, Heat& mass transfer are vital operations required in main production processes as heat source and a sink. Different types of equipment and sub-systems are in use to undertake the stated operations. Without these, continuous operation of main process is next to impossible. This section is included to have a discussion on the sub-systems: cooling tower, heat exchangers, condenser & evaporator.

1.1 Cooling Tower

The Cooling Tower is an important utility equipment which plays a vital role by providing cold water which is obtained by utilizing atmospheric air to cool the incoming hot water. Cooling water is required for cooling purposes and is used for process cooling, condensation of vapour, to take away heat from heated part of any equipment etc. Return hot cooling water is again used for the cooling purpose after getting cooled in an equipment with the help of air, called cooling tower.

1.2 Cooling Tower Basics - Principles of Operation

Warm water from the heat source is pumped to the water distribution system at the top of the tower. The water is distributed over the wet-deck filled by spray nozzles. Simultaneously, air is drawn through air-inlet louvers and through the wet-deck surface causing a small portion of the water to evaporate. The evaporative process removes heat from the water. The warm, moist air is drawn out of the top of the tower. The resulting cold water is then recirculated back through the heat source in a continuous cycle.

1.3 Types of cooling tower

Different types of Cooling towers are available in industry for the purpose of cooling the hot water. The types are mainly based on interaction between air and water and are discussed below.

1.3.1 Cross flow Cooling Tower

In cross flow towers, Fig-9.1(a), air flow is directed across the water flow. Air enters through the vertical faces of the tower to meet the fill. Hot water is distributed to the fill, perpendicular to the air flow, by gravity through perforated basins. The air passes through the fill, interact with the water and gets discharged out of tower with the help of induced draft fan. The turbulent air flows through the fill structure is to maximize the contact with the water thus drawing heat out of the water. Air moves horizontally across the path of flowing water in a **cross-flow cooling tower**. A cold water basin provided at bottom of the tower to collect the cooled water after its interaction with the air flow.

1.3.2 Counter flow Cooling Tower

In counter flow towers, air flows opposite to the water flow, Fig-9.1(b). Air enters at the tower below the fill and is drawn up vertically into the tower. Above the fill, hot water is sprayed through nozzles in fine droplets over the surface of the fill. Heat of water is drawn away by the cooling air as water progresses to the bottom of the tower. The drift eliminator above the spray nozzles captures water droplets and returns the water to the circulating system. A cold-water basin collects the water after its interaction with the air flow. In a counter flow tower, air flows in the opposite direction (counter) to the falling water. The cross-flow cooling tower occupies less area than a crossflow cooling tower.

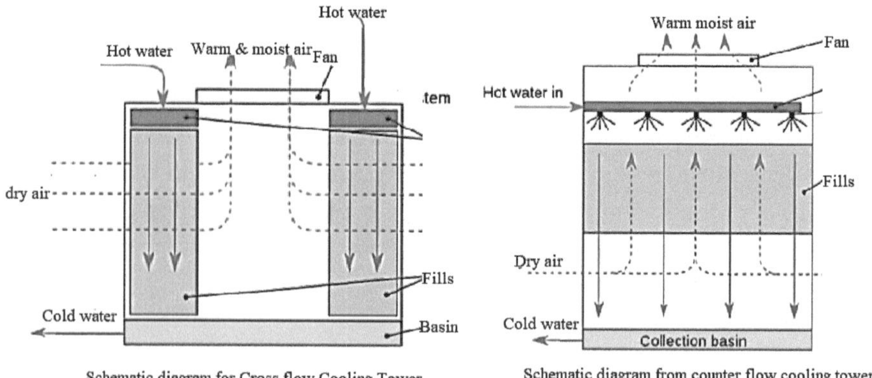

▲ **Fig-9.1 (a) & (b): Cross flow cooling tower & Counter flow cooling tower**

1.3.3 Induced Draft Cooling Towers

In induced draft cooling towers, Fig-9.2(a), incoming cooling water is injected throughout the cooling tower by a spray distribution header. Air enters the cooling tower's interior through appropriately built 'louvres.' Big fans pull air through the baffled region of cooling the water.

1.3.4 Forced Draft Cooling tower

Forced draft fans are provided at the bottom of the tower to supply cold air inside the tower, Fig-9.2(b). Hot water is sprayed on wet deck surfaces, made of vertical fills, through nozzles. While ascending, air comes in close contact with the hot water to take way heat from it.

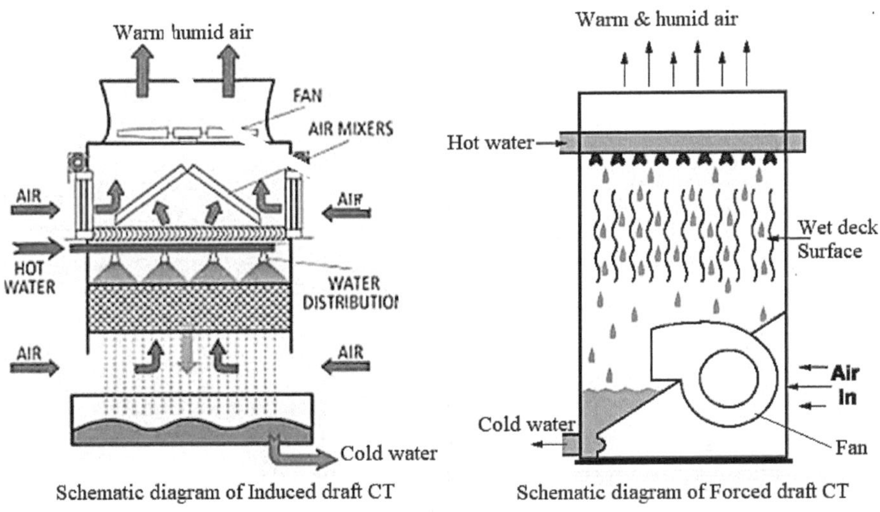

▲ **Fig-9.2 (a) & (b): Induced draft cooling tower & Forced draft cooling tower**

1.3.5 Natural Draft Cooling Tower

Natural drafting cooling tower generates airflow by chimney effect due to its shape, which is a hyperbolic structure, Fig-9.3(a). As warm and wet air is less dense than cold and dry air, it ascends out of the cooling tower and in turn drawing in denser fresh air from the surrounding environment. The temperature difference between the warm air within the tower and the colder air outside, generates the ideal airflow inside the tower. Natural draught cooling towers are the best option for cold and humid locations as well as for high winter loads.

1.3.6 Dry Cooling Towers

Dry cooling towers are among the other cooling towers intended to remove heat from hot circulating water without water evaporation rather by heat exchanging, Fig-9.3(b). In such cooling tower air is used as coolant for cooling the return hot water. Conventionally, in wet cooling tower water gets cooled by evaporation with the help of air where as, in dry cooling tower air takes away heat from hot water by heat exchanging which is circulated through the finned tube. Such type of tower is commonly used in those area where there is the scarcity of water.

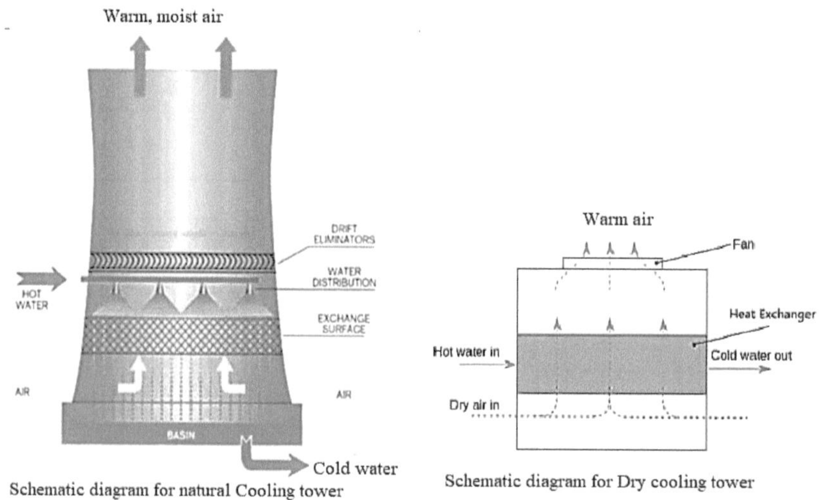

▲ **Fig-9.3(a)&(b): Natural cooling tower & Dry cooling tower**

1.4 Performance assessment of Cooling tower

1.4.1 Definitions related to CT performance assessment

Following definitions need to be discussed before assessing the performance of Cooling Tower.

i. **Range (R)**- It is the temperature difference between inlet hot water & cooled outlet water from cooling tower.

ii. **Ambient WBT**- Temperature of ambient measured when the bulb of the thermometer is made wet. Cooling Tower performance is majorly driven by ambient WBT.

iii. **Approach (A)**-Temperature difference between Cooled water temperature & ambient WBT.

1.4.2 Equations related to cooling tower Performance assessment

i. **Effectiveness**

This is the ratio between the range and the ideal range (in percentage), i.e. difference between cooling water inlet temperature and ambient wet bulb temperature.

Or, CT Effectiveness (%) = Range / (Range + Approach). =R/(R+A) x100.

Or, CT Effectiveness (%) = 100 x (CW inlet temperature - CW out temp) / (CW in temperature -ambient WB temperature).

The higher this ratio, the higher the cooling tower effectiveness.

ii. **Cooling capacity**

Cooling capacity is derived as the heat rejected in k Cal/hr. or TR, given as product of mass flow rate of water, specific heat and temperature difference. The capacity of Cooling Tower can be calculated from water side heat and material balance as per following formula:

Heat capacity = (m x C_p x ΔT) Kcal/hr.

or, TR = (m x C_p x ΔT)/ 3024.

When,

a) m: mass flow rate of circulating water (Kg/hr)

b) C_p: Specific heat of circulating water (Kcal/ Kg °C)

c) Δ T: Temperature difference (I/L water - O/L water) (°C).

iii. **Evaporation loss**

Evaporation loss is the water quantity that gets evaporated for cooling effect.

a) Theoretically, the evaporation quantity works out to 1.8 m^3 for every 1,000,000 kcal heat rejected (1,000,000/540x1000 =1.8 m^3, Latent heat of evaporation of water at atm. pressure is 540 Kcal/kg)

b) The following formula can be used (Perry): Evaporation loss (m^3 / hr.) = 0.00085 x 1.8 x circulation rate (m^3 /hr.) x (T1-T2) [T1 - T2 = temperature difference between inlet and outlet water].

c) Or, Evaporation loss= Heat load (Kcal/hr.) /540 x Water circulation rate(kg/hr.).

d) For every 5.5°C of water cooling in the tower, there will be 1 percent of water mass lost due to evaporation.

iv. **Drift loss**

The loss depends on the efficiency drift eliminator. The typical losses for different type of CT can be as below.

a) Natural draft CT - 0.3 to 1.0% of circulating water.

b) Induced draft CT- 0.1 to 0.2% of circulating water

c) High efficiency drift eliminator- 0.0005 to 0.001% of circulating water

v. **Cycles of Concentration (C.O.C)**

As water evaporates during the normal operations of the cooling tower, dissolved solids (TDS), such as magnesium, silica, chloride, and calcium, remain in the water cycle that recirculates through the system. This concentration of solids can become too high over a course of time, which can cause both scale and corrosion to form.

The highly concentrated water is replaced by draining, called Blow Down, with makeup water. Blowdown loss is directly related to the cooling tower's cycle of concentration (COC), which is the ratio of chloride/TDS/ Silica content in circulation water and in makeup water.

This is the ratio of dissolved solids in circulating water to the dissolved solids in makeup water.

COC = Chloride or TDS or silica in Circulating water / Chloride or TDS or silica in make-up water

vi. **Blow down loss**

Blow down losses depend upon cycles of concentration and the evaporation losses and is given by formula: Blow down quantity (kg/hr.) = Evaporation loss (kg/hr.) / (C.O.C. - 1).

vii. **Total Water Loss in Cooling tower**

Total water loss can be derived and this is the quantity of make water to be added to make up loss.

M = D + E + B

When,

a) M = Makeup water

b) D = Drift loss

c) E = Evaporation loss

d) B = Blowdown loss

viii. **Liquid/Gas (L/G) ratio**

The L/G ratio of a cooling tower is the ratio between the water and the air mass flow rates.

Therefore, the following formulae can be used:

L (T1 - T2) = G (h2 - h1)

Therefore, L/G = (h2 - h1) / (T1 - T2).

When:

a) L/G = liquid to gas mass flow ratio (kg/kg)

b) T1 = hot water temperature (°C)

c) T2 = cold-water temperature (°C)

d) h2 = enthalpy of air-water vapor mixture at exhaust wet bulb temperature (kcal/kg).

e) h1 = enthalpy of air-water vapor mixture at inlet wet-bulb temperature (kcal/kg).

Cooling towers have certain design values, but seasonal variations require adjustment and tuning of water and air flow rates to get the best cooling tower effectiveness. Adjustments can be made by water box loading changes or blade angle adjustments. Thermodynamic rules also dictate that the heat removed from the water must be equal to the heat absorbed by the surrounding air.

1.4.3 Factors affecting Cooling tower effectiveness

Table- 9.1 shows CT effectiveness in different operating conditions.

▼ **Table-9.1: Factors affecting CT effectiveness**

Parameter	**Unit**	**Design condition**	**Condition-1**	**Condition-2**
WBT	^{0}C	28	28	26
CW inlet temperature	^{0}C	37	39	40
CW outlet temperature	^{0}C	32	32	32
CT range	^{0}C	5	7	8
CT approach	^{0}C	4	4	6
CT effectiveness	%	5/(5+4)=55.5	7/(7+4)=63.6	8/(8+6)=57.1

Inference from the chart:

a) Higher the range, higher the effectiveness

b) Higher the approach, lower the effectiveness

c) CT Effectiveness is higher with: higher Range & Lower Approach

d) Lower CT O/L water temp, better the performance of Cooling Tower

I. **O/L Water temperature:** Since, the heat transfer is in between water and air, the approach practically could not be zero & hence, the cooling tower effectiveness can't be 100%. Lower CT O/L water temp, better the performance of Cooling Tower. But the CT O/L temperature gets largely affected by following five factors:

   a) Ambient WBT

   b) Hot Water Temperature (CT Inlet Water Temperature)

   c) Cold Water Temperature (CT outlet/Sump Water Temperature)

   d) Air Flow Rate [L-Liquid)

   e) Water Flow Rate [G-Gas)

II. **Effect of L/G ratio:** Cooling Tower Performance is not to be decided based on the value of % effectiveness. But the same is also to be evaluated from L/G ratio (Liquid to Gas ratio) & then the same is to be compared with designed conditions for particular cooling tower. It is important to maintain the required L/G ratio as per designed conditions.

   a) If L/G ratio < than the rated: It means either circulating water flow rate is lower or the air flow rate is higher.

   b) Other condition of higher L/G > than the rated: It means, either circulating water flow rate is higher or the air flow rate is lower.

The corresponding adjustments can be done either in liquid or gas flow rates, as per convenience by adjusting the VFD frequency to get the corresponding saving in energy consumption.

1.4.4 Cooling tower size estimation

Cooling tower size is affected by the heat load, range, approach & Wet Bulb temperature in following ways:

a) Higher is the Cooling tower size, higher the cooling tower can take load [Range, approach &WBT are constant]

b) Higher is the size, lower is the range achieved [load, approach & WBT are constant]

c) Higher is the size, lower is the approach (can be achieved) [load, range & WBT are constant]

d) Higher is the size required, if WBT is low [load, range& approach are constant].

2.0 Heat-Exchanger

Heat exchangers are equipment that transfer heat from one medium to another. Proper design, operation and maintenance of heat exchangers make the energy-efficient process and minimize energy losses.

2.1 Types of Heat exchanger

Different types of HEs are available in industry but most commonly used HEs are a) shell & tube type b) Plate heat exchanger.

2.1.1 Tube & shell type heat exchanger

A shell and tube heat exchanger uses multiple tubes installed in a cylindrical shell to exchange heat between two working fluids through thermal contact. Shell and tube heat exchanger is also called a tube and tube heat exchanger.

2.1.2 Types of Shell and Tube Heat Exchangers

The shell and tube heat exchanger has the following major types:

a) **Fixed tube sheet exchanger-** In fixed tube sheet exchangers, the tube sheet is welded with the shell. The fixed tube sheet heat exchanger has a simple and inexpensive structure that enables mechanical or chemical cleaning of the tube holes. Except for chemical cleaning, the tube's outer surface of this exchanger is not accessible, fig-9.4(a).

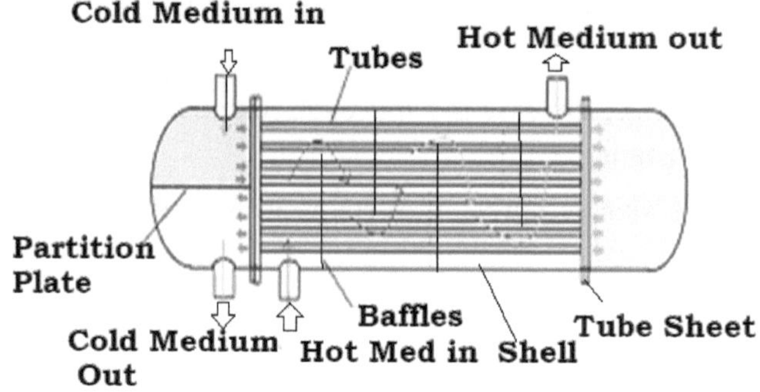

▲ **Fig-9.4 (a): typical shell-tube heat exchanger**

b) **Floating head Heat Exchanger-** In the floating heat exchanger, the tube sheet at the end of the rear header is not welded with the shell but can float or move. The diameter of the front header tube sheet is more than the shell, and the sealing method is the same as the fixed tube sheet exchanger, Fig-9.4(b).

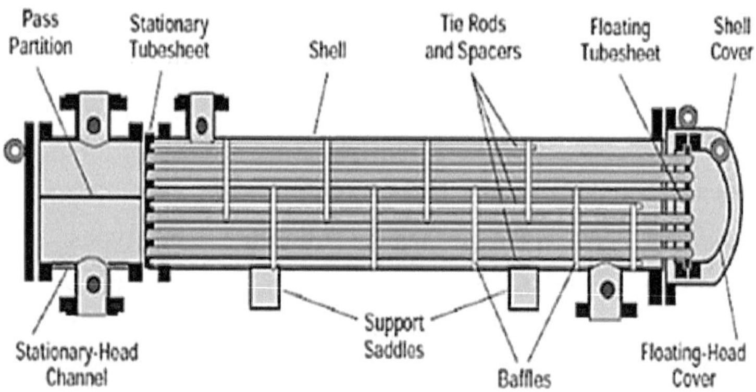

▲ **Fig-9.4 (b): Floating heat exchanger**

As the tube sheet at the shell's rear header end has somewhat less diameter than that of the shell, so a tube bundle can easily pass by the shell. By using a floating head exchanger, the thermal expansion can be taken into account, and the tube bundle can be removed and cleaned easily.

c) **U-tube Heat Exchanger-** In this type of shell and tube heat exchanger, the tubes are bent into a U-shape. Therefore, this heat exchanger is known as a U-tube heat exchanger. This heat exchanger uses an M-type rear head and any type of front header, fig-9.4(c).

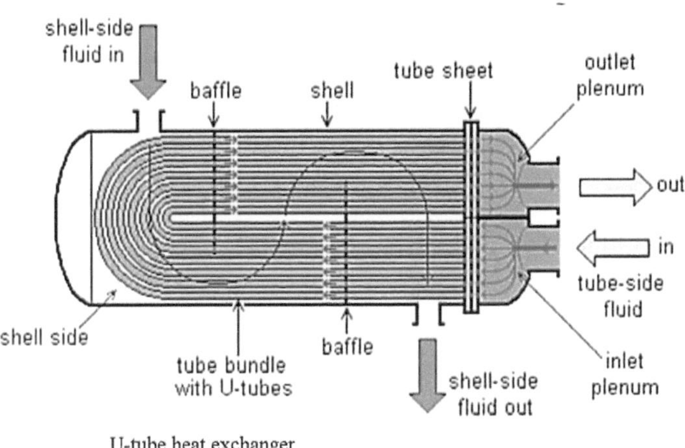

▲ **Fig-9.4 (c): U-tube heat exchanger**

2.1.3 Advantages and Disadvantages of Shell and Tube Heat Exchanger

i. Advantages of Shell and Tube Heat Exchanger

a) These are the best choice for high-temperature and pressure applications instead of plate heat exchangers.

b) The tube heat exchangers have a lower pressure drop than the plate heat exchangers.

c) The shell and tube heat exchanger doesn't foul so easily like a plate heat exchanger.

d) These exchangers have steady and flexible designs.

e) There are no dimensional limitations.

f) These exchangers can be easily assembled and disassembled for cleaning, repair, and maintenance.

g) These types of heat exchangers have more versatility than other types.

h) The shell and tube heat exchanger has lower cost than the plate heat exchanger.

ii. Disadvantages of Shell and Tube Heat Exchanger

  a) This heat exchanger has low efficiency than a plate exchanger.

  b) It requires more floor space for installation.

  c) The cooling capacity of these exchangers can't be improved.

2.1.4 Application of Shell and Tube Heat Exchangers

The shell and tube heat exchangers are most commonly used in the following major applications.

1. Power plant
2. Chemical industries
3. Refrigeration
4. Metals and mining
5. Oil and Gas
6. Fertilizer industries
7. Cooling of engines, compressors, and turbines
8. Pharmaceutical
9. Refinery and Petrochemical
10. HVAC systems

2.1.5 Plate Heat exchanger

The design of a plate heat exchanger (PHE) comprises several heat transfer plates. Held by a fixed plate and a loose pressure plate to form a complete unit, Fig-9.5(a). Each heat transfer plate has a gasket arrangement, providing two separate channel systems.

The arrangement of the gaskets allows through flow in single channels. This enables the primary and secondary media in a counter-current flow, Fig-9.5(b). The mediums are not mixed because of the gasket design.

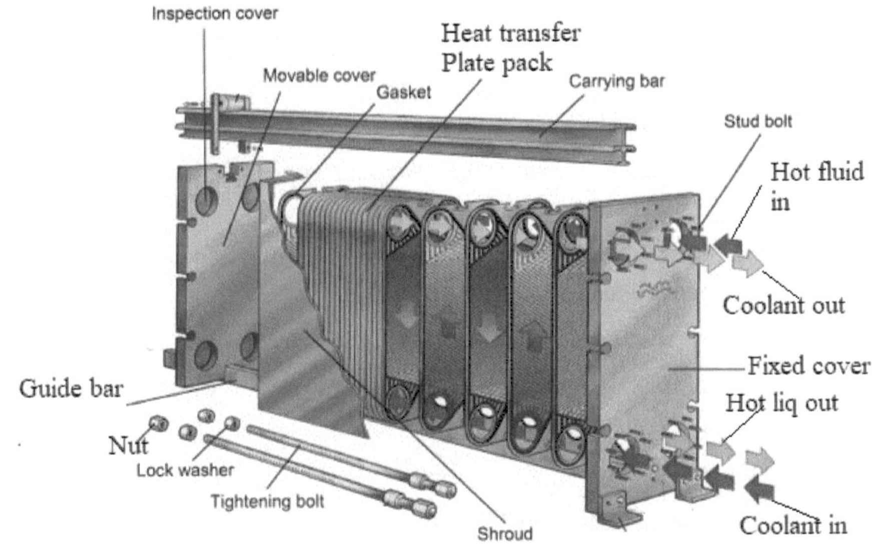

▲ **Fig-9.5(a) Different parts of PHE**

The corrugated plates create turbulence in the fluids as they flow through the unit. This turbulence gives an effective heat transfer coefficient.

Plate heat exchanger optimizes heat transfer. The corrugated plates provide easy heat transfer from one gas or liquid to the other.

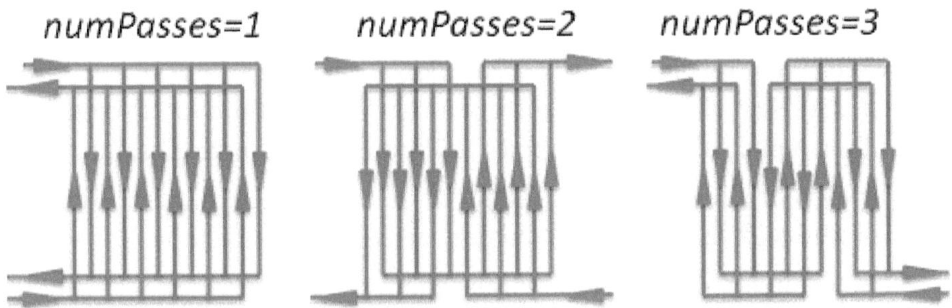

▲ **Fig-9.5(b): Schematic diagram showing flow path in Plate Heat exchanger**

Fluids run counter-current through the heat exchanger. This gives the most efficient thermal performance. It also enables a very close temperature approach.

i. **Advantages of plate heat exchanger**

a) Heat transfer – Improved temperature approach, true counter-current flow, 80-90% less hold-up volume.

b) Low cost - Low capital investment, installation costs, limited maintenance and operating costs.

c) Greatest reliability - less fouling, stress, wear, and corrosion.

d) Responsible - least energy consumption for most process effect, reduced cleaning.

e) Easy to expand capacity – adjustable plates on existing frames.

ii. **Disadvantages of plate heat exchangers:**

a) Poor sealing would cause leakage which will be a replacement hassle.

b) Limited pressure use, generally not more than 1.5MPa.

c) Limited operating temperature due to temperature resistance of the gasket material.

d) Small flow path, and not suited for gas-to-gas heat exchange or steam condensation.

e) High blockage occurrence especially with suspended solids in fluids.

f) The flow resistance is larger than the shell and tube.

2.2 Working Principle of Heat exchanger

Heat exchanger functions by transferring heat from higher to lower temperatures. Heat can thus be transferred from the hot fluid to the cold fluid if a hot fluid and a cold fluid are separated by a heat-conducting surface.

The operation of a heat exchanger is governed by thermodynamics. Heat can be transferred with the help of conduction, convection, or radiation.

a) Conduction is the transfer of thermal energy from one material to another through the separating body.

b) Convection is the transfer of thermal energy from one surface to another through the motion of a fluid such as heated air or water.

c) Thermal radiation is a heat energy transfer mechanism characterised by the emission of electromagnetic waves from a heated surface or object.

Mostly heat exchangers transfer heat through convection & conduction heat transfer phenomena.

2.3 HEs type based on fluid flow direction

i. **Counter Flow-**Exchangers when in the fluid flow direction of the cold and exchanger hot fluids are opposite. Normally preferred flow in exchanger.

ii. **Co- Current Flow-** An exchanger when in the fluid flow direction of the exchanger cold and hot fluids are the same.

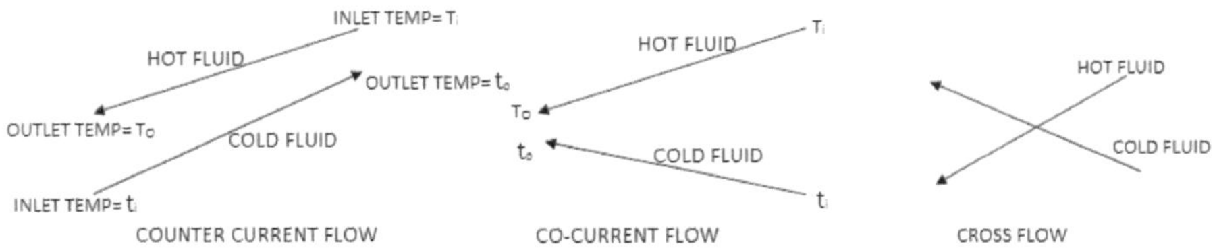

▲ **Fig-9.6: Fluid flow directions with heat exchange**

iii. **Cross flow** -An exchanger when in the fluid flow direction of the cold and hot fluids are in cross.

2.4 Performance assessment for heat exchanger

Performance of heat exchanger can deteriorate with time due to:

a) Operation beyond design
b) Fouling inside tube
c) Scaling on heat transferring surface
d) Insulation failure.

Hence, it is important to assess the heat exchanger performance periodically in order to maintain its efficiency. For the purpose of performance assessment, test is undertaken to derive performance related parameter and to compare with the design values for assessing the causes of deviation.

2.4.1 Derivation of Parameters to assess performance of a Heat Exchanger

1. **Overall heat transfer coefficient, U**

   i. **Equation for U- value**

   Overall heat transfer coefficient U is derived from the standard heat transfer equation:

   Q= UA LMTD; Then, U = Q/ (A x LMTD)

   When,

   a) U= Overall heat transfer coefficient Kcal/ (m^2-hr-°C) or W/ ($m^{2o}C$) [1W/m^2-°C = 0.85984 kcal/(hr- m^2- °C).

   b) Q= heat transferred (Kcal/ hr.)

   c) A= Surface area (m^2)

   d) LMTD= Log Mean Temperature Difference (°C).

   ii. **Calculation of U-value**

   U can be evaluated from the equation, fig-9.7:

   1/U= 1/ h1 +L/k+1/h2

   When,

   h1&h2= convective heat transfer coefficient, W/($m^{2o}C$)

   L = thickness of the wall, m

   k = thermal conductivity, W/(m °C)

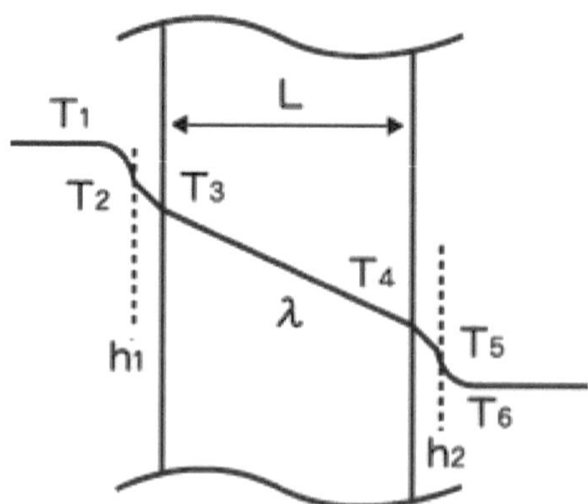

▲ **Fig-9.7: Diagram of Heat exchange**

Table-1 shows typical values for water & steam for h as reference for the convective heat transfer coefficients:

▼ **Table-1: Convective heat transfer coefficient(h) of few fluids**

Fluid	**Convective heat transfer coefficient (h)**	
	W/(m² °C)	Kcal/(hr-m² -°C)
Water	about 1000	860
Hot Water	1000 – 6000	860-5160
Steam	6000 – 15000	5160-12900

Example: A carbon steel (k = 50 W/(m °C) jacketed kettles with an inner wall thickness of 15mm are used to heat water. One uses hot water as the heat source. Assuming heat transfer coefficients of 1000 W/m² °C for the water being heated, 3000 W/m² °C for hot water, let's calculate the U values for heating processes.

Solution:

1/U= 1/h1+k/L+1/h2=1/3000+0.015/50+1/1000

U=612 W/ (m²°C) =526 Kcal/(hr-m²-°C).

▼ **Table-2: Typical Overall heat transfer Coefficient (U)**

Type of Heat Exchanger	Over all heat transfer coefficient (U)	
	Watt/(m²-°C)	Kcal/(m²-hr-°C)
Free convection gas to free convection gas	1-2	0.86-1.72
Flowing gas to flowing gas (laminar)	10-15	8.6-13
Water (forced) to oil (forced)	100-400	86-344
Water (forced) to water (forced)	850-2000	730-1700
Steam Condenser (water in tube) forced to Condensing	1000-5000	860-4300
Ammonia condenser (water in tube)	800-1400	680-1200
Alcohol condenser (water in tube)	250-700	210-600
Feed water heater (water in tube) Condensing to forced	1000-6000	860-5100
Plate heat exchanger (liquid to liquid)	1000-4000	860-3440

2. Derivation of LMTD

LMTD is Log Mean Temperature Difference.

Equation: LMTD = (ΔT1- ΔT2)/ Ln (ΔT1 /ΔT2)

When, ΔT1& ΔT2 are terminal temperature difference, fig-9.8(a).

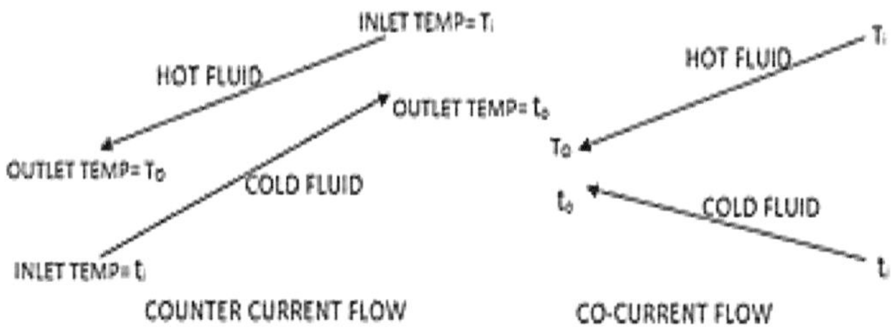

▲ **Fig-9.8 (a): Co-current & Counter current heat exchanging**

Terminal temp difference (°C)	Counter current flow	Co-Current flow
ΔT1	T_o-T_i	T_i-t_i
ΔT2	T_i-t_o	To-t_o

3. Heat duty of a heat exchanger

It is defined as the quantity of heat the heat exchanger can transfer per unit time. Heat duty of the exchanger can be calculated either on the hot side fluid or cold side fluid, Fig-9.8(b) as per below equations.

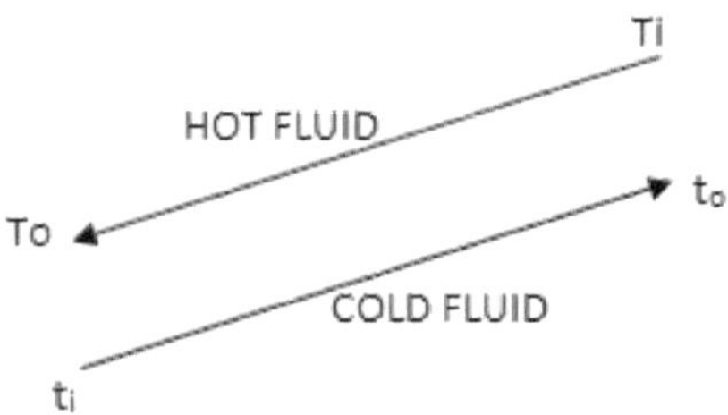

▲ **Fig: 9.8(b): Counter current heat exchanging**

a) Heat Duty for Hot fluid, Q_h = W x C_{ph} x (Ti –To) x 4.18 KJ/ s= kW

b) Heat Duty for Cold fluid, Qc = w x C_{pc} x (t_o–t_i) x 4.18 KJ/s= kW

c) Heat duty for evaporation = W x h_e x 4.18 kW

d) Heat duty for condensation= W x h_c x4.18 kW

When,

a) W, w = Quantity of hot and cold fluid respectively (Kg/s.)

b) C_{ph}, C_{pc} = Specific of hot & cold fluid respectively (Kcal/kg-°C)

c) h_e, h_c = Heat transfer coefficient during evaporation & condensation (Kcal/ (m^2-hr-°C).

If the operating heat duty is less than design heat duty, it may be due to:

a) Heat losses

b) Fouling in tubes

c) Reduced flow rate (hot or cold) etc.

4. **Capacity Ratio (R)**

Capacity ratio can be calculated as:

a) Heat load in hot fluid ÷ heat load in cold fluid.

b) Or, Temperature range of the hot fluid ÷ Temperature range of the cold fluid.

Higher the ratio, greater will be size of the exchanger.

5. **Effectiveness (S)**

It is defined as: Actual heat transfer rate to cold fluid/maximum possible heat transfer rate

Equation: S = Increase in temperature of cold fluid (t_o-t_i)/[Hot fluid inlet temperature(T_i)-Cold fluid inlet temperature(t_i)].

Higher the ratio, lesser will be requirement of heat transfer surface.

6. **Correction factor(F) (counter flow & cross Flow)**

Correction factor can be derived with help of S&R from the equation:

F= [√(R^2+1) x Ln{(1-S)/(1-RS)}] ÷ (R-1) Ln[2-S{R+1-√(R^2+1)}/{2-S(R+1) +√(R2+1)}].

7. **Temperature Range**

The difference in the temperature between the inlet and outlet of a hot/cold fluid in a heat exchanger is called temperature range.

a) $\Delta T = (T_i - T_o)$ is the temperature range for hot fluid

b) $\Delta t = (t_o - t_i)$ is the temperature range for cold fluid.

8. **Corrected LMTD & Heat flow**

a) Corrected LMTD = **LMTD x F (correction Factor)**

b) Corrected Heat Flow Equation: **Q= U x A x F x LMTD**

9. **Cleanliness**

   Cleanliness (C) = (Design LMTD/ Actual LMTD) x (Q actual/Q Design)

2.4.2 Other derived parameters related to HE performance

1. **Heat Flux** -The rate of heat transfer per unit surface of a heat exchanger is called Heat flux J/(s-m^2) =W/ m^2.
2. **Heat Duty**-The capacity of the heat exchanger equipment expressed in terms of heat transfer rate J/s = W.
3. **Pressure drops across heat Exchanger**- It is the pressure difference between inlet to outlet
   a. **Equation**: P_i- P_o. When, P_i- Inlet pressure, Po–Outlet pressure.
   b. It can also be calculated as Pressure drop α (Flow)^2.
4. **Fouling & Fouling factor**
   a) **Fouling** is the phenomenon of formation and development of scales and deposits over the heat transfer surface diminishing the heat flux. The process of fouling gets indicated by the increase in pressure drop.
   b) **Fouling Factor** -The reciprocal of heat transfer coefficient is called fouling factor. It indicates of the dirt deposition formed in the heat exchange process. Higher the factor lesser will be the overall heat transfer coefficient. (m^2-°C)/W.

3.0 Condenser

A condenser is an essential component in industry and work as heat sink to condense any fluid from its vapour state to liquid state. It works at positive pressure also at negative pressure depends of vapour getting condensed.

3.1 Working of a condenser

The concept behind working of a condenser is heat transfer. That is, the thermodynamic principle describing that heat always moves from warmer to cooler environments.

Three different phases may happen in condenser.

a) The first phase is called the de-superheating. The vapour entering in the condenser is superheated and super pressurized. De- superheating means to remove sensible heat from the vapour and it is the initial cooling process.

b) The second phase is condensation when latent heat of evaporation is removed. Condensation involves the transformation of the gaseous phase into its liquid state.

c) The third and last phase is a sub-cooling state. The sub-cooling state is to cool the liquid below saturation temperature.

3.2 Types of Condensers

There are basically **three major types** of condensers used-

a) Air Cooled

b) Water Cooled

c) Evaporative type

d) Jet Condenser

3.2.1 Air Cooled Condensers

These **types of condensers exchange heat through the air, known as Air Cooled Condensers (ACC).** Air-cooled types use air to remove heat from vapour or steam inside the tube. When high-temperature Steam or vapour enters the tube, Steam transfers heat to the tube and hence, transfers heat into the surroundings due to the connection of air to the tube. The air flow can be natural or forced.

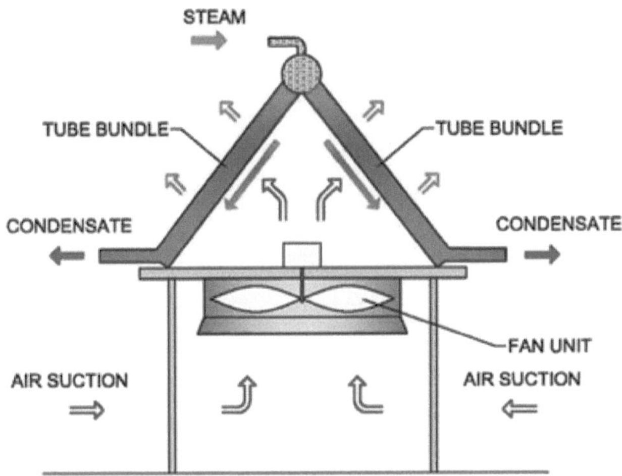

Schematic diagram of Air cooled condenser- forced draft type

▲ **Fig-9.9: Air cooled condenser-forced draft type**

In this Air-cooled type, there are **two Sub-types** of condensers. These are:

a) **Natural Convection**–When atmospheric air is passed through the condenser, vapour or gas inside the tube releases its heat to the surrounding atmosphere, resulting in the cooling of steam and turning to a liquid state. In Natural convection, Fins are used to increase the overall efficiency of the condenser. Fins raised the surface area of the condenser and therefore increased Heat rejection.

   Example of Natural convection is **Home Refrigerator**, As you see, there is no external fan to blow air to the condenser. It uses surrounding air as a coolant.

b) **Forced Convection**–This type of condenser needs a fan or blower to circulate the air over the condenser tube to cool. This condenser is much more efficient than natural convection because the mounted fan pressurized the natural air and passes through the condenser tubes and fins. Fig-9.9 shows forced convection air cooled condenser.

   Example: Forced air-cooled condensers are used in domestic deep freezers, water coolers, window air conditioners, power plant etc.

3.2.2 Water Cooled Condensers

Water-cooled type condensers use water as a cooling medium. These types of condensers are used in large refrigerating industrials needs and power industries.

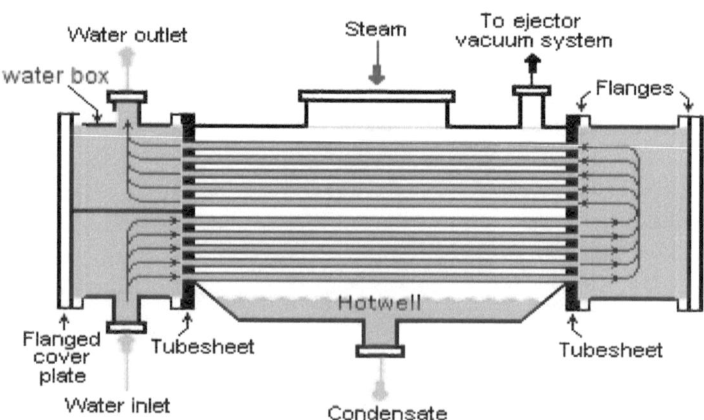

Schematic diagram for surface condenser

▲ **Fig-9.10: Surface condenser**

3.3 Main components- of a shell & tube type water cooled condenser

The main components of the condenser, fig: 9.10, are given below.

i. **Shell-** Shell is the outer surface and base of the condenser.

ii. **Tubes**– Tubes are the main component of the condenser.

iii. **Coolant**– To create temperature difference and to absorb heat from vapour, a coolant is needed. Normally water or is used as coolant.

iv. **Coolant inlet & outlet**– Whether coolant is to be flown through tube or shell depends on working pressure. Normally higher pressure fluid passes through tube as a design consideration.

v. **Vapour Inlet**– The connected flange, where high-temperature vapour enters into the condenser.

vi. **Condensate Outlet**– Exit flange where condensed liquid is taken out from condenser.

vii. **Non-condensing vapour outlet**– non-condensing gases or vapour is taken out for further process.

The best example of this type is surface condenser used in power plant.

3.3.1 Evaporative Type Condenser

This type of condenser has the features of both air and water-cooled condensing facilities.

Air and water are both used as a cooling medium.

Water is pumped from the sump of the evaporative condenser to a spray header sprayed over the condenser coil and at the same time, the fan draws air from the bottom side of the condenser and discharges it out at the top of the condenser.

The spray water comes in contact with the condenser tube surface, then it evaporates into the air streams.

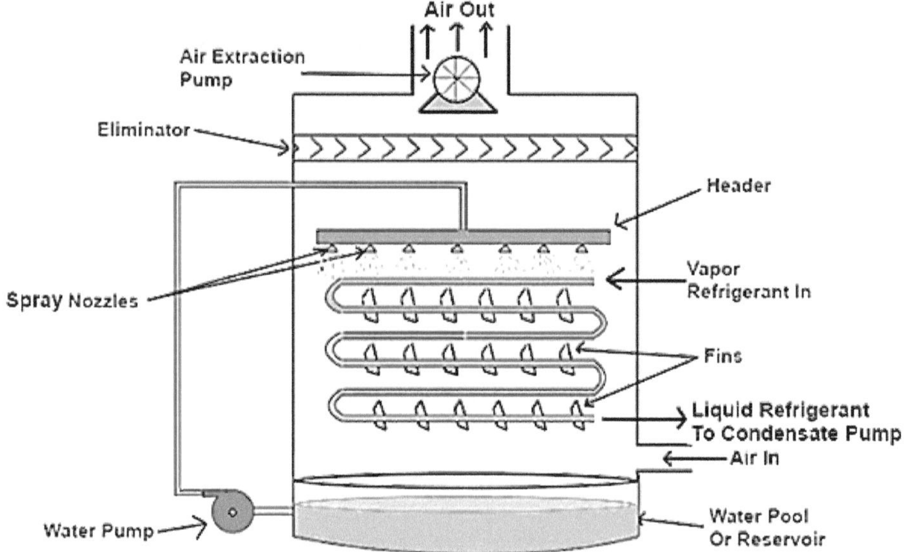

▲ **Fig-9.11: Evaporative condenser**

The source of heat for vaporizing the water is taken from the refrigerant, thereby condensing the gas.

Evaporative types consume less power than water-cooled types.

3.3.2 Jet type condenser

Jet condenser is a mixing type condenser where exhaust steam is condensed mix up with cooling water. In a jet condenser, high power is required for condensation. To reuse the water for boiler, condensate is used in **jet condenser** for condensation.

3.3.3 Different Types of Jet Condenser

According to the direction of flow and arrangement of tubing system, mainly four **types of jet condenser** is used.

1. Parallel flow jet condenser
2. Counter flow or low-level jet condenser
3. Barometric or high-level jet condenser
4. Ejector condenser

i. **Parallel-Flow Jet Condenser:** Exhaust steam, and cooling water enter the top of the parallel jet condenser and mix up together in the condenser shell. This mixture is removed from the bottom of the condenser. The condensate is removed by pump and uncondensed gas by air pump. At the time of separation, a vacuum is created in the condenser chamber. Condensate extraction pump delivers the condensate to the hot well where an overflow pipe is connected to the cooling water tank when surplus water of hot well is stored, Fig-9.12 (a).

ii. **Counter-Flow or Low-level Jet Condenser:** In counter flow or low-level jet condenser, exhaust steam enters lower side and cooling water from upper-side of the condenser chamber, Fig-9.12 (b). The flow direction of exhaust steam and cooling water is counter-current. Air pump which is placed at the top of condenser shell creates a vacuum; resulting drawing of the cooling water. A perforated conical plate stores this falling water, from which it escapes as jets and meets the exhaust steam entering at the bottom. Condensate and cooling water released for this rapid condensation to discharge the hot well through a vertical pipe by condensate pump.

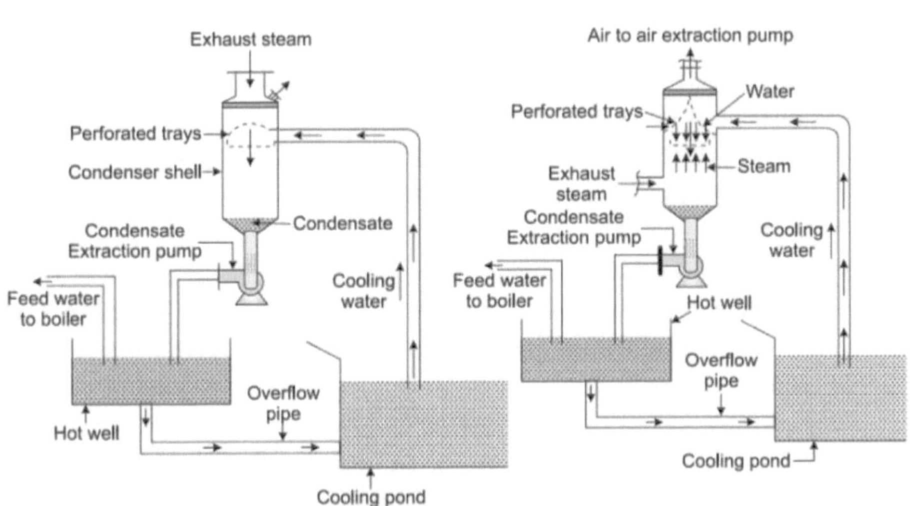

▲ **Fig-9.12 (a) & (b): Parallel flow jet condenser & Counter flow jet condenser**

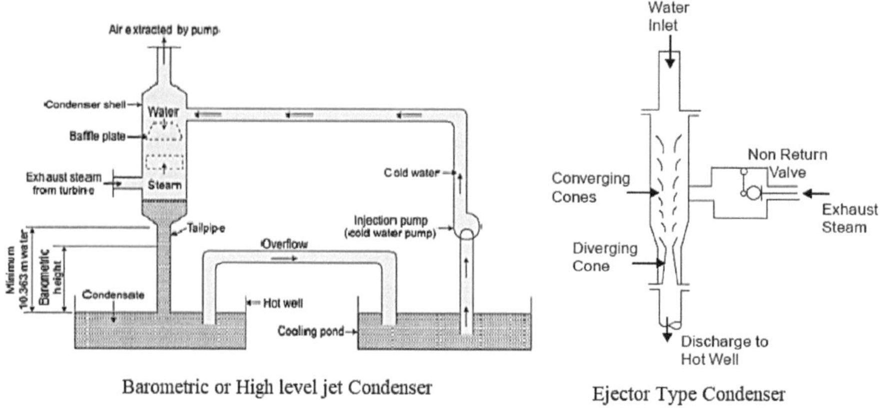

▲ **Fig-9.13 (a) &(b): High level jet condenser & Ejector type condenser**

iii. **Barometric or High-Level Jet Condenser:** In barometric or high-level jet condenser, a long discharge pipe is attached to the lower side of the condenser shell, called tailpipe, fig-9.13 (b). Condenser shell arrangement of this type is same as the low-level jet condenser. Exhaust steam enters to the lower side of the condenser and direction of steam flow is upward direction. Cooling water enters upwards from cooling pond and stored at perforated conical plate. Vacuum is created by air pump, placed at the top of the condenser shell. Now steam and cooling water mix up together and discharge hot well through a discharge pipe placed at the bottom of the condenser shell. There is no pump attached in between hot well and discharge pipe. The surplus water store in a cooling tank through an overflow pipe.

iv. **Ejector Condenser:** In ejector-type condenser, steam and water mix up simultaneously and pass through a series of metal cones. A non-return valve arrangement is arranged for entering the exhaust steam in the condenser shell. Water enters at the top through a number of guide cones. Steam and air pass through the hollow truncated cones. In the cone, kinetic energy is partly transformed to pressure energy. The condensate and discharge to the hot well, Fig: 9.13 (b).

3.3.4 Advantages and Disadvantages of Jet Condenser

i. **Advantages of Jet Condenser**

   a) Design of the jet condenser is simple.

   b) Less floor area is required as compared to the other condenser.

   c) Price is cheap of this condenser.

ii. **Disadvantages of Jet Condenser**

   a) Vacuum efficiency is low.

   b) Pure condensation is not possible. So it can't be reused.

3.4 Materials of Construction

MOC for condenser are:

a) Copper

b) Brass

c) Steel

d) Titanium

e) Nickel

3.5 Advantages of Condenser

Here are some major Advantages of Condensers:

a) It increases the expansion ratio of steam, and hence, increases the efficiency of the plant.

b) It reduces the backpressure of the steam, and thus more work can be obtained.

c) It reduces the temperature of the exhaust steam, and its increasing work efficiency.

d) The reuse of condensate as feed water for the boilers reduces the cost of power generation.

e) The temperature of the condensate is higher than that of the freshwater, therefore the amount of heat supplied per kg of steam is reduced.

f) Useful in Obtaining the better vacuum.

g) Less space is required.

3.6 Disadvantages of Condenser

Some disadvantages of condensers:

a) Needs Adequate supply of freshwater.

b) Difficult to repair.

c) Corrosion/ scale deposition.

d) Requires more space.

e) High maintenance cost.

3.7 Applications

Applications of condensers: Power plant, Chemical Processes, Air conditioning etc.

3.8 Condenser efficiency and Vacuum Efficiencies

Condenser efficiency is assessed by following two ways:

3.8.1 Condenser Efficiency

In an ideal condenser, the steam should give only its latent heat to the circulating water so that the temperature of the condensate becomes equal to the saturation temperature corresponding to the condenser pressure.

It means there should be no undercooling of the condensate, Maximum temperature to which the cooling water can be raised is the condensate temperature.

The condenser efficiency is then defined as:

The ratio of the actual rise in the temperature of the cooling water to the maximum possible rise.

Equation: $\eta_{cond} = (t_2 - t_1)/(t_3 - t_1)$

Example:

The Outlet & inlet temperatures of cooling water to a condenser are 37.5°C and 30°C respectively. If the vacuum is 706mm of Hg with a barometer reading of 760mm of Hg. Determine the condenser efficiency.

Solution:

Given: t_o = 37.5°C, t_i = 30°C, Actual vacuum, Pv = 706mm of Hg, Standard atmospheric pressure, Patm= 760mm.

We need to find saturation temperature (t_s) and to calculate absolute pressure in the condenser, i.e. (Pc).

The absolute pressure of the condenser, Pc = P_{atm} + (– Pv) = 760-706 =54 mm of Hg.

To convert into the bar (so that, the saturation temperature can be determined using steam table).

54mm of Hg= 54×0.001333=0.072 bar [1mm of Hg= 1.01325/760= 0.001333]

From the steam table, the saturation temperature corresponding to 0.072bar is 40°C.

So, from the formula: η_{cond}= (37.5-30)/(40-30)= 75%

Hence, the **efficiency of the condenser** will be 75%, at the given condition.

3.8.2 Vacuum Efficiency of condenser

The volume of water is highly reduced hence, in a vessel due to the reduction of volume, a vacuum is created. It is a measure of the degree of perfection of maintaining the desired vacuum in the condenser.

The ideal vacuum means the vacuum due to steam alone when air is absent. Under such conditions, the total pressure in the condenser will reach the pressure of steam corresponding to the saturation temperature of the steam.

If there is no air leakage into the condenser, then partial pressure of air (Pa) =0 and hence, the vacuum efficiency becomes 100%.

The vacuum efficiency depends upon the effectiveness of the air cooling and the rate at which the air is removed by the air pump. Generally, the vacuum efficiency is about 98 to 99%.

a) **Vacuum efficiency=Actual vacuum (gauge)/ (atmospheric pressure or barometric pressure-Absolute pressure). [Absolute pressure = gauge pressure+ atmospheric pressure]**

b) **Or, Vacuum efficiency= Actual vacuum/ maximum obtainable vacuum**

Example:

Exhaust steam entering into condenser: 42 °C, Vacuum Gauge reading= -0.89 kg/cm²(g), find vacuum efficiency.

Solution: From steam table, Exhaust pressure corresponding to 42°C is 0.084 kg/cm²(a).

Maximum vacuum obtainable= 0.084-1.033 = -0.95 kg/cm²(g)

Vacuum efficiency= -0.89/-0.95=93.6%.

3.9 Definition of condenser parameters and fault diagnosis of condenser

3.9.1 Parameters

a) **Terminal temperature difference (T.T.D.)**-The temperature difference between the exhaust steam and the cooling water is least at the top of the condenser when the cooling water leaves. Here the cooling water has its highest temperature. This particular temperature difference is very important and is given a special name. It is called the terminal temperature difference. The important point is that any increase in this terminal difference leads directly to increase in the saturation temperature of the exhaust steam and a higher back pressure.

T.T.D. = Condensing Steam saturation temperature- Cooling Water outlet temperature

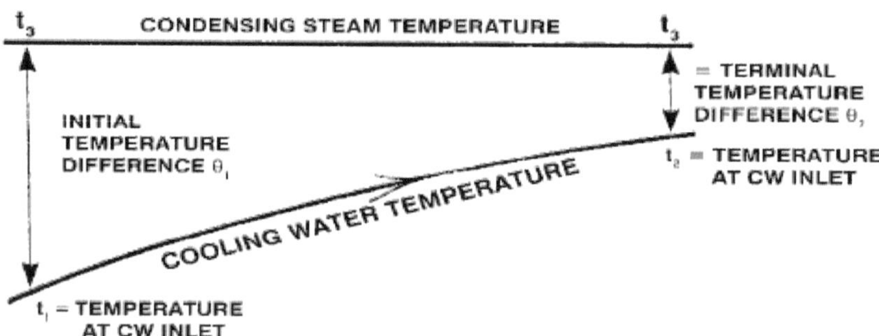

Diagram showing the temperature profile of condensing steam & CW in a condenser

▲ **Fig-9.14: Temperature profile of condensing steam & CW in a condenser**

b) **Initial temperature difference (I.T.D.)**-It is defined as the difference between saturation temperature of the condensate water in the condenser and temperature of cooling water coming into the condenser.

I.T.D. = Condensing Steam saturation temperature – Cooling Water inlet temperature

c) **LMTD**-The logarithmic mean temperature difference (LMTD) is used to determine the temperature driving force for heat transfer in flow systems. Mathematically it can be given as,

$LMTD = (\theta_1 - \theta_2)/Ln\ (\theta_1 / \theta_2)$

d) **Saturation temperature**- The saturation temperature is the temperature for a corresponding saturation pressure at which a liquid boils into its vapour phase. The liquid can be said to be saturated with thermal energy. Any addition of thermal energy results in a phase transition.

3.9.2 Fault diagnosis

a) **High CW temperature rise** – An indication of cooling water flow low or, condenser is over loaded than design.

b) **TTD is more than design**- An indication of scaling or fouling in condenser tube.

c) **High difference between saturation temperature & condensate temperature**- 2°C temperature difference is allowed, but if it is more than that, an indication of air ingress in condenser.

d) **Exhaust temperature more than that of corresponding exhaust pressure**- Turbine is running at part load.

e) **LMTD vs Flow rate** - LMTD indicates the flow rate of coolant and hot fluid. If the inner tube flow rate increases at constant outer flow rate, the LMTD increases. As the outer tube flow rate increases with constant inner tube flow rate, the LMTD decreases.

4.0 Evaporator- its working O&M practices and performance assessment

Evaporator is a tube-shell heat exchange that absorb the heat from the supplied source to vaporize the liquid flowing through it. The flow of supplied heat or the feed can be through tube or shell depends on the process requirement and design. Evaporators are manufactured in different shapes, types and designs to suit a diverse nature of process requirements. There is a difference between drying and evaporation also differs from distillation. Evaporation is used mainly to concentrate a solution containing a non-volatile solute and a volatile solvent.

4.1 Classification of evaporator

Evaporator can be classified based on Effect:

a) **Single effect**- Such type of evaporator works in a single heat exchanger when the heat is supplied in the HE at one end and product is achieved from another end of the HE maintain constant pressure in it.

b) **Multiple effect**- In multiple effect evaporator the heat source is one and supplied to HE, but the vapour exited from one HE is used in subsequent HEs in a staggered way at varied lower pressure as the heat source for them. The raw feed is supplied in end of the HEs and getting treated in in sequence to get product from the last HE.

4.2 Single effect evaporator

Single effect evaporator tubular evaporator can be sub divided as:

4.2.1 Long tube vertical evaporator

It can be sub-classified as described below.

i. **Upward flow (climbing Film)** – In climbing Film Evaporators (RFE), 9.15 (a), vaporisation occurs in the heating tubes and a separator called evaporator body which is used to separate vapour and liquid. The heat exchanger is mounted vertically with feed introduced at the bottom of the heated tube bundle. The evaporator liquid flows in the upward direction in long tubes in the form of annular or film flow for the substantial portion of the tube length. The water in the evaporator liquid boils as the liquid rises in the tubes and forces the evaporator liquid out of the tubes to the separator. it works based on Thermo-Siphon Principle which helps in circulation of the fluid in a natural manner, eliminating the need of a centrifugal pump. The heat exchange is based on natural convection.

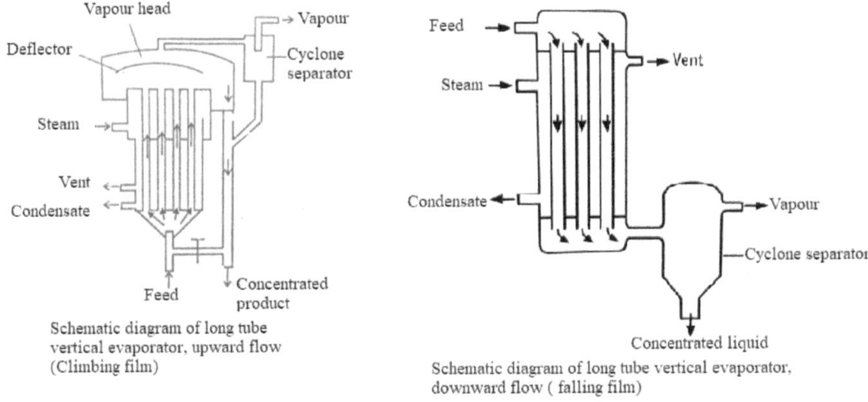

▲ **Fig-9.15 (a) & (b): Long tube vertical evaporator, upward flow & falling film**

ii. **Downward flow (falling film)-** In falling film evaporators, Fig-9.15 (b), the liquid feed usually enters the evaporator at the head of the evaporator. In the head, the product is evenly distributed into the heating tubes. The liquid enters the heating tube and forms a thin film on the tube wall when it flows downwards at boiling temperature and is partially evaporated. In most cases, steam is used for heating the evaporator. The product and the vapour both flow downwards in a parallel flow. The separation of the concentrated product from its vapour takes place in the lower part of the heat exchanger and the cyclone separator.

iii. **Forced circulation-** If a liquid boils on a heating surface, its fouling characteristics causes problems of easy flowing which is mitigated by forced circulation evaporator mechanism by raising the boiling point of the liquid. In forced circulation evaporator, Fig-9.16(a), the liquor is pumped through tubes at high velocity from the bottom of the cone of the vapour body, through the tubes and then back into the vapour body. When the liquid comes out of the tubes and enters the vapour body, there is a sudden fall in pressure, resulting the flashing of superheated liquor. Hence, evaporation takes place in the vapour body. Forced circulation evaporator working also creates a degree of agitation.

The liquid is maintained at a certain height above the heat exchanger tubes as well as above the vapour body to prevent mass boiling at the tube surface and at the inlet. Boiling in heat exchanger would cause fouling of the heat transfer surface due to precipitation in the tubes. Adequate tube velocity is ensured by providing a high circulation rate for the liquid through pumping. This in turn helps achieve the desired heat transfer.

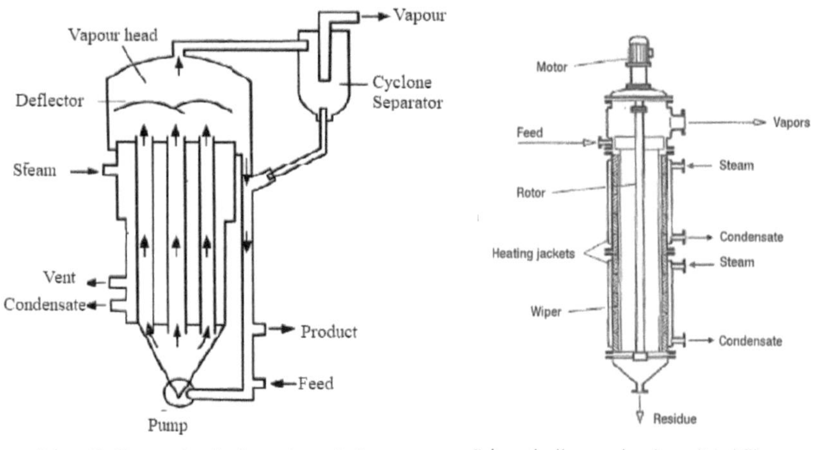

▲ **Fig-9.16 (a) & (b): Long tube vertical evaporator forced circulation & Agitated film evaporator**

4.2.2 Agitated film evaporator

Agitated Thin Film Evaporators (ATFE), Fig-9.16 (b) are mechanically aided with turbulent flows and constitutes of vapour body assembly and a rotor. The process liquid that enters the evaporator is spread across the thermal surface of a tube, with the help of mechanical blades. The feed enters the evaporator by means of a nozzle which transports the liquid through the vessel by means of gravity. The rotor blades spread the liquid in the form of thin film due to the turbulent flow generated by rotation of rotor. This helps in easily disengaging the liquid and vapours.

Principally, in ATFE thermal separation of a liquid feed occurs by producing a thin film at the heated wall of the evaporator by generating high turbulence with intense heat and mass transfer using a rotor. The intense agitation of the liquid film reduces the scale formation on the heat transfer surfaces thereby reducing fouling.

4.3 Multi Effect evaporator

The multi-effect evaporation system is a heat exchange system composed of multiple evaporators in series. Connect multiple evaporators in series, the secondary steam of the previous evaporator is used as the heating steam of the next evaporator, the heating steam of the next evaporator, and the heating chamber of the next evaporator is the condenser of the previous evaporator.

4.3.1 Types of multi effect evaporator

The multi-effect evaporation process is divided into four types based on method of feeding:

i. **Multi-effect evaporation co-current flow:** The flow direction of solution and steam are the same, and both flow sequentially from one effect to the last effect, which is called co-current feeding method, Fig-9.17. Since the pre-effect pressure is higher than the post-effect, the material liquid can flow by the pressure difference. However, the concentration of the final solution is high, the temperature is low, and the viscosity of the solution is large, so the heat transfer coefficient is low. For example, the flow sequence of materials and steam in the evaporator in the three-effect co-current state is one-effect, two-effect, and three-effect.

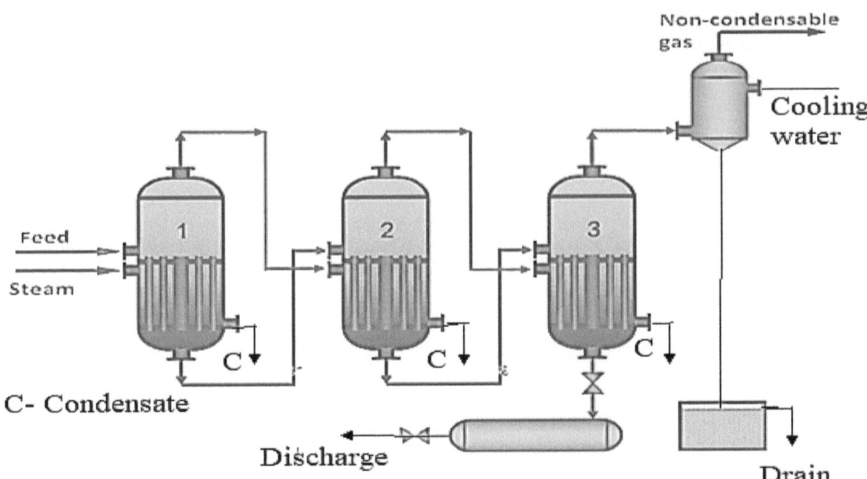

▲ **Fig-9.17: Co-Current multi effect evaporator**

ii. **Multi-effect counter-current flow evaporator:** The direction of the solution and the secondary steam are opposite, Fig-9.18. It is necessary to use a pump to send the solution to the previous effect with higher pressure. The effect of the concentration and temperature of each effect solution on the viscosity is roughly offset, and the heat transfer conditions for each effect are basically the same. For example, the flow sequence of the material in the evaporator under the three-effect counter current state is three-effect, two-effect, and one-effect. The order of steam is opposite to that of one-effect, two-effect, and three-effect. This method is suitable for processing solutions whose viscosity changes greatly with temperature and composition, but not for processing heat-sensitive solutions.

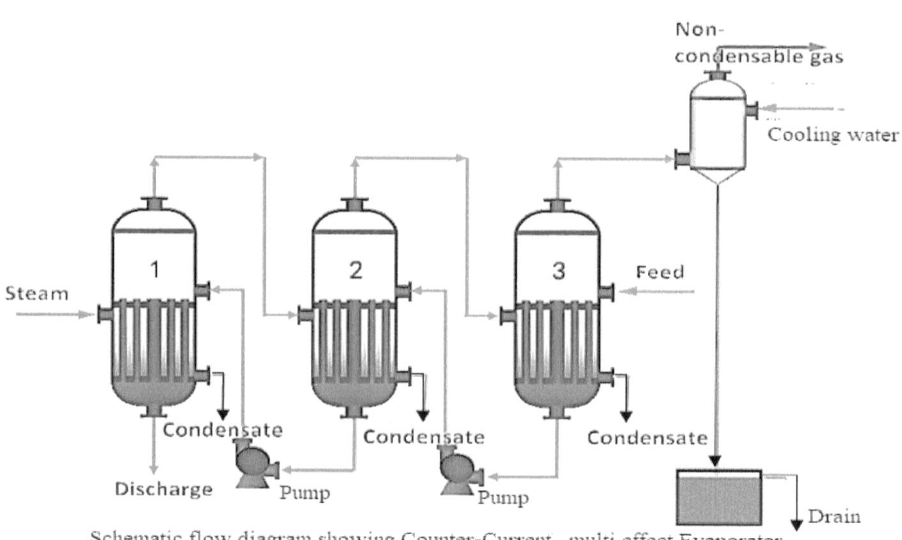

▲ **Fig-9.18: Counter Current multi effect evaporator**

iii. **Multi-effect evaporator with parallel feed:** The raw material liquid is added to each effect evaporator in parallel, and the finished liquid is also discharged from each effect separator, fig-9.19. The flow of steam is still from the first effect to the last effect. It is more suitable for materials with high viscosity and easy to crystallize. It can also be used for the simultaneous evaporation of two or more different liquids.

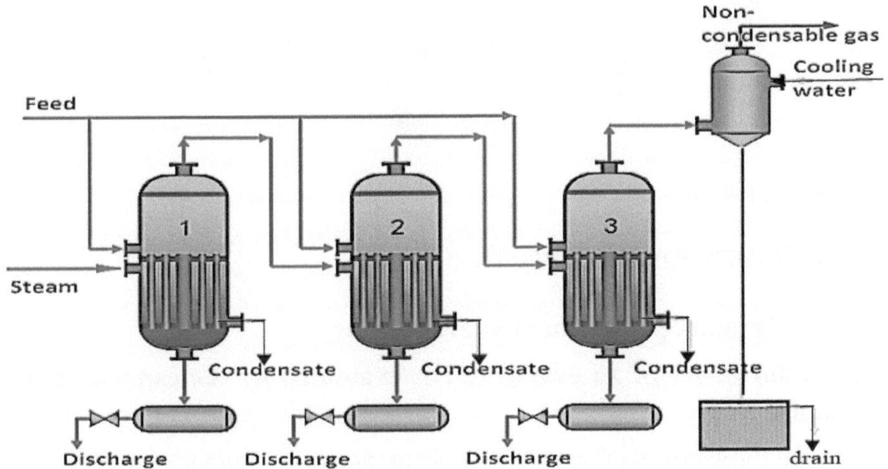

▲ **Fig-9.19: Parallel feed multi effect evaporator**

iv. **Multi-effect evaporation cross-flow process:** In a multi-effect evaporation process, the feeding method can have parallel flow or counter flow, which becomes a cross-flow method, Fig-9.20. For example, the order of the evaporation process in the three-effect cross-flow state can be two-effect, one-effect, three-effect or other combinations. The operation of the cross-flow process is more complicated, and each effect of the feed liquid is entered and exited separately. This process is suitable for the feed liquid with crystal precipitation.

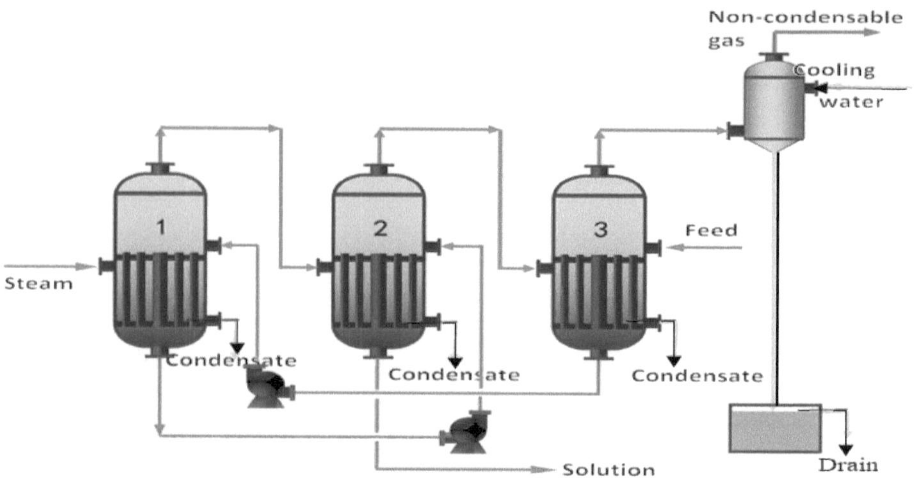

▲ **Fig-9.20: Cross flow multi effect evaporator**

4.4 Performance assessment of evaporator

The performance of an evaporator is evaluated by the capacity and the economy.

There are three criteria that affect performance of an evaporator,

1. Capacity (kg vaporized / time)
2. Economy (kg vaporized / kg steam input)
3. Steam Consumption (kg / hr.)

When, Consumption = Capacity / Economy.

4.4.1 Economy

Economy is defined as: kg of water vaporized / kg of steam fed to the unit.

a) Economy calculations are determined using enthalpy balances.

b) The key factor in determining the economy of an evaporator is the number of effects.

c) The economy of a single effect evaporator is always less than 1.

d) Multiple effect evaporators have higher economy but lower capacity than single effect.

e) The thermal condition of the evaporator feed has an important impact on economy and performance.

4.4.2 Capacity

Capacity is defined as: kilograms of water vaporized per hour.

Factors affecting the capacity:

a) Feed temperature at boiling point- All the heat transferred through the heating surface is available for evaporation and the capacity is proportional to Q.

b) Feed temperature is cold- Additional heat required to heat it to its boiling point and the capacity for a given value of Q is reduced accordingly.

c) Feed temperature is above the boiling point- A portion of the feed evaporates spontaneously by adiabatic equilibration with the vapor-space pressure and the capacity is greater than that corresponding to Q, this process is called flash evaporation.

4.4.3 Steam consumption

Steam consumption is very important to know, and can be estimated by the ratio of capacity divided by the economy.

Steam consumption = Capacity/Economy.

=Kg of water vaporized/hr ÷ kg of water vaporized/kg of steam

=Kg steam/hr

4.5 Single effect vs Multi effect Evaporator

For single effect evaporator, practically the steam economy is 0.8 maximum-1, which translates to 0.8/1 kg of steam needed to evaporate 1 kg of water.

So to decrease the evaporator steam economy, the multiple-effect design uses the exhaust vapours from the product to heat the downstream evaporation effect and reduce the steam consumption.

The capacity of a multiple effect evaporator (n effects) is n x single effect evaporator capacity and the economy is about 0.8 x n. Following are the differences between single effect & multi effect evaporators.

i. **Single Effect (SE) Evaporator**

   a) Small capacity but wasteful energy (1 kg steam vaporize 1 kg water)

   b) Overall temperature drop for single effect is somewhat equal to multiple effect

ii. **Multiple Effect (ME)**

a) Each individual effect will have a smaller temperature difference, thus high area of heating surfaces to compensate temperature difference.

b) Capital cost- Higher than single effect.

c) Operating cost- Steam economy, only required for the first effect (1 kg steam vaporizes 3 kg water).

Example: A Single effect evaporator is fed with a NaOH solution at 37 ^{0}C with a flow rate 9000kg/hr with a concentration 20% to produce 50% concentration. Steam at a pressure of 2 kg/cm^2 (g) is used for the purpose. The vapour space pressure is 0.13 kg/cm^2(g), condensate temperature. Find amount of water to be evaporated, capacity & Economy of the evaporator.

Solution:

Given:

a) Steam Pressure in (ata) = 2.0 +1.032 =3.032

b) Enthalpy of saturated steam (from steam table) =651 kcal/kg

c) Condensate enthalpy=132 kcal/kg

d) Vapour space pressure (ata)= 0.13+1.032 =1.16

e) At 1.16 ata, Enthalpy of vapour =640 kcal/g, Boiling temp=103°C.

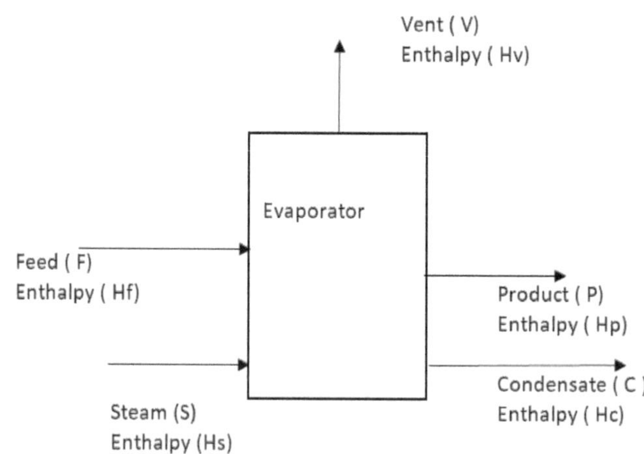

▲ **Fig-9.21: Evaporator-Mass & Heat Balance**

Mass balance:

We know, Feed (F)= Product (P)+ Vent (V) [Fig-9.21]

By NaOH balance: 9000 x 0.20= Px0.5; P= 9000 x0.2/0.5=3600 kg/hr.

Then steam vented = 9000-3600=5400 kg/hr.

Heat required to evaporate by Heat balance, Fig-9.21

Steam quantity (s) x Hs + Feed quantity (F) x Hf =Vent quantity (Hv)+ Product quantity(P)x Hp + Condensate quantity x Hc

Sx651 + 9000x1x37=5400x640+3600 x 103+C x132

S x (651-132) =3,826,800-333000=3493800 [since steam & quantity are same, S=C]

S=3493800/519=6732 kg/hr.

Answer:

a) Quantity of water evaporated or capacity =5400 kg/hr

b) Steam consumption= 6732 kg/ hr.

c) Economy = Kg of water evaporated/Kg of steam= 5400/6732=0.8

4.6 Factors that affect Evaporator Efficiency

Following are the key factors which affect evaporator efficiency

a) **Temperature**- Temperature is a critical factor, as it directly affects the rate of evaporation. Higher temperatures can increase the rate of evaporation, but they can also cause issues with product quality or safety if they are too high.

b) **Pressure-** Pressure can impact the boiling point of the liquid being evaporated. If the pressure is too low, the liquid may not boil efficiently, while if the pressure is too high, it can cause damage to the evaporator or other equipment.

c) **Flow rate-** Flow rates are another important factor to consider when optimizing your evaporator. The rate at which liquid is fed into the evaporator can affect the efficiency of the process, as well as the quality of the final product. If the flow rate is too low, the evaporator may not be able to handle the volume of liquid being fed into it, while if the flow rate is too high, it can cause issues with product quality or safety.

d) Equipment design is also critical when it comes to evaporator efficiency. The design of the evaporator can impact how efficiently it operates, as well as how easy it is to maintain. Some key design considerations include the type of heat exchanger used, the shape and size of the evaporator, and the materials used in construction.

4.7 Best Practices for Evaporator Optimization

a) To maintain proper temperature and pressure throughout the process which can be achieved through careful monitoring and control of the evaporator and regular maintenance and cleaning of the equipment

b) To keep the evaporator and other equipment clean and well-maintained to avoid build-up or damage

c) Addition of appropriate chemicals and additives such as anti-scaling agents or surfactants to the liquid can also help improve evaporator efficiency. Addition of such agent prevent build-up or foaming, which reduce efficiency.

Questions:

1. What are the different types of cooling tower?
2. Define the terms related to cooling tower-a) range b) approach) effectiveness.
3. What are components responsible for total water loss- discuss?
4. What are the factors that affect CT performance?
5. Explain tube & shell type heat exchanger.
6. Discuss about advantages & disadvantages of tube & shell type HE.
7. Derive the equation for overall heat transfer coefficient, U.
8. What is LMTD? Write equation LMTD for counter flow HE.
9. Explain the working of a surface condenser.
10. Deduce condenser thermal efficiency and vacuum efficiency.
11. Explain working of single effect long tube climbing film evaporator.
12. What are the different types of multi-effect evaporator?
13. What are the criteria for assessment of evaporators?

SECTION 10 RAW WATER TREATMENT, WASTE WATER TREATMENT, DM PLANT, AND WATER SOFTENER PLANT

The performance evaluation of a water treatment plant is a procedure to measure the functioning efficiencies based on some established performance indicators such as degree of removal of pollutants like turbidity, colour, suspended impurities etc. Equipment, systems related to water treatment for industrial purpose and waste water treatment are discussed in this section.

1.0 Clarifier, its working and performance assessment

1.1 Working of a clarifier

Clarifiers work on the principle of gravity settling. The heavier suspended solids settle in the clarifier due to the motionless conditions provided in the Clarification Zone. The settled solids are swept to the centre-well provided for collection of sludge with help of moving scraper blades. Clarification occurs following series of steps that include:

i. **Coagulation-C**oagulation means coming together, or clattering of particles. Colloidal substances are either commonly negatively charged particle, chemicals like Alum & ferric chloride supplies positive charge to clatter the charged particles. This process is called Coagulation.

ii. **Flocculation**- Coagulation process results in flock formation to make it heavier to settle down. This process is called flocculation.

iii. **Sedimentation**- Heavier particles generated due to flocculation process gets settle down due to gravity, called settling or sedimentation.

iv. **pH booster**- During coagulation process, pH decreases and to maintain the same, lime solution is also dosed.

Clarifiers are built for continuous removal of solids being deposited by sedimentation. It is generally used to remove solid particulates or suspended solids from liquid for clarification and (or) thickening. Concentrated impurities, discharged from the bottom of the tank are known as sludge.

1.2 Clarifier Technology

Although sedimentation might occur in tanks of any shapes, following points are taken care for clarifier design.

a) Mechanical solids removal devices move as slowly as practical to minimize resuspension of settled solids.

b) Tanks are sized to give water an optimal residence time within the tank.

c) Considerable attention is focused on reducing water inlet and outlet velocities to minimize turbulence and promote effective settling throughout available tank volume.

d) Baffles are used to prevent fluid velocities at the tank entrance from extending into the tank; and overflow weirs are used to uniformly distribute flow from liquid leaving the tank over a wide area of the surface to minimize resuspension of settling particles.

1.3 Types of clarifiers

Following are the main two types of clarifiers used in water treatment plant.

i. **Conventional clarifiers-** These are gravity sedimentation tanks (settlers) whose primary end product is a clarified liquid. They are available in rectangular and circular configurations. Such clarifiers are used in waste water treatment plant.

ii. **Sludge blanket clarifiers-** These are up-flow system, meaning that water flows upward from the bottom of the clarifier through a blanket of suspended solids that acts as a filter. These clarifiers contain an inverted cone that produces an increasing cross-sectional area from the bottom to the top. The upward velocity or the water decreases as it flows towards top.

The upward velocity of the water balances the downward velocity of the solid particle, and the particle becomes suspended. The heavier particles are suspended more closely to the bottom while the lighter ones are suspended more closely to the surface. As the water containing flocculated solids passes upwards through this blanket, the particles are absorbed onto the larger floc, and drops to a lower level. Eventually, it falls to the bottom of the clarifier to be recirculated or drawn off.

1.4 Different Parts of a clarifier

Fig-10.1 shows all major components associated with a clarifier.

a) Raw water Inlet
b) Clarified water Discharge
c) Sludge Discharge
d) Rake/Plough
e) Skimmer
f) Flash mixer
g) Scum Trough

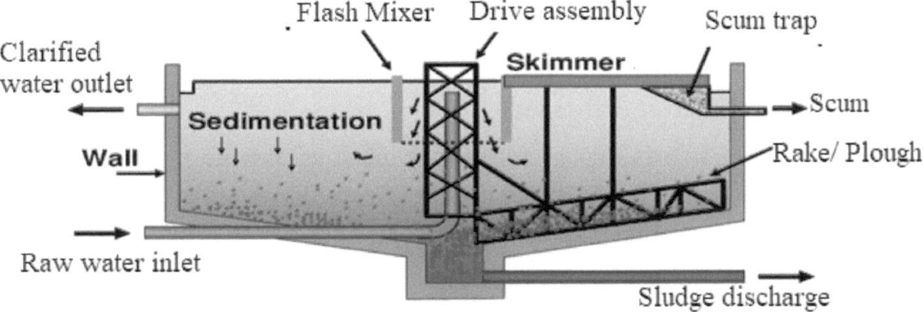

▲ **Fig-10.1: Clarifier**

1.5 Applications

i. **Potable water treatment-** Sedimentation in potable water treatment generally follows a step of chemical coagulation and flocculation, which allows grouping particles together into flocs of a bigger size. This increases the settling speed of suspended solids and allows settling colloids. Polyelectrolytes and ferric sulphate cause fine and suspended particles in the water to bunch together. Water may be treated with these coagulant chemicals prior to entering the clarifier. Clarified water may then be passed through granular filters to remove any residual particles remaining in the water. Filtered water then flows to a pumping station for storage and use.

ii. **Waste water Treatment-**Sedimentation tanks have been used to treat wastewater for ages. Primary treatment of sewage is removal of floating and settling solids through sedimentation. Primary clarifiers reduce the content of suspended solids and pollutants embedded in those suspended solids. Because of the large amount of reagent necessary to treat domestic wastewater, preliminary chemical coagulation and flocculation are generally not used, remaining suspended solids being reduced by following stages of the system. Clarifiers remove flocs of biological growth created in some methods of secondary treatment including activated sludge, trickling filters, and rotating biological contactors.

iii. **Mining-** Methods used to treat suspended solids in mining wastewater include sedimentation and floc blanket clarification and filtration. After dissolving the ore, the saturated borate solution (solution of boric acid) is pumped into a large settling tank. Borates float on top of the liquor while rock and clay settle to the bottom.

1.6 Performance assessment for a clarifier

1.6.1 Factors affecting Efficiency

There are several factors that affect proper clarifier operation. Temperature, detention time, short circuits, weir overflow rate, surface settling rate and, solids loading as described below.

a) The nature of solids in the wastewater and their source.

b) The transit time and temperature of the wastewater stream have a large, negative impact on the solids.

c) Hydraulic loading rate on the clarifier- The important performance factor of a circular primary clarifier is the flow rate and velocity of influent through the clarifier. The surface overflow rate (SOR) compares the *total influent flow rate* to the surface area of the tank. The weir overflow rate (WOR) measures the *flow over the weir* compared to the *length of the weir*. Both of these ratios (SOR & WOR) are useful in determining clarifier performance.

d) Maintenance status of the clarifier.

1.6.2 Performance parameter of a clarifier

A primary clarifier is expected to remove:

a) Settable solid: 90% to 95%

b) Suspended solid: 40% to 60% (scum)

c) Bacteria: 25% to 75% of bacteria

d) Total Oxygen Demand (TOD): 25% to 50% of total BOD.

2.0 Reverse Osmosis (RO) Plant

RO plant is a water treatment plant that is used for treatment of water to be used for drinking and other domestic purposes. It is also used for pre- treatment of water that goes for final treatment to achieve high quality of water to be used for boiler or pharmaceutical process. RO plant is also provided to get **zero discharge** from industry.

2.1 Working of RO Plant

RO plant is a membrane based physical process for the separation of salt molecules from water. RO membrane is a permeable sheet securely wrapped on the porous tube. Raw water is forced to pass through the semi-permeable membrane and the microporous membrane allows the passage of small molecules whose molecular size is smaller than the membrane pores.

Driving force of RO is water pressure, which separates the feed line to further two lines with varying concentrations by passing through the membrane and is effected by the temperature of the water. The schematic diagram for RO plant is shown in Fig-10.2.

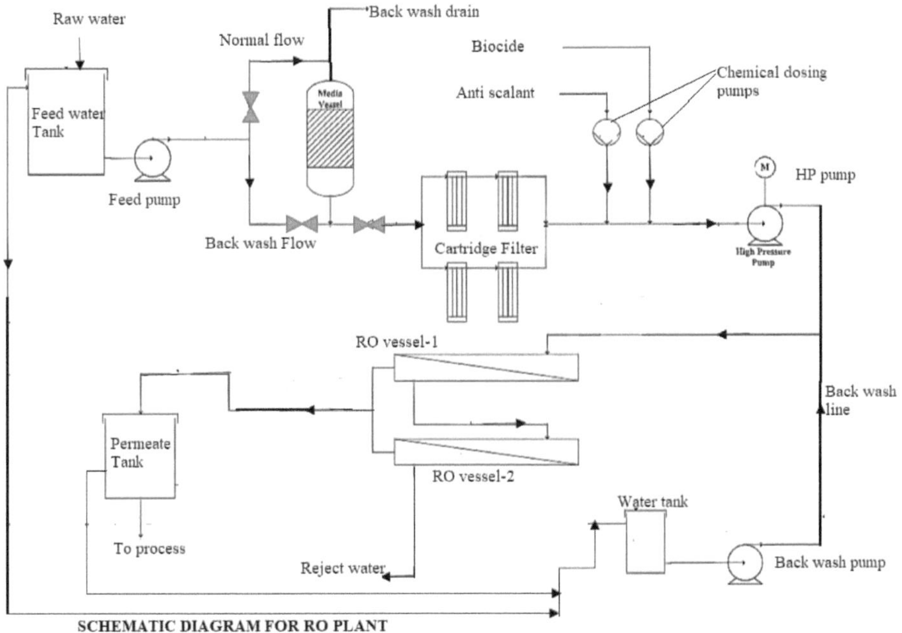

▲ **Fig-10.2:Reverse Osmosis plant**

The velocity, by which raw water moves across the membrane module, is equivalent to the differential pressure that exceeds the natural osmotic pressure differential.

2.2 Hindrances in RO Plant and remedy

RO system designed precisely and accurately according to the raw water properties so that, every element of the system performs its normal function. Product quantity and quality can be changed with the variation in the feed properties and by a change in operating parameters. For smooth operation, hindrances faced by RO plant are:

i. **Membrane fouling-** Its causes, phenomenon and mechanism are to be identified and corrected.

ii. Corrective and preventive controls are required according to the source water properties as well as treated water properties. These measures may be implemented in pre-treatment techniques, intra-treatment modification and post-treatment requirements. By using periodic measurement, the performance can be analysed.

2.3 Performance monitoring parameter and corrective measures

Key factors to analyse the performance of the Reverse Osmosis plants are:

a) The salt rejection

b) Permeate flux

c) Chemical oxygen demand (COD) rejection

Above parameters demonstrate working performance of membrane system. To achieve above performance, operating parameter to be monitored strictly are describe below.

2.3.1 Differential pressure of RO system (ΔP)

Differential pressure between the membrane inlet and outlet pressure changes due to the reasons as follows:

a) High SDI (more colloidal and suspended particles)

b) Biological fouling

c) Salt concentration

d) Boundary conditions of membrane wall

e) Viscosity

f) Density of the fluid.

Increase of 10–15% of pressure drops in feed and reject line is normal. Typically, pressure drop increases with the passage of time because salts are going to accumulate over the membrane surface.

Maximum pressure drop in one membrane module should not be more than 15% of feed pressure. Membrane may undergo changes in structure if the pressure is higher than the designed pressure of membrane.

2.3.2 Percent salt rejection

Water and salt transportation from an identical membrane module is distinguished by three coefficients of transport; a) Permeability of solute (signifies permeability of solute) b) Coefficient of reflection (signifies selectivity of the membrane) & c) Hydraulic permeability (water permeation followed by pressure applied).

Per cent rejection is a measure of quantity of TDS removed from the RO feed water or TDS present in the permeate.

Salt rejection can be affected if:

a) Feed pump pressure is relatively higher than the designed pressure of plant

b) Increase in temperature.

Salt rejection or percent rejection can be calculated by:

$(C_{av}$-$C_p)/C_{av}$ x100%. [When, C_{av}-average conductivity at inlet, C_p- Conductivity of permeate.]

2.3.3 Monitoring of transmembrane pressure

The pressure difference between the feed side and the permeate side of the membrane is called transmembrane pressure. It can be calculated among a single membrane module or more than one module of the membranes connected by the spacer in a single membrane casing.

Transmembrane pressure (TMP) = $(P_f + P_c)/ 2 - P_p$ [P_f- feed pressure, P_c- Reject line pressure, P_p-Permeate pressure.]

Calculating this pressure, the internal performance of the membrane can be analysed.

It can only be increased by fouling and the scale formation on the membrane element. Cleaning of the membranes with chemical is recommended when transmembrane pressure (in operation) reaches near 1.0 kg/cm^2.

2.3.4 Silt Density Index (SDI)

Silt Density Index is a measure of fouling potential of source water. It is the onsite calculation of colloidal and suspended particles. SDI is basically a performance analysis test for equipment used for pre-treatment. SDI measurements should be performed with pre-post cartridge filters and pre-post multimedia filters. SDI value should be as minimum as possible which can be done by operating the filter media at its design differential pressure (ΔP) and by the proper backwash of the multimedia filters.

The test involves below steps:

a) Collect a 500 ml sample through the filter (clean filter) at the start of the test and measure the time taken to collect the sample. Let it is Ti

b) Allow flow water through filter for 15 minutes, denoted by T(T=15 min)

c) Collect another 500ml sample and measure time to collect the sample. Let the time taken to collect sample is T_f.

d) The sample times are applied to the formula below to obtain the SDI_{15}.

Calculation:

$SDI_{15} = \% P_{30}/T$; When, $\% P_{30}$ = Percentage plugging at 30 Psi pressure

$\% P_{30} = [(T_f - T_i)/T_f \times 100]$

Hence, $SDI_{15} = [(T_f - T_i)/T_f \times 100]/T$.

2.3.5 Normalized permeate flux (J)

It is a calculation of the amount of product water generated per unit area of membrane and time. It can be calculated for the singular membrane modules as well as, for multiple modules of membrane joined by a spacer enclosed in a single membrane casing. Flux can be increased or decreased depending on the operating parameters of the plant.

Equation: Flux (J) = Q_p/Am m^3/m^2-hr. [J – Permeate flux, Q_p-permeate flow (m^3/hr, Am-surface area of membrane (m^2)].

2.3.6 Rate of Recovery

Rate of recovery is the ratio between permeate flow rate to feed flow rate. Monitoring the rate of recovery is to measure the performance characteristics of the membrane. Recovery can decrease with the passage of time due to many factors such as pressure, flow, salt rejection and few others.

Equation: Recovery rate (Z) = Q_p /Q_f x 100%. [Q_f-feed flow rate m^3/hr, Q_p-Permeate flow rate m^3/hr]

Recovery of first pass is always be less than the recovery of second pass, because the load of salt concentration on the first pass is much higher than the second pass.

2.3.7 Flow Rate

Flow rate is an important factor to be monitored in the performance of the RO plant as it is the main parameter which is directly concerned. Once the intermediate blocking of membrane starts, flow rate decreases gradually as flux J decreases.

Flow rate is a function of filtrate volume. It decreases exponentially as a function of time.

Feed water stream is divided into two main streams

a) permeate

b) reject.

The flow rate of feed stream remains constant as it is summation of permeate & reject. Permeate streams can be reduced in flow, as a result of fouling or scaling factors, but total flow remains constant with increased flow in reject. Flow rate can be maintained to membrane's design capacity by backwashing technique of membrane.

2.3.8 Back wash

Backwash is performed for the membrane itself and it is carried out by using the permeate. By using permeate water for backwash purposes, ensures slightly less scale formation and biological growth over the membrane element.

2.4 Water Conditioning before feeding RO membrane

Standard recommendation for water conditioning before feeding to RO membranes:

i. Reducing Agents: Sodium bisulphite is used as a preservative of Reverse Osmosis membrane. Sodium meta bisulphite (SMBS) is used to remove naturally occurring oxygen from water.

ii. Flocculation: Calculated amount of non-ionic or anionic flocculating polymers is used for the removal of algae, because most of the membranes have negative charge on its surface.

iii. Coagulation: Coagulants with aluminium basis should be avoided to use in pre-treatment technologies because they are more soluble, residual aluminium can cause precipitative scaling over RO membranes.

iv. Pre-Chlorination: Shock dose of chlorine and intermittent chlorination is far good than the continuous dosage of chlorine to avoid production of polysaccharides.

v. Scale Inhibitor: Anti-scalants and mineral acids are used to inhibit the scale. Sulphuric acid is mostly preferred or hydrochloric acid due to safety and cost effectiveness.

2.5 Treatment technologies

Different types of problems are encountered while operating the RO plants. If these problems are eliminated at early stages, then RO can perform accurately. RO plant can be divided into three main zones where treatment is undertaken as discussed below:

2.5.1 Pre-Treatment

Natural source of water contains two type of solids in it, dissolved solids (TDS) (mineral ions like sodium, calcium, magnesium etc.) and suspended solids (TSS) (colloids, particulates, silt etc.). In pre-treatment of raw water, it is desired to remove all the TSS and some of the salts. TSS are necessarily removed from the raw water because it can affect the porous structure of membrane. Membrane productivity is lost with the passage of conventional treatment techniques used in pre-treatment. Hence, non-conventional techniques are also adopted along with conventional treatment in membrane technologies for better performance as described below.

a) **Conventional pre-Treatment**: Conventional pre-treatment refers to media filtration which is one of the main techniques used as a pre-treatment technology to remove suspended solids in the raw feed of the RO plants. Media filtration principally involves:

   1. Carbon and Sand Filtration - Carbon filters is basically an activated carbon bed rested on the gravels. Whereas, in the case of Sand filters sand is used instead of the carbon bed.

   2. Other techniques can also be used which depends on the purity required. For example, green sand or activated alumina is commonly used as arsenic removal media only if the permeate is used for drinking and pharmaceutical applications.

3. Dissolved air flotation is also a conventional pre-treatment technique in which air is introduced in the raw water which assists to remove hydrophobic contents present in water. Air to solid ratio is very analytical to remove 90–99% of algal cells.

b) **Non-Conventional Pre-Treatment**: Micro, Ultra or Nano filtration membranes are used in the Non-Conventional Pre-Treatment techniques. These membranes can remove particulate matters, colloids and some organic foulants easily. These membranes have comparatively high efficiency and life as they can be back washed periodically to get the effective results and to increase the life of membrane. Selection of membrane depends on the purity of feed water required or according to the quality of feed water.

c) **Chemical pre-treatment**

1. Coagulation chemicals are frequently used for the coagulation of the particles. These chemicals combine particles together and settle them down as a result of gravity or being removed by filtration. Chemicals are also added to remove metals.
2. Chlorine is commonly used as disinfectant of feed water. But the excess amount of residual chlorine can cause the distortion of membrane's porous structure.
3. Maintaining pH of feed water is very important in the rejection of salts as below:
    a. At elevated pH about 10% rejection of sulphate and of sodium is 95%.
    b. Removal of arsenate is much higher at pH 3 to 10. Arsenite removal increase at pH rages 7 to 10.
    c. Flux of water is not dependent on pH, but rejection of salts decreases at extreme values of $pH > 11 < 5$.

2.5.2 Intra-treatment

Bio-foulants deposited over the membrane surface, start growing in the form of colonies. Once the growth is started, it is very difficult to get rid of these foulants. Biological growth rate is dependent on factors like pressure, temperature etc. and favourable at a temperature between10 to 30 ^{0} C.

a. Saline coated membranes are having a strong tendency against the foulants growth over the membrane element. Commonly saline solution is used for preservation of membrane.

b. Different biocide chemicals are used to avoid the bio growth which is also used continuously on daily bases at the time of operation.

c. Anti-scalants are used as a preventive action to remove or avoid scaling of salts over the membranes. Usually anti-scalants are of phosphonates and poly-phosphonates.

d. Activity of carbon media is slowed down over a period of time and this media shapes like mud and passes with the feed water to the inlet of membrane and forms a layer. This carbon layer is removed by cleaning. Cleaning is based on low pH (commonly HCI) and high pH (commonly caustic and EDTA) based chemicals.

e. Sodium hypochlorite with water is also used to increase membrane permeability. Chemical cleaning of plant is undertaken after3-4 months depending on pressure drop and flow rate of the plant.

f. Membranes are connected to each other by inter-connectors and end caps of casing are connected with the membrane by adapters. O-rings of inter-connectors and adapters are basically isolating seals which help restrict intrusion of concentrated water into permeate. It is a very rare case that the seals get damaged but if such circumstances occur, probable reasons are:

1. With time, the large generation of back pressure would affect the seals.
2. Workmanship on the inter-connectors in the membranes casing can damage rings.
3. RO plants if established without shed on, the constant exposure to sunlight can harden the seals with time and cause malfunctioning.

2.5.3 Post-treatment for membrane permeate water

Post-treatment of membrane permeate water are:

a) **pH adjustment-** According to the regulatory requirements, pH can be increased or decreased by adding chemicals.

b) **Corrosion inhibitor-** Corrosion inhibitors mostly used are-polyphosphates, phosphonate and silicate based. Inhibitors control corrosion in many ways like generation of film coating inside the pipes to avoid direct contact of water with walls of pipes. Another way is the use of chemicals to that specific ions which are causing corrosion.

c) **Oxygen scavengers** – These are mostly used in boilers, that react with oxygen and reduce the probability of corrosion by oxygen. Oxygen scavengers are mostly sulphite base chemical and hydrazine.

d) **Blending or mineral dosing** -Blending or mineral dosing is recommended when permeate water is required either for drinking or for industrial purposes. Blending is suggested only if the quality of water is good enough to drink, there shouldn't be micro-organisms present in the water that is blended with a specified ratio with permeate water to get desired TDS level.

e) **Disinfection** -Disinfection is required at last after all the processes have been completed to remove all the bacteria from water. Disinfection can be done by chemical dosing or by UV ray. Disinfection by chemicals includes chlorine dosing or ozone gas dosing. In ultra violet lamp, treated water is passed over the lamp which kills bacteria.

2.5.4 Post treatment for membrane reject water

a) Reject water of membrane can also be utilized. It can be used for flushing in bath rooms, watering the gardens, vehicle washing, floor mopping etc. For industries, it can be used by making some treatments like microfiltration of reject water. Microfiltration removes traces of particles but salts remains same.

b) In industrial sector, ion exchange or softener can be used to soften the water for reuse in the cooling towers, fire hydrants or in bathrooms. If zero discharge is concerned, then open pits are recommended when water is evaporated by leaving salts behind. Rejection of RO water is disposed according to the rules of state or local regulations.

3.0 Sewage Water Treatment Plant

3.1 Introduction

Waste water discharge from industries and domestic area, called sewage water, contains several contaminants which result in water pollution. Hence, it is important to treat the waste water and to maintain contaminants with the limit before it is discharged.

3.2 Working of sewage water treatment plant

Sewage water contains sludge containing organic & inorganic contaminants and treatment basically consists of two phases a) primary treatment b) secondary treatment.

In primary treatment the inert like stone, clay and metal pieces are removed. Secondary stage is the important stage where biological reactions happen to remove the organic contaminants. A typical sludge treatment scheme is shown in Fig-10.3.

Discussion in this section is limited to the commonly used sewage treatment processes.

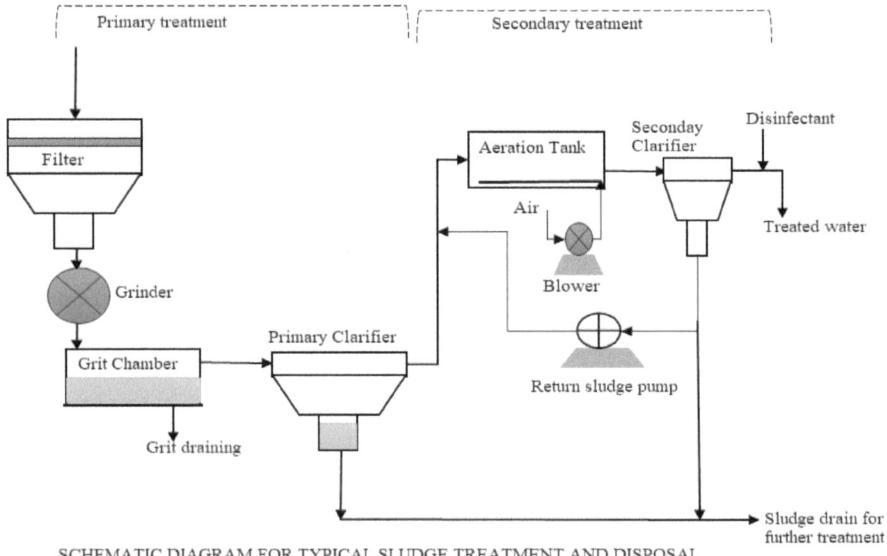

▲ **Fig-10.3: Typical Treatment and Disposal**

Some of the commonly sewage water treatment processes are described below.

3.3 Activated Sludge Process (ASP)

3.3.1 Working of ASP

ASP process flow is shown in Fig-10.4.

This process is an old process and is used for the treatment of sewage and industrial wastewaters.

Though there are many different designs, but in general, all ASP have three main components:

a) An aeration tank that serves as a bio reactor

b) A settling tank for separating as solids

c) Treated waste water.

d) A Return Activated Sludge (RAS) equipment that transfers settled activated sludge from the clarifier to the aeration tank's influent.

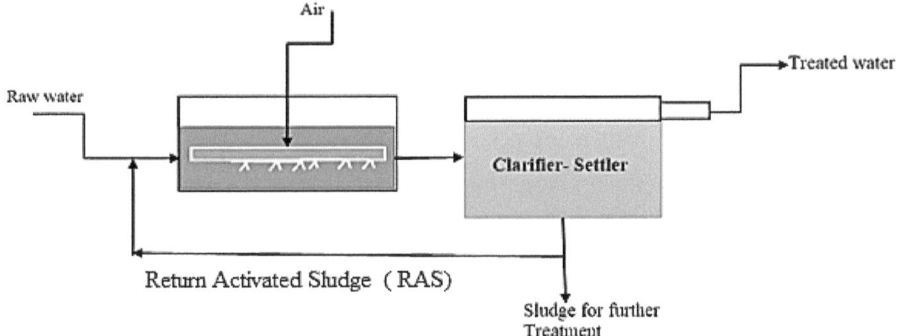

▲ **Fig-10.4: Activated Sludge Process**

The concentration of biodegradable components in the influent is reduced in all activated sludge plants due to biological (and sometimes chemical) processes in the aeration tank.

The **activated sludge** process in wastewater treatment involves activities as:

i. Injecting oxygen or air into raw, unsettled sewage where the solids are smashed during this process.

ii. The sewage is bubbled, and the sewage liquor is discharged into another chamber with activated sludge.

iii. Live bacteria sink to the bottom of the tank, while dead bacteria float to the surface.

iv. The live bacteria return to the digestion chamber; clean water is discharged into a soak away or watercourse.

It is important to note:

a) What **Activated Sludge** is and how the entire activated sludge system operates-is an important understanding for the process. As the particles are actively flooded with beneficial bacteria that digest the sewage, the sludge is considered activated.

b) **Hydraulic Residence Time** (HRT) in the aeration tank, which is defined as the aeration tank volume divided by the flow rate, controls the removal efficiency.

3.3.2 Advantages & disadvantages of ASP

i. **Advantages of activated sludge treatment:** Activated sludge treatment process has following advantages over other alternatives.

   a) The quantity of unwanted sludge is reduced.

   b) Beneficial bacteria reseed themselves in sewage treatment plants.

   c) The procedure is extremely dependable.

   d) The procedure is less complicated.

   e) The method can be odourless.

ii. **Disadvantages of the activated sludge treatment process**

   Some disadvantages of activated sludge are:

   a) ASP incurs initial high capital and operating costs.

   b) Skilled personnel are required to operate and maintain the treatment of activated sludge.

   c) Electricity must be used continuously, which increases the energy consumption of wastewater treatment.

   d) Sludge and effluent may necessitate additional treatment before discharge.

3.5 Sequencing Batch Reactor (SBR)

SBR is an activated sludge process, fig-10.5, used for waste-water treatment. To reduce the organic matters, oxygen is bubbled through the mixture of wastewater and activated sludge. A SBR is used in small package plants. The SBR system consists of a single complete mix reactor in which all the steps of the activated sludge process occur as per following sequence.

1. The reactor basin is filled for a duration of 3 h.
2. Then aerated for a certain period of time, usually 2 h.
3. After the aeration cycle is complete, the reactor is allowed to settle for a duration of 0.5 h.
4. Effluent is decanted from the top of the unit, which takes approximately 0.5 h.
5. Decanting of supernatant is carried out by either a fixed or floating decanter mechanism.
6. When the decanting cycle is complete, the reactor is again filled with raw sewage, and the process is repeated.

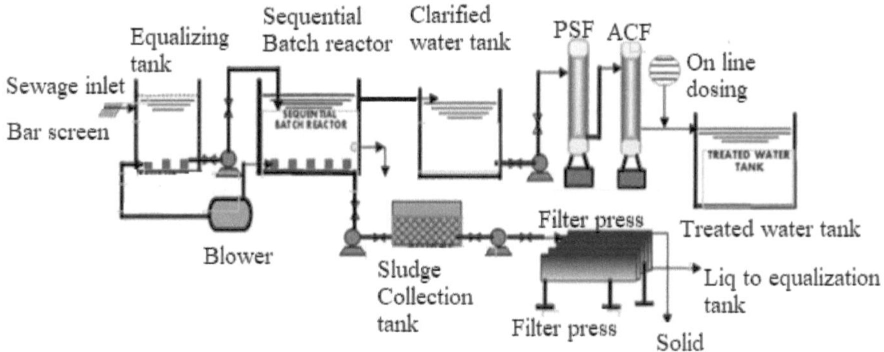

Schematic diagram for Sequencing Batch Reactor (SBR) Process

▲ **Fig-10.5: Sequencing Batch Reactor Reactor**

3.6 Moving Bed Biofilm Reactor (MBBR)

The MBBR system, Fig-10.6 (a) consists of an activated sludge aeration system where the sludge is collected on recycled plastic carriers. These carriers have an internal large surface for optimal contact water, air and bacteria.

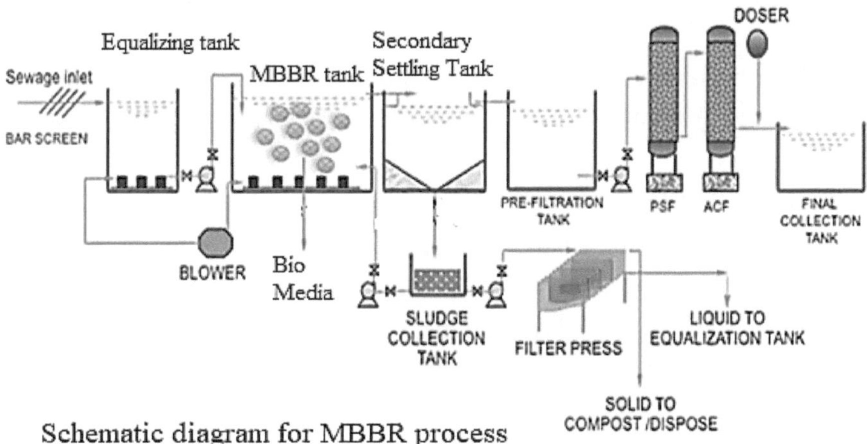

Schematic diagram for MBBR process

▲ **Fig-10.6 (a): MBBR process**

The bacteria/activated sludge grows on the internal surface of the carriers. The aeration system keeps the carriers with activated sludge in motion. The excess sludge gets separated and flows from the carriers with the treated water towards the final separator.

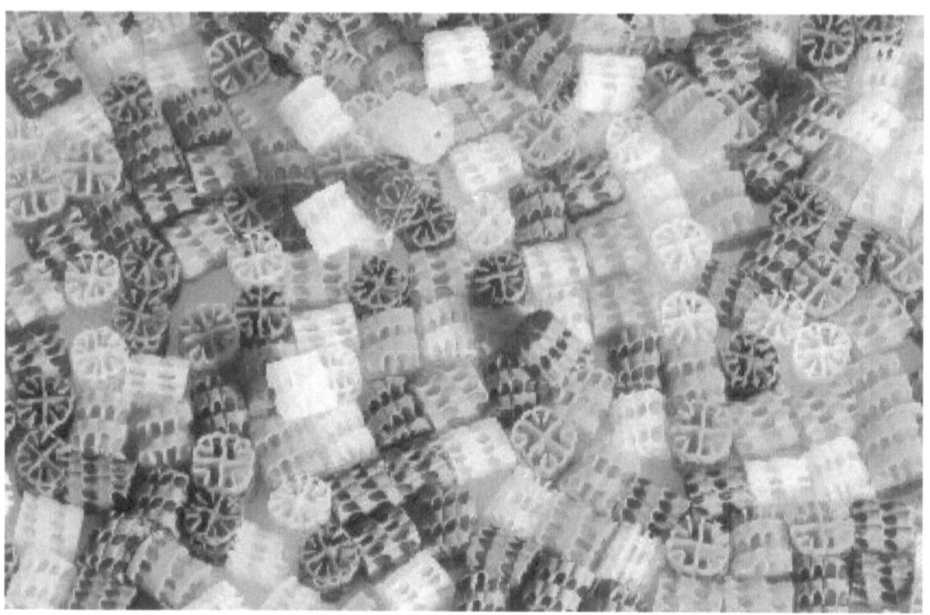

▲ **Fig-10.6 (b):The carrier material used inside an MBBR system**

The specific bacteria remain in the duty tank as the carriers remain in only 1 tank, protected by screens.

3.7 Membrane Bio Reactor (MBR)

MBR is the combination of the membrane process and a biological wastewater treatment process. It is one of the most widely used processes.

In an MBR system, Fig: 10.7 (a), the membranes are submerged in an aerated biological reactor. The membranes have porosities ranging from 0.035 microns to 0.4 microns, which is considered between micro and ultrafiltration.

This level of filtration allows for high quality effluent to be drawn through the membranes and eliminates the sedimentation and filtration processes typically used for wastewater treatment. Because the need for sedimentation is eliminated, the biological process can operate at a much higher mixed liquor concentration.

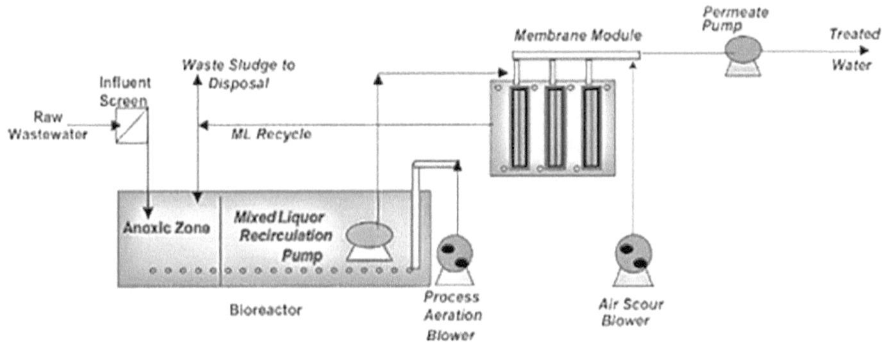

Schematic Diagram for Membrane Bio- Reacror (MBR) Process

▲ **Fig-10.7(a): Membrane Bio-Reactor (MBR) Process**

This dramatically reduces the size of the process tank required and allows many existing plants to be upgraded without adding new tanks. To provide optimal aeration and scour around the membranes, the mixed liquor is typically kept in the 1.0-1.2% solids range, which is 4 times that of a conventional plant.

If air is supplied in the membrane, it is called a Membrane Air Bio Reactor (MABR) as shown in Fig-10.7(b).

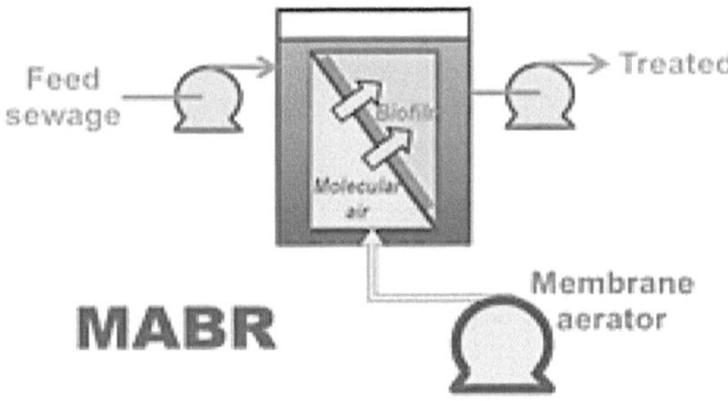

▲ **Fig-10.7(b): Member Air Bio Reactor**

3.8 Differences among different processes

The differences among SBR, MBBR, MBR and MABR is shown as below Table 1.

▼ Table-1:The differences among SBR,MBBR,MBR and MABR.

Item	SBR	MBBR	MBR	MABR
Definition	Sequencing batch Reactor	Moving bed biofilm reactor	Membrane bio reactor	Membrane aerated-biofilm reactor
Principle	Phase separation is gravity settling method	Biofilm process	Activated sludge & membrane separation	Using a membrane both as the biofilm substrate and for molecular aeration
Effluent quality	Satisfactory	MMBR cannot remove SS alone. In addition to TP (total phosphate) and SS (Suspended Solid), it can meet the requirements	In addition to TP, it can meet the requirements	Satisfactory
Daily operation management	Difficult	Simple MBBR packing does not need to be replaced	More difficult MBR membrane module needs to be replaced once every 5 years	Difficult
Cost	It is cost-effective and takes care of the process required.	The initial investment is large, and the later operation is small	The initial investment is large, and the later operation is large. The cost of MBR component replacement is high	As like MBR

3.9 Electro Coagulation

Electrocoagulation, Fig-10.8, is performed by applying an electric current across metal plates that are submerged in water. Heavy metals, organics, and inorganics are primarily held in water by electrical charges. By applying another electrical charge to the contaminated water, the charges that hold the particles together are destabilized and separate from the clean water. The particles then coagulate to form a mass, which can be easily removed.

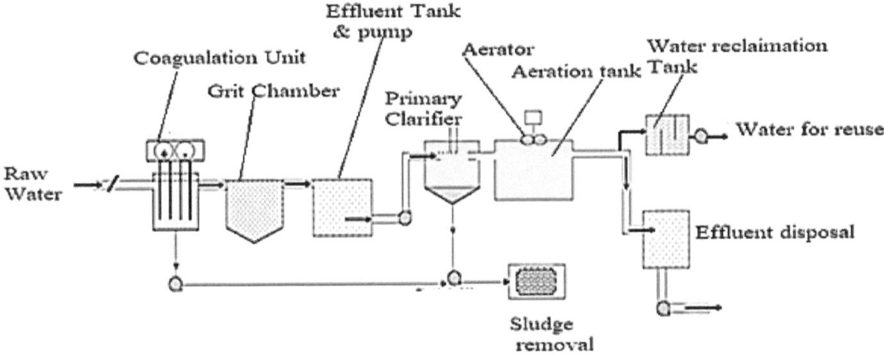

▲ **Fig-10.8: Electro-coagulation process**

3.10 Performance assessment for waste water treatment Plant

In usual practice performance of wastewater treatment plant is assessed by percentage removal of Chemical Oxygen Demand (COD), Biochemical Oxygen Demand (BOD) and Total Suspended Solids (TSS) from wastewater after processing.

3.10.1 Definition related to performance parameter

a) Biodegradable organics- The contaminants which is organic in nature and a food for microbes. Microbes consume to make water free from bio-organics.

b) Biological Oxygen demand (BOD)-Oxygen required for seed (microbes) to oxidize/consume biodegradable organics (food). Quantity of oxygen required by microbes to consume total organics present in effluent, is called BOD.

c) Initial DO (Dissolved Oxygen)- Oxygen content in inflow water is Initial DO.

d) Final DO- Effluent sample is kept in incubator for 5 days at a temp of 20°C. After 5 days the DO in sample is reduced due to consumption by microbes and it is again tested, called Final DO.

e) BOD_5-As the sample is tested after 5 days, calculated BOD is also called $BOD_{5.}$

f) Seed- Seed is simply a solution that contains a sufficient population of bacteria. The term "seed" refers to microbes that consume the biodegradable organic matter (food) in samples for measurement of Biochemical Oxygen Demand (BOD).

g) BOD (unseeded)-BOD calculated without adding seed is called BOD (unseeded).

h) BOD (seeded)-BOD calculated after adding seed is called BOD (seeded).

i) BOD Bottle- Standard capacity of BOD bottle is 300 mL. Hence, calculations are based on 300 mL.

j) Seeding- The BOD test relies on the presence of healthy organisms. If the samples tested contain materials which could kill or injure the microbes, the condition must be corrected and healthy, active organisms are added. This process is known as seeding.

k) MLSS- The suspended solids in an aeration tank are called **Mixed Liquor Suspended Solids** (MLSS).

l) MLVSS (**Mixed Liquor Volatile Suspended Solids**)-If the filtrate used for MLSS testing is ignited at 550°C for 30 minutes, the weight lost on ignition of the solids represents the **Volatile Solids in the Sample**. MLSS is the suspended solids in the aeration tank include both organics and inorganics. MLVSS, on the other hand, **is the volatile portion only**, which are microbes.

m) **The mean cell residence time (MCRT)**-It is an average measure of time for how long the microorganisms remain in contact with the substrate (food source). MCRT is also known as solids retention time (SRT).

n) **RAS (Return Activated Sludge)**-To calculate how many Kgs of microbes required to be added to the aeration basin over the span of a day. Using mixture of sludge and microbes from the clarifier this requirement can be fulfilled. This mixture is known as return activated sludge (RAS).

o) **Sludge volume index (SVI)** – This calculation tells about whether the mixed liquor suspended solids (MLSS) in the aeration tank are settling at the right rate, or if they are hindering the performance of your facility.

3.10.2 BOD Determination

When determining BOD, it is necessary to have a population of microbes that can oxidize, or consume, the biodegradable organic matter present in the sample. If there is too little seed present in the sample, complete consumption of the organic matter may not occur, resulting in inaccurate results. However, in certain sample types, such as some industrial wastes, high temperature wastes, and treated effluent, there is not enough bacterial activity to consume the material that is present. In these cases, seed must be added. Seed is simply a solution that contains a sufficient population of bacteria and is usually provided at the plant.

BOD determination of a seeded and unseeded sample.

a) BOD(unseeded) mg/L= (Initial Dissolved Oxygen(DO)- Final DO)/mL sample]x300ml

b) BOD (seeded), mg/L= (Initial DO, mg/L-Final DO, mg/L-seed correction, mg/L)/mL sample x300 ml/ml of sample.

Example:1

A BOD test was done on a 5 mL sample. The initial DO of the sample is 8.3 mg/L. The final DO of the sample is 4.5 mg/L. Determine the BOD in mg/L for this sample.

Solution: BOD (Unseeded)= (8.3-4.5)/5x300 = 228 mg/L (mg/Lx mL/mL= mg/L).

Example:2

A BOD test was done on a 5 mL sample. It was determined that the initial DO for the sample was 9.2 mg/L and the final DO was 3.8 mg/L. A seed correction of 0.59 mg/L is needed for the sample. Determine the BOD, in mg/L for the sample.

Solution: BOD (seeded)= [(9.2-3.8)-0.59]/5x300=288.6 mg/L.

3.10.3 BOD Loading

When calculating BOD or suspended solids loading on an aeration process (or any other treatment process), loading on that process is usually calculated as kg/day.

Equation: BOD Loading (Kg/day) = BOD concentration x inflow rate = mg/Lx L/day x1/10^{6}kg/day

Example -1

The BOD concentration of the wastewater entering an aerator is 180 mg/L. If the flow to the aerator is $8.5x10^6$ L/day, what is the BOD loading in lb/day (kg/day?

Solution: BOD, kg/day =$8.5x10^6x180/10^6$ kg/day = 1530 kg/ day.

Example-2

The flow to an aeration tank is 14,500 L/min). If the BOD concentration is 155 mg/L, how much Kg/day of BOD are applied to the aeration tank daily?

Solution: BOD applied=$14500x155x60x24/10^6$ kg/day=3236 kg/day.

3.10.4 Solids Inventory

In the activated sludge process, it is important to control the amount of solids under aeration. The suspended solids in an aeration tank are called **Mixed Liquor Suspended Solids (MLSS).**

To calculate the quantity (kg) of solids in the aeration tank, MLSS concentration and the aeration tank volume are required. MLSS can be calculated using the following equation:

MLSS, (kg) = MLSS (mg/L) x Volume (L)x$1/10^6$ kg

Example:

If the MLSS concentration is 850 mg/L and the aeration tank volume is 200 m^3, how much kg of suspended solids are in the aeration tank?

Solution: MLSS, Kg= 850x200 x$1000/10^6$=170 kg

Some calculations need the mixed liquor volatile suspended solids (MLVSS) measurement. That can be determined from the MLSS measurement:

MLVSS = MLSS x% (decimal) volatile matter (VM).

3.10.5 Food-to-Microorganism Ratio (F/M Ratio)

The number of microorganisms which are used to seed the aeration chamber is carefully controlled and is based on the food to microorganism ratio (F/M ratio). The microbes efficiently break down the organic matter in water if they are present in the right proportion. The food value in the F/M ratio for computing loading can be either BOD or COD (Chemical Oxygen Demand).

The typical F/M ratio for ASP are shown below in Table-2:

▼ **Table-2: The typical F/M ratio for ASP**

Process	BOD Kg per day/kg MLVSS	COD Kg per day/kg MLVSS
Conventional	0.2- 0.4	0.5- 1.0
Contact stabilization	0.2- 0.6	0.5- 1.0
Extended aeration	0.05- 0.15	0.2- 0.5
Pure oxygen	0.25- 1.0	0.5- 2.0

Equation to calculate the proper amount of microbes to be added:

F/M ratio = [BOD (kg/day)]/MLVSS (kg)

kg/day is BOD going into the aeration and MLVSS is the amount of microbes inside the aeration tank that consumes the incoming BOD. The F/M ratio determines how many kg/ day of BOD is available for each kg of microbe in the tank.

Example:

A wastewater treatment plant has 84,000kg of MLVSS in the aeration tank. If the primary effluent has a BOD of 185 mg/L and a flowrate is 17,000,000 L/ day, what is the F/M ratio of the tank?

Solution:

BOD in kg/day: $17{,}000{,}000 \times 185/10^6 = 3145$ kg/day

The amount of microbes ready to consume the incoming BOD. (MLVSS = 84000kg)

F/M ratio: 3145/84000=0.037

3.10.6 Mean Cell Residence Time (MCRT)

MCRT is a measure of number of days, microbes are kept in the activated sludge process before being wasted. As a rule of thumb, approximately 0.5 kg of new solids is produced per kg of BOD removed. So, if 100 kg of BOD is removed, approximately 50 kg of new solids are produced. If fails to remove the new solids produced, treatment system suffers in performance. A longer MCRT- age, yields less sludge production than a younger MCRT- age. This is because BOD (food) is used for both staying alive and growing. Equation for MCRT:

MCRT (Days)= [Aeration Tank TSS (Kg) + Clarifier TSS (Kg)]/(TSS Wasted kg/day+ Effluent TSS kg/day) i.e. "solids in the aeration system, kg" divided by "solids leaving the system, kg/day".

Example:

Given the following data, calculate the MCRT:

- MLSS = 2250 mg/L
- Aeration tank + clarifier volume =5.7 x10^6 Litre
- Effluent TSS = 11 mg/L
- Flow = 17x10^6 L/day
- WAS flow = 1.13 x10^6 L/day
- Wasted TSS = 4850 mg/L

Solution:

TSS in Aeration tank= 5.7x10^6 x2250/10^6=12825 kg.

Effluent TSS, kg/day=17x10^6x11 /10^6 =187 kg/day

Wasted TSS, Kg/day=1.13x10^6 x4850 /10^6kg/day=5480 kg/day

MCRT (days)= 12825/(5480+187) =2.26 days.

3.10.7 Return Activated Sludge (RAS)

A mixture of sludge and microbes from the clarifier is fed back in digestion chamber for BOD reaction, this mixture is known as return activated sludge (RAS). The return sludge is only made up of approximately two percent microbes.

Example:

A plant has a flow of 3.21x10^6 L/day. The BOD of the wastewater is 185 mg/L. To achieve a F/M ratio of 0.6 for optimum performance, determine the amount of microbes needed per day.

BOD required (kg/day) =3.21x10^6 x 185 /10^6 =594 kg/ day.

F/M= BOD (kg/day)/MLVSS (kg)

Or, MLVSS =BOD/(F/M) =594/0.6=990 kg/day

RAS flow= 990/24 =41.25 kg/hr.

3.10.8 Return Sludge Rate-Solids Balance (Bioreactor)

In order to determine if your system is balanced properly you take into consideration the MLSS, RAS and the flow to the basin. The balance can be determined with the formula:

Return sludge & solid balance= MLSS mg/Lx Flow, L/day/(RAS L/day-MLSS, mg/L)

3.10.9 Sludge Volume Index (SVI)

Sludge volume index (SVI) calculations reveal whether the mixed liquor suspended solids (MLSS) in the aeration tank are settling at the right rate, or they are hindering the performance of your facility. To calculate the SVI, a sample collected from the aeration tank is allowed to settle for 30 minutes before beginning analysis. Analyse the sample and find out the concentration of suspended solids. this is MLSS concentration, represented in grams per litre (g/L). Divide the wet volume of the settled sludge (represented in mL/L) by the MLSS value from the last step. This calculation gives SVI value (represented in mL/g).

The typical SVI for a system should be between 50 and 150 mL/g. If SVI is outside this range, system level needs to adjust as given in Table-3.

▼ **Table-3: SVI Value and the indications**

SVI Value, mL/g	Indications
Less than 100	Old bio-solids, possible pin flow. Effluent turbidity increasing.
100 - 250	Normal operation, good settling. Low effluent turbidity.
Greater than 250	Bulking bio-solids, poor settling. High effluent turbidity.

To increase the SVI, it is required to increase the waste sludge rate. This results in a slower rate of settling, which in turn traps more of the suspended solids in the mixed liquor, leading to a clearer effluent.

To decrease the SVI, do the reverse- reduce the waste rate. This results in a thicker sludge with heavier particles. As the density increases, so does the rate of settling, making the process more efficient.

The sludge volume index can be determined with the following equation:

SVI (ml/g)= [Settled sludge volume (ml/L)/MLSS(mg/L)] x1000 (mg/g)

Example:

A sample is tested and it is determined the SSV is 420 mL/L and the MLSS is 2850 mg/L. What is the SVI? What does this indicate about the plant?

SVI (mg/L) =420/2850x100=147.37 mL/g

This indicates the plant is running in normal operation, with good settling and a low effluent turbidity.

3.10.10 SDI (Sludge Density Index)

To determine the sludge density index of the sample:

Equation: SDI=100/SVI

This means the sludge density index of the sample above would be: 100/147.37=0.68.

4.0 DM Plant

Demineralization is a method of purifying water. The term **demineralization** can refer to any treatment technique that removes minerals from water. It is done by **ion exchange** processes that remove ionic mineral pollutants almost completely. The terminology deionization and demineralization are frequently used, but are the same.

Minerals and salts dissociate into their constituent ions in the presence of water. These dissolved solids are made up of anions (negatively charged ions) and cations (positively charged ions), both of which are attracted to counter ions of resin.

DM plant is installed in water treatment plant to remove dissolved solids so that salt free water can be used in boiler to prevent any kind of scaling and corrosion.

4.1 DM Process

Schematic diagram Fig-10.9 shows the process about DM Plant. Clarified water initially passed through sand filter to remove any floating particle from it. Water is then passed through the Strong Acid Cation (SAC) resin bed where cationic part of salt is removed. As cationic part is removed, pH of water after SAC is at the range of 3-4.

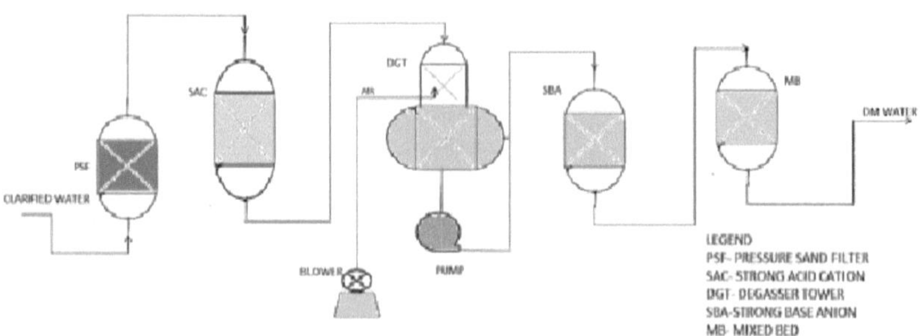

Typical DM plant flow scheme

▲ **Fig- 10.9: Demineralization Plant**

Water coming out from SAC is fed at the top of degasser tower. Water contains carbonic acid which is removed in degasser tower by using air fed counter currently. Purpose of removing CO_2 before feeding to Strong Base Anion (SBA) is to reduce load in SBA.

Degassed water is then fed to SBA where all the anionic part (acid) is removed from water. As the anionic part is removed, water gains it pH up to 5-5.5. TDS is reduced at the level of 1 ppm and silica is 0.1-0.2 ppm.

Water at the outlet of SAC is passed through Mixed Bed (MB) to get DM water having TDS at the level of 0.5 ppm, silica <0.02 ppm and pH 6-6.5.MB contains both cation and anion resin and are mixed to have an intimate contact with water for better performance.

4.2 Reactions in exchangers

4.2.1 SAC Reaction

De-mineralization reaction: $R\text{-}H + CaCO_3 = R\text{-}Ca + H_2CO_3$; RH is cation resin

Regeneration Reaction: $R\text{-}Ca + HCl = RH + CaCl_2$

4.2.2 SBA Reaction

De-mineralization reaction: $R\text{-}OH + HCI = RCI + H_20$; R-OH is anion resin

Regeneration Reaction: RCI + NaOH= R-OH + NaCl

4.2.3 Mixed Bed (MB)

Silica in water remains as silicic acid, sodium silicate and they are feebly ionized while dissolved. Other salts get easily ionized and do not allow silica to de- mineralization reaction. So, SAC & SBA are unable to remove silica. MB inlet water is free from other salt so silica can be easily removed from water, so is the reason to have MB in DM plant. High pressure boiler requires feed water free from silica. MB out silica is < 0.02 ppm.

4.3 DM plant monitoring and corrective measures

Despite the many benefits of using ion exchange resins in DM plant challenging issues are faced. Some **common problems in DM plant are described below:**

4.3.1 Resin fouling

Ion exchange treatment system loses some percentage of operating ability over time of operation due to fouling

Some of the most **common resin foulants are:**

1) Suspended solids such as silica, iron, and manganese as particles or colloidal
2) Oils and greases
3) Bacteria and algae
4) Organic substances

4.3.2 Remedy for resin fouling

a) In general, caustics are used to remove foulants from anion resins, while acids or strong reducing agents are used to remove foulants from cation resins.
b) Similarly, surfactants are typically used to clean oil from fouled resins, though it is necessary to use care in selecting a surfactant that should not itself foul the resin, and sometimes an aggressive backwash with air scour helps.
c) Organic fouling is common and can be difficult to correct, although using a brine squeeze on anion resin at elevated temperatures may be effective.
d) Preventative measures for organic fouling are: pre-chlorination and clarification, activated carbon filtration.
e) Generally, the best way to avoid resin fouling is to ensure proper pre-treatment that removes the foulants before it becomes an issue in addition to using appropriate cleaning, storage, and **regeneration measures** in the day-to-day operation of the ion exchange system to make sure no problematic foulants accumulate over time.

4.3.3 Oxidation of Resin

If oxidizing agents such as **chlorine, chlorine dioxide, chloramine and ozone** come into contact with both cation and anion resins under certain conditions, they can damage the resins, leading to **capacity loss and inhibited performance**. This compaction obstructs the flow of liquids through the resin bed, which affects the overall effectiveness of the unit, and lead to inconsistent effluent quality due to channelling in the resin bed.

While oxidation damage to resins **cannot be reversed, it can be prevented** through various pre-treatment measures. Common preventative measures for oxidation degradation include application of:

a) Activated carbon filtration

b) Ultraviolet irradiation

c) Chemical pre-treatment through the application of a reducing agent.

4.3.4 Thermal degradation of resin

Extremely high or low temperatures can permanently reduce the effectiveness of resins. Over time, thermal degradation alters the resin's molecular structure such that it is no longer able to bind with the functional groups of ions resulting in decrease in operational performance and shorter product life.

Resin capacity has an inverse relationship with temperature, so **it is important to consider the recommended operational temperatures and other process conditions to minimize thermal degradation** over time. Cation resins are more resistant to thermal degradation than are anion resins, though both can generally withstand brief applications of high heat for occasional sterilization or other purposes. Prolonged exposure to extreme temperatures usually means a shorter useable life for resins.

4.3.5 Improper regenerant and inadequate regeneration

Suboptimal system function can result when regenerant solutions are administered incorrectly. Sometimes regeneration methods yield varying results, even when they are implemented with the same procedure. The outcome from resin regeneration depends on the overall resin condition, regeneration process water quality, regenerant chemical concentration, flow rate, temperature, and contact time. Both cation and anion resin can be scaled from improper regeneration.

For example, regenerating with a too-high concentration of sulphuric acid can cause calcium sulphate scale on the resin. With some anion resins, silica can precipitate with improper caustic concentrations.

Following the resin manufacturer's guidelines for regenerant concentration, application time, and flow control can help prevent issues and is advisable.

i. **Common issues on Regeneration**

Once regeneration is initiated, it is vital:

1. Regenerant is the right strength
2. Regenerant is injected at the correct flow rate and over correct time to provide sufficient contact time of regenerant
3. Slow (displacement) rinse is at the correct flow rate and time
4. Fast rinse is at the correct flow rate and time.

ii. **Common issues on Regenerant quality**

Important to use best quality regenerant to protect resin life from:

1. Problem with long term degradation and/or fouling
2. Normal regenerant strength for acid or caustic is around 4%

iii. **Usage of sulphuric acid**

Normally HCI solution is used for regeneration. If at all used sulphuric acid, it is advisable to use stepwise regeneration to avoid calcium sulphate precipitation

a) Two steps – at 0.75 to 1.5% then at 2 to 2.5%

b) Determined by% Ca in raw water

iv. **High concentration of regenerant**

Too high a regenerant strength (>8%) irreversibly damages the resin through osmotic shock.

4.3.6 Channelling

Channelling occurs when liquids pass through the resin unevenly, carving pathways that result in the uneven exhaustion of the resin and breakthrough of untreated solution into the effluent stream. Channelling can be caused by:

a) Incorrect flow rates

b) Failure of the distributor mechanism

c) Inadequate backwashing

d) Blockages by dissolved solids

e) Damaged resin beads.

4.3.7 Resin loss or migration

Resin loss occurs when resin beads flow out of a column, or flow from one vessel to another. There are multiple causes for resin loss such as:

a) Excessive backwashing

b) Mechanical failures in underdrain screening or other resin retention equipment.

c) Fragmentation of resin beads due to exposure to high temperatures, chlorine, and/or osmotic shock.

Resin loss and migration reduces overall system capacity and efficiency. In demineralization systems, for example, the migration of cation resin into the anion unit can result in sodium leakage and excess rinse time.

4.3.8 Input Water quality

Changes in raw water quality have a major impact or plant performance – both capacity and treated water quality

a) May be day to day or seasonal

b) May be due to rainfall patterns or supply management

It is common practice to analyse raw water quality routinely off-line for multiple parameters, but the most valuable parameter to support day-to-day operation is continuous on-line conductivity measurement on a flowing sample.

5.0 Water softener plant- working and performance

A water softener is a filtration system that removes hardness-causing calcium and magnesium minerals from water through a process called ion exchange. A water softener changes hard water to soft water which builds scale in water system used in plant for different purposes.

5.1 Working of a Softener

A water softener removes calcium and magnesium from water through a process called ion exchange. When the hard water enters into the mineral tank, it flows through a bed of spherical resin, also called zeolite, beads. Resin are charged with a sodium ion with negative charge. The calcium and magnesium minerals have positive charges are attracted by the negative charge of the resin. As the hard water passes through the resin, hold the mineral ions and remove them from the water releasing the sodium ion. The column of resin removes all the hardness out of the water as it passes through the bed and softened water flows out to soft water tank.

Equation Involved: $Na_2Z + CaSO_4 = Na_2SO_4 + CaZ$

5.2 Components of a water softener plant

A water softener is made up of three components Fig-10.11, a control valve, a mineral tank, and a brine tank. These three work in conjunction to remove the minerals from hard water, monitor the flow of water, and periodically clean the system through a regeneration process.

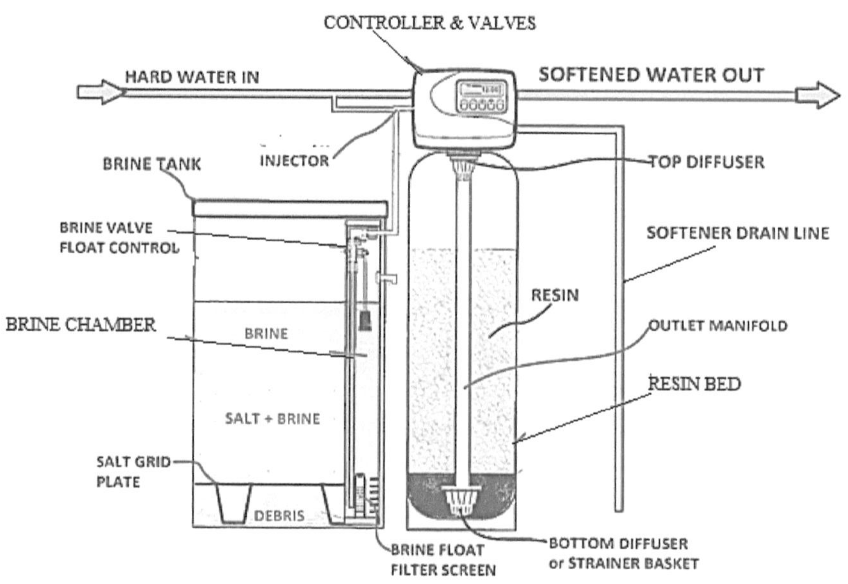

▲ **Fig-10.11: Softener system**

a) Resin Bed-The mineral tank is the chamber where the hard water is softened. The water supply line feeds the hard water into the tank. The water seeps through the bed of resin beads, depositing the water-hardening calcium and magnesium ions. The water exits the tank soft and flows through your pipes and out to your household appliances.

b) The control valve -The control valve measures the amount of water passing through the mineral tank. The valve houses a meter that tracks the volume of water entering the mineral tank. As hard water flows through the resin bed, the resin beads exchange their sodium ions for hardness ions. Over time, this depletes the capacity of the resin to continue to effectively soften water. Before the beads become too burdened with mineral content to continue removing calcium and magnesium ions, the control valve automatically initiates a regeneration cycle. This maximum capacity is pre-programmed into the control valve's on-board computer. Control valves are operated through the controllers.

c) Brine tank-The brine tank holds a highly concentrated solution of salt and is placed adjacent to softener bed. The brine solution is used for bed regeneration. Salt is manually added to the brine tank which dissolve in the water to prepare brine solution. When the control valve registers the softening capacity of the resin is reducing, the heavy brine solution is drawn out of the tank and flushed through the resin in the bed.

5.3 Regeneration process

Water softener regeneration cycles fills the resin beads with a highly concentrated brine solution, washing off the hardness minerals and draining them out of the system. The resin beads are recharged and ready remove the hardness minerals. Water softeners regenerate by one of two methods: co-current or counter-current regeneration.

Equation involved: $Ca\,Z + 2NaCl = Na_2Z + CaCl_2$

a) Co-current regeneration cycle-In a co-current regeneration cycle, the brine solution enters the mineral tank in the same direction as the service flow. The brine solution flows down the depth of the bed of resin beads and an ion exchange process occurs again, in reverse. Co-current regeneration uses more water and salt to complete the regeneration process than counter-current.

b) Counter-current regeneration cycle-In a counter-current regeneration cycle, brine enters the tank through the bottom of the mineral tank, where the water usually exits. The counter current cycle runs the brine up the resin bed, beginning at the bottom where the resin beads are usually the least depleted. A counter-current cycling water softener uses 75% less salt and 65% less water than co-current cycling.

5.4 Minerals removed by softener

Water softeners are designed to remove calcium(Ca^{+2}) and magnesium(Mg^{+2}) ions from hard water which are hardness-causing minerals. The ion exchange process can also eliminate any positively charged ion (also known as a cation) such as iron(Fe^{+3}) and manganese(Mn^{+4}).

5.5 Water Softener Preventative Maintenance

Below are several essential steps of the industrial water softener inspection and maintenance process.

i. Evaluation of Valve Wear and Tear-This maintenance involves pulling internal parts to verify wear and tear and replacing them if necessary.

ii. Brine Study-A brine study uses an instrument called a Salometer or Salinometer to measure the brine concentration at the water softener's drain. During the brine cycle, the salometer records or graphs the percentage of salt traveling through the drain. Ideally, it is 30% salt moving through the drain for 30 minutes. If not enough salt is getting into the water, need adjustments to increase the brine concentration and provide effective softening. Too much salt use, on the other hand, can indicate that the softener's resin requires replacement.

iii. Solenoid Replacement or Rebuild-The solenoid valve serves as a water meter and a flow restrictor for the water softener. It regulates the water flow by turning the water off and on in response to predetermined flow rates.

   Eventually, the solenoid valve may wear out and require replacement or rebuilding. The calcium present in hard water can build up in the valve over time as lime scale, eventually restricting the movement of the solenoid armature and reducing its ability to regulate water flow.

iv. Cleaning Pilot Screens-Over time, the pilot screens in industrial water softeners may become clogged with build up or valve corrosion chips. This clogging can cause issues along the pilot line and in the multiport valve. The brine tank may overflow and the cycles may seem shorter. If controller system is running hydraulically, cleaning the screens can help. Removing build-up from the screens allows the water to flow more freely and mitigates the performance issues associated with a clogged pilot screen.

v. Lubrication- Industrial water softener design includes many moving parts, Lubrication on these essential components helps keep them gliding smoothly to provide optimal softener performance.

vi. Visual Inspections-Visual inspection of a water softener usually includes several evaluative elements:

a) **Checking components for leaks:** Visual inspections of tanks, external pipes and valve pistons can verify leaking points.

b) **Checking the brine tank level:** The brine tank level can provide valuable information about leaks. A high water level in the brine tank often means the brine valve is leaking.

c) **Assessing pressure gauges:** The pressure gauges on the industrial water softener's inlet and outlet should be checked for proper functioning.

d) **Inspecting the drain:** The drain provides clues about leaks and resin degradation. The drain itself may be leaking, or if water flows through the drain while the softener isn't regenerating, the valves may be failing. The drain may also contain bits of resin if the resin is deteriorating.

vii. Resin Testing-Resin testing can help determine its remaining lifespan and condition and identify foulants. Analysis of these samples provides valuable information about capacity, moisture and percentage of broken resin.

viii. Checking Backwash Flow Controllers-The backwash flow controllers are the components of softener that send water flowing upward over the resin. Backwash is essential for removing debris and broken resin beads and resetting the resin for further ion exchange. Checking and cleaning backwash flow controllers can ensure enough flow for backwash and no restrictions for brine.

ix. Assessing Resin Level-Checking the resin level can help determine if enough resin is in the softener. A certain amount of resin loss is natural over the softener's life span and by topping it up to increase capacity. Eventually, the entire resin need replacement after its life span.

x. Monitoring Pressure Gauges-Knowing pressure drop across the softener can help in understanding what additional problems may occur. Minimal pressure is often optimal, high-pressure loss can indicate degraded resins, improper valve function or clogged distributors.

xi. Cleaning the Brine Tank-Cleaning the brine tank annually can help with proper brine injection. Cleaning the tank generally involves disconnecting all lines and hoses, draining the tank, removing any remaining salt and then scrubbing the tank with a cleaner such as a bleach solution.

6.0 Quality Standard of water

6.1 Service water

▼ Table-4: Class of water, its criteria for designated-best-use

Designated-best-use	Class of water	Criteria
Drinking water source without conventional treatment but after disinfection	A	- Total Coliforms < 50 MPN/100 ml - pH between 6.5 and 8.5 - Dissolved Oxygen > 6 mg/l - BOD_5 days 20°C 2 mg/l or less
Outdoor bathing (organized)	B	- Total Coliforms < 500 MPN/100 ml - pH between 6.5 and 8.5 - Dissolved Oxygen > 5 mg/l - BOD_5 <3 mg/l or less
Drinking water source after conventional treatment and disinfection	C	- Total Coliforms < 5000 MPN/100 ml - pH between 6 to 9 - Dissolved Oxygen > 4 mg/l - BOD_5 < 3 mg/l
Propagation of wildlife and fisheries	D	- pH between 6.5 to 8.5 - Dissolved Oxygen > 4mg/l - Free Ammonia (as N) < 1.2 mg/l
Irrigation, industrial cooling, controlled waste disposal	E	- pH between 6.0 to 8.5 - Electrical conductivity at 25°C micro mhos/cm max. 2250 - Sodium absorption ratio max. 26 - Boron max. 2 mg/l
	Below-E	Not meeting A, B, C, D, & E criteria

6.2 Waste water standard

Table-5: Standard Parameters of waste water

	Parameters	General norms[g] 1986				Draft norms Nov. 2015**	MoEF & CC notification, Oct. 2017**	NGT order 2019**
		Inland surface water	Public sewers	Land irrigation	Marine coastal areas			
1	BOD [mg/l]	30	350	100	100	10	30 20 (metro cities)[h]	10
2	COD [mg/l]	250	–	–	250	50	–	50
3	TSS[i] [mg/l]	100	600	200	100 (process water)	20	100 50 (metro cities)	20
4	pH	5.5–9	5.5–9	5.5–9	5.5–9	6.5–9	6.5–9	5.5–9
5	TN[j] [mg/l]	100	–	–	100	10	–	10
6	Ammonical Nitrogen as N [mg/l]	50		–	50	5[k]	–	–
7	Free NH3 [mg/l]	5			5	–	–	–
8	Nitrate [mg/l]	10			20	–	–	–
9	Diss. PO4 as P [mg/l]	5	–	–	–	–	–	1[l]
10	Fecal Coliform [MPN/100ml]	–	–	–	–	<100	<1,000	<230

[g] Standards set in 1986 cover in total 40 parameters, which are not depicted in this illustration. NOTE: industrial wastewater standards are regulated under CETP (Common Effluent Treatment Plant) set, which is not focus of this this study.

*[h] Metro Cities, all state capitals except in the state of Arunachal Pradesh, Assam, Manipur, Meghalaya Mizoram, Nagaland, Tripura Sikkim, Himachal Pradesh, Uttarakhand, Jammu and Kashmir and Union Territory of Andaman and Nicobar Islands, Dadar and Nagar Haveli Daman and Diu and Lakshadweep Areas/Regions. **Standards applicable for discharge into water bodies and land disposal/applications, while reuse is encouraged.*

[i] As SS in [mg/l] in General Norms, 1986.

[j] As Total Kjedahl Nitrogen in General Norms, 1986.

[k] As NH_4-N.

[l] Valid for Phosphorus Total (for discharge into ponds and lakes).

6.3 Standard of water for miscellaneous purposes:

Water used for miscellaneous purposes should be of specific parameters. In case of toilet flushing, fire protection, vehicle exterior washing are almost same and TDS & pH are not so important as the pH remains near to neutral. For different use of water in landscaping, horticulture and agriculture slightly concentration are allowed which are shown in Table-6:

▼ Table-6: Standard parameters of water for miscellaneous uses.

Parameter	Toilet flushing	Fire protection	Vehicle exterior washing	Non-contact impound-ments	Landscaping, horticulture & agriculture			
					horticulture, golf courses	Crops		
						Non-edible crops	Edible crops	
							Raw	Cooked
Turbidity (NTU)	<2	<2	<2	<2	<2	AA	<2	AA
SS	nil	nil	nil	nil	nil	30	nil	30
TDS				2100				
pH				6.5 to 8.3				
Temp. (°C)				Ambient				
Oil and Grease	10	nil	nil	nil	10	10	nil	nil
Minimum Residual Chlorine	1	1	1	0.5	1	nil	nil	nil
Total Kjeldal Nitrogen	10	10	10	10	10	10	10	10
BOD	10	10	10	10	10	20	10	20
COD	AA	AA	AA	AA	AA	30	AA	30
Dissolved Phosphorus as P	1	1	1	1	2	5	2	5
Nitrate	10	10	10	5	10	10	10	10
Fecal Coliform/ 100 ml	nil	nil	nil	nil	nil	230	nil	230
Helminthic eggs/liter	AA[m]	AA	AA	AA	AA	<1	<1	<1
Color	Colorless	Colorless	Colorless	Colorless	Colorless	AA	Colorless	Colorless
Odor				Aseptic (Not septic and no foul odor)				

[m] as arising when other parameters are satisfied.

6.4 Standard Effluent discharge water after treatment

Several standard biological treatments and advanced treatment technologies are presently in use for the treatment of waste water before safe discharge or to reuse for different purposes. The results of different water parameters after application of different treatment technologies are given in Table-7 below.

▼ Table-7: Different biological treatment technologies and the characteristics of treated water.

Assessment parameter/technology	ASP	MBBR	SBR	UASB+EA	MBR	WSP	DEWATS[n]
Performance after Secondary Treatment							
BOD (mg/l)	<20	<30	<10	<20	<5	<40	
SS (mg/l)	<30	<30	<10	<30	<5	<100	
Fecal Coliform, Log unit	Upto 2<3	Upto 2<3	Upto 3<4	Upto 2<3	Upto 5<6	Upto 2<3	
T-N removal efficiency (%)	10–20	10–20	70–80	10–20	70–80	10–20	
Performance after Tertiary Treatment							
BOD (mg/l)	<10	<10	<10	<10	<10	<10	<20
SS (mg/l)	<5	<5	<5	<5	<5	<5	<40
TN							<10
NH_3N (mg/l)	<1	<1	<1	<1	<1	<1	
Total Coliforms, MPN/100 ml	10	10	10	10	10	10	

[n] DEWATS technology serves as comparative for nature-based solutions due to lack in data availability for other systems.

Questions:

1. What do you understand by primary process and secondary process of sewage treatment purification?
2. Enumerate the various tests for water. Explain the significance of those tests which are performed.
3. What characteristics of water are important?
4. What is clarifier? What are the types of clarifiers used in water treatment plant?
5. During application of clarifier, mention the factors affecting the efficiency of a clarifier.
6. What are the difference between Osmosis and Reverse Osmosis? Why R.O. process is needed for water treatment?
7. Define the term Silt Density Index (SDI).
8. What are the standard recommendations of water conditioning before feeding to R.O. membranes?
9. Mention the post-treatment undertaken for membrane permeate water.
10. Compare the advantages and disadvantages of activated sludge process (ASP) with tricking filter process (TFP).
11. What is importance of the sequential batch reactor in a STP plant?
12. When electro-coagulation process is applied in water treatment process?
13. Define the terms BOD and COD and their importance in waste water treatment process.
14. How to determine the BOD and COD of a waste water sample?
15. Briefly describe the working principle of a demineralized water plant.
16. Mention the term 'Fouling of Resin' of a DM process? What are the remedies?
17. Explain the regeneration process in DM plant?

SECTION 11

CYCLONE SEPARATOR, BAG FILTER, SONIC CLEANING, ESP, FGD & SCR

1.0 Introduction

Separation of solid from particle-laden gas is a vital process in industry. The process is implemented in many ways, out of those, separation of product from product-laden gas is the main process. The other important function of solid separation from the gas or air to remove the particles before releasing in atmosphere. The exit gas/ air which gets released out from industrial activities, contains particles having size < 10μm other than toxic gasses.

Particles of this size goes into the body by inhalation and poses the hazard if inhaled. It can reach the lower regions of the lungs and prolonged exposure to such dusts can cause permanent damage to the lung tissues (pneumoconiosis) with a symptom of shortness of breath and respiratory infection. The released gas, which contains oxides of sulphur and nitrogen, causes health hazard and acid rain resulting damage to plant and vegetation.

Hence, it is mandatory to maintain the hazardous components within limit before gases are released in the atmosphere.

This section is for discussion of solid-gas separation process by using different systems such as cyclone separator, bag filter, Sonic system, Electro Static Precipitator (ESP). FGD, SCR systems have also been added which are at present being implemented to capture toxic gases like SOx & NOx from exit gas.

1.1 Cyclone Separator- Operation and Performance Assessment

Cyclonic separation is a process of separating particles or liquid from a gas stream or, separating different liquid phases based on different liquid densities. A cyclone separator is referred to several nomenclatures such as- **'dust separator'**, **'dust collector'**, **'dust extractor'**, **'cyclone extractor'** and **'cyclone separator'**. Generally, smaller units are referred to as 'dust separators or extractors', while large scale industrial separators are referred to as 'cyclone separators'.

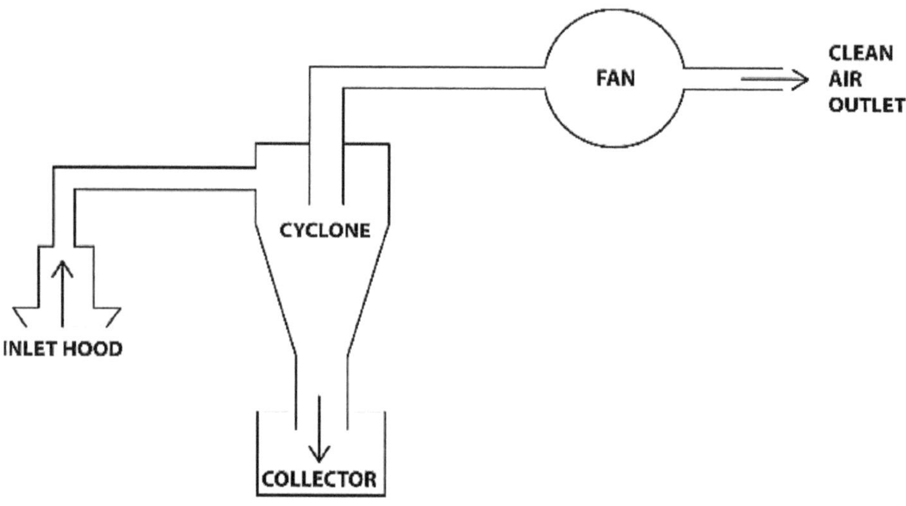

▲ **Fig-11.1: Typical Gas Cyclone Installation**

Cyclone separators often form part of a pre-cleaning stage prior to a gas or liquid being discharged. Cyclone separators can be installed as single units, or in multiples, known as **multi-cyclones**. It is also possible to install cyclones in series, or, in parallel. Separators can be installed with a horizontal or vertical orientation. Fig -11.1 shows a typical installation scheme in industry.

There are two main designs of cyclone separator, these are the **gas cyclone** and **hydro-cyclone**.

a) Gas cyclones are used to remove entrained particles from a gas stream.

b) Hydro-cyclones are used for separating fluids of different densities.

1.2 Gas Cyclone Separators

Gas cyclone separators are sub-grouped into two main categories, **reverse-flow** and **axial-flow**. This categorization is mainly based on direction of input flow and exit flow.

1.2.1 Reverse flow cyclone separator

These are cone shaped. Gas enters the top of the separator body tangentially, flows downwards, then flows back upwards and is discharged as shown in Fig-11.2 (a). As the gas flows back in opposite direction of gas flow, such design is called as reverse flow cyclone separator.

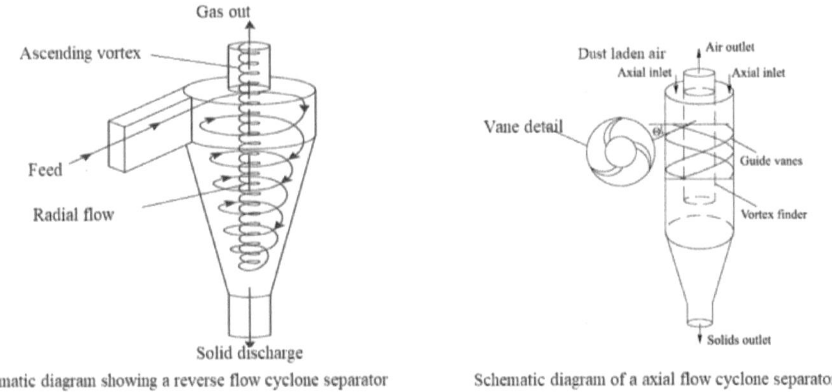

Schematic diagram showing a reverse flow cyclone separator

Schematic diagram of a axial flow cyclone separator

▲ **Fig-11.2 (a) & (b): Reverse flow cyclone separator & Axial flow cyclone separator**

1.2.2 Axial Flow Cyclone Separator

For **axial flow** cyclone separators, gas enters axially at one end and is discharged at the opposite end or same side. Axial flow separators are not as common as reverse flow separators. Vanes are provided on the vortex finder (deep tube) to impart cyclonic action on the inlet gas, as shown in Fig-11.2(b). As the gas enters into the cyclone axially, so is called axial flow cyclone separator.

1.3 Working of a cyclone separator

The main cylindrical part of the cyclone separator is known as the **body** or **barrel**. The gradually narrowing conical section is known as the **cone**.

Un-treated gas enters tangentially through the inlet at the side of the separator. Entrained particles within the gas stream are separated from the gas stream and discharged through the reject port at the base of the separator. Cleaned gas exits through the accept port at the top of the separator, Fig-11.3 (a).

Gas containing entrained particles enters at high velocity through the tangential inlet at the top of the cyclone. The gas flows into the cyclone body/barrel at a tangent and begins to flow in a circular downward spiral towards the lower reject port; this downward flowing spiral is referred to as a spiral vortex.

Gradually decreased cone diameter causes the gas velocity to increase. The outer vortex creates an additional inner vortex closer to the centre of the separator body and this inner vortex flows spirally upwards towards the accept port as shown fig-11.3 (b).

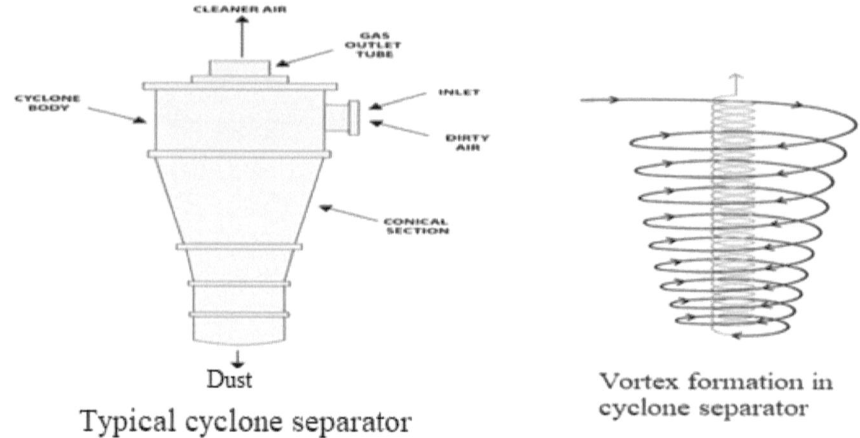

▲ **Fig-11.3(a) & (b): Cyclone separator & Vortex formation**

Particles with more inertia impact with the side of the cyclone whilst particles with lower inertia remain within the gas stream. When an external force is applied, such as by the cyclonic vortex, the particles with low inertia do not continue to travel in a straight line, instead travel spirally as they are swept along by the gas stream.

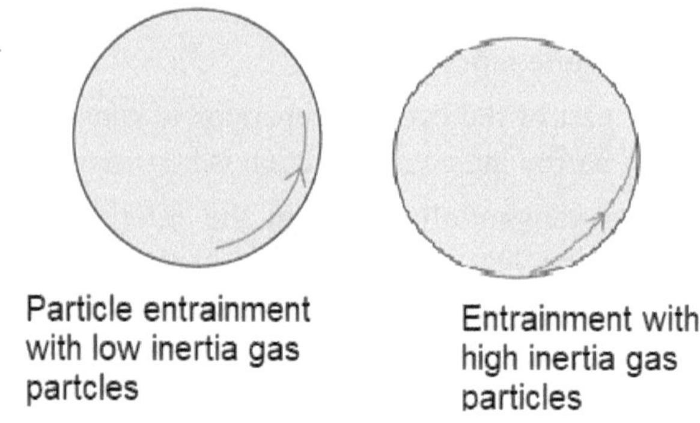

▲ **Fig-11.4: Trajectory of particles**

Particles with greater inertia, less affected by the vortex, continue travelling in a straight line. This straight-line trajectory causes the high inertia particles to move out of the gas stream and impact with the cyclone separator body, Fig-11.4. These particles then fall to the base of the cyclone separator and get out through the reject port. In this way, entrained particles of a certain size can be separated from the gas stream.

Another way it can be explained that higher density particles colliding with the cyclone body whilst less dense particles are retained within the gas stream.

1.4 Hydro-cyclone separator

1.4.1 Introduction

A hydro-cyclone is a cone-shaped cylindrical equipment without any moving parts. It is used to separate solid particles in a liquid stream or liquid in gas stream by size and density.

1.4.2 Description of hydro-cyclone

A hydro-cyclone consists of a cylindrical top with an offset feed inlet pipe, an overflow outlet (top flow) on top and a tapered cone leading down to a small outlet on the bottom. The conventional hydro-cyclone consists of a cylindrical chamber connected to a conical body, which leads to the bottom outlet at the apex of the cone. This discharges the "underflow". The reverse flow is located by a pipe which projects axially into the top of the chamber, is termed the "vortex finder", and discharges the "overflow"/ top flow. Fig-11.5(a) shows reverse flow liquid-solid separation & fig-11.5(b) shows axial flow gas-liquid separation.

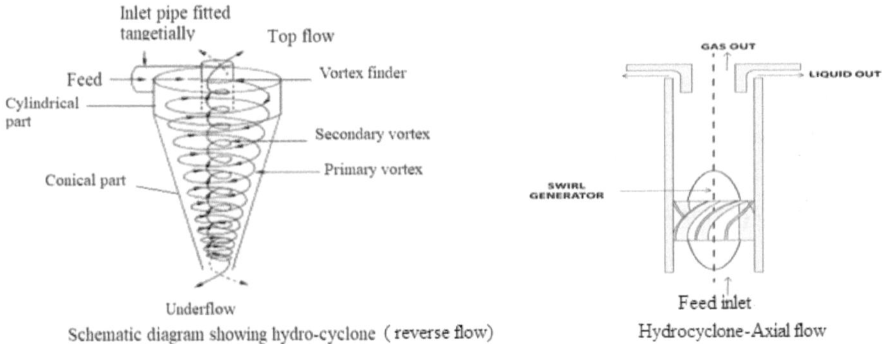

▲ **Fig-11.5 (a) & (b): Hydro-cyclones (Reverse flow & Axial flow)**

1.4.3 Working of a hydro-cyclone

The hydro-cyclone uses fluid pressure to generate centrifugal force and a flow pattern which can separate particles or droplets from a liquid medium. These particles or droplets must have a sufficiently different density relative to the medium in order to achieve separation.

The flow pattern in a hydro-cyclone is cyclonic. This is induced by tangential injection of the liquid into a cylindrical chamber, which causes the development of a vortex. Some of the liquid has to reverse its path and flow counter currently to an axial top outlet. This reverse flow continues to rotate and an air core develops due to lower pressure at the axis of rotation.

For axial flow hydro-cyclone input is from bottom and the discharge axially for gas and liquid.

Solids and water are pumped in under pressure, creating a swirling motion inside the cyclone cylinder which throws the solids to the outside of the cylinder by centrifugal force. The larger particles move downwards under gravity and leave through the bottom outlet with a small quantity of water. A small amount of the fine solids remains near the centre of the cyclone with most of the water and then spirals upwards around an air core and is removed via a central pipe (the vortex finder) and out the top overflow.

1.5 Efficiency assessment for cyclone separator

As reverse flow gas cyclone separator is the commonly used, efficiency of such cyclone is discussed.

1.5.1 Collection efficiency

The collection efficiency - is a measurement of a cyclone's ability to separate particles from the flowing gas stream.

The efficiency of a cyclone represents the proportion of particles that the cyclone can retain:

Efficiency = (Inlet loading - Outlet loading)/Inlet loading x 100%

When,

inlet loading = the concentration of solids in the gas entering the cyclone (kg/m^3 or mg/m^3)

outlet loading = the concentration of solids in the gas leaving the cyclone (kg/m^3 or mg/m^3)

1.6 Efficiency assessment calculation using design data

Efficiency of a cyclone separator can be assed using the design data as explained below.

1.6.1 Data

Following data are required:

a) Cyclone diameter
b) Cyclone gas inlet width
c) Number of effective turns in the cyclone
d) Inlet gas loading in particles
e) Particles diameter
f) Density of the particles
g) Gas inlet temperature
h) Gas inlet viscosity
i) Gas inlet velocity

1.6.2 1st step- cut diameter calculation

The cut diameter is the diameter of particles for which 50% of those particles are collected by the cyclone (50% efficiency). Bigger particles (more than cut diameter) are collected at >50% efficiency; smaller particles (less than cut diameter) are collected at <50% efficiency. Cut diameter can be calculated with the following formula [Lapple]:

$$\mathbf{d_{pc} = \sqrt{[9\mu B_c/\{2\pi n_t v_i(\rho_p - \rho)\}]}}$$

When,

d_{pc} = particle cut diameter (microns)

μ = gas viscosity (Pa-s)

B_c = cyclone inlet width (m)

n_t = number of effective turns in the cyclone

v_i = inlet gas velocity (m/s)

ρ_p = particle density (kg/m^3)

ρ = gas density (kg/m^3).

1.6.3 2nd step- to calculate the ratio of average particle diameter to the cut diameter

The ratio is equal to: Particle size ratio = d_p/d_{pc}

When,

d_p = mean particle diameter of the particles in the inlet flow (microns)
d_{pc} = cut diameter (microns)

1.6.4 3rd step- Efficiency Calculation

Cyclone separator efficiency = $\eta = 1/[1+1/(d_p / d_{pc})^2]$x100%

Or, $\eta = 1/[1+(d_{pc} / d_p)^2]$x100%

1.6.5 Estimation of the outlet flow particle loading

Once the efficiency of the cyclone is estimated, it is also possible to estimate the particle loading in the gas flow out of the cyclone.

Outlet loading = Inlet loading x (1- η /100).

1.6.6 Example

A cyclone separator is provided on the workshop air exhaust flow, having a diameter of 0.6 m, an inlet width of 0.15 m and is rated for 5 effective turns. Find the efficiency of the cyclone with air exhaust flow having a load of 17.65 mg / m^3, with solids particles of 2500 kg/m^3 and mean particle diameter of 10 microns. Conditions considered, the inlet velocity of the air flow is 15 m/s and the air viscosity is 1.8×10^{-5} Pa-s and the air density is 1.2 kg/m^3.

Solution:

Given data-

μ = gas viscosity (Pa-s) = 1.8×10^{-5} Pa-s

B_c = cyclone inlet width (m) = 0.15 m

n_t = number of effective turns in the cyclone = 5

v_i = inlet gas velocity (m/s) = 15 m/s

ρ_p = particle density (kg/m^3) = 2500 kg/m^3

ρ = gas density (kg/m^3) = 1.2 kg/m^3

dp = mean particle diameter of the particles in the inlet flow (microns) = 10 microns

i. **Calculation of cut diameter**

   a) The following formula can be used to calculate the cut diameter of particle:
   b) $d_{pc} = \sqrt{[9 x \mu x B_c / \{(2\pi \times n_t x v_i) x(\rho_p - \rho)\}]}$
   c) or, d_{pc} = $[9x1.8x10^{-5}x0.15/ \{2x3.14x5x15 \times (2500-1.2)\}]^{0.5}$ = 4.54 micron

ii. **To calculate the particle size ratio**

The particle size ratio= inlet particle size diameter/Cut diameter = d_p/d_{pc} = 10/4.54=2.2; d_{pc}/d_p=0.45

iii. **Cyclone collection efficiency estimation**

The efficiency can be estimate with the following formula:

a) Efficiency = E = $1/[1+(d_{pc}/d_p)^2] = 1/(1+0.45^2) = 0.829$

b) The efficiency is thus estimated for this application at 82.9%

It is advisable to double check with Lapple's curve, fig-11.6.

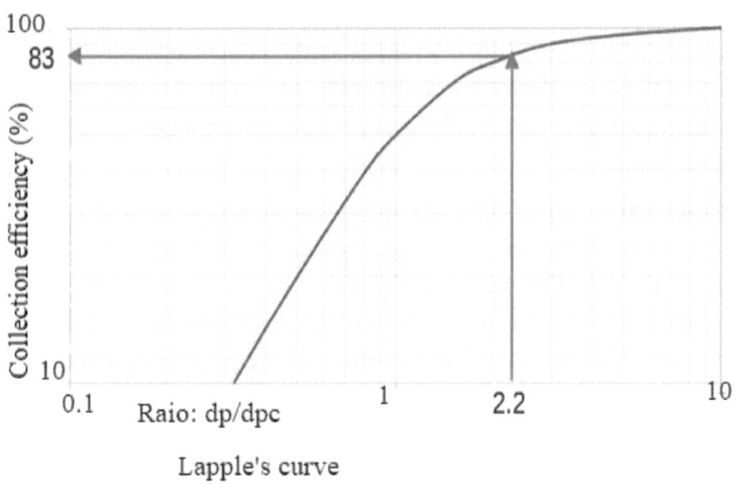

▲ **Fig-11.6: Lapple's curve**

iv. **To Estimate the outlet flow particle loading**

Efficiency of 0.829, the loading of the air flow leaving the cyclone can be calculated as:

Outlet loading = Inlet loading x(1-η/100) = 17.65 x (1-0.829) = 3.02 mg/m^3.

1.7 Factors Effecting Efficiency

There are several factors that can affect a cyclone separators efficiency as discussed below.

Following are important aspects which affects cyclone efficiency.

i. **Particle Density-It** is one of the most deciding factors affecting a cyclone's ability to remove entrained particles. Dense particulates such as ferrous oxides can be separated with an efficiency of 99% or more, irrespective of particle size. When the particle density decreases, the efficiency decreases (assuming no other system changes occur).

ii. **Particle size** -Larger particles can be more easily separated than smaller particles. Particles smaller than 5μ are difficult to separate without using very small separators. Particle exceeding 200 μ can often be separated using other means such as gravity-settling chambers. A reduction in particle size gives a corresponding reduction in efficiency.

iii. **Geometry of separator**- A separator's geometry greatly impacts the efficiency of the unit. A larger diameter cyclone separator does not able to separate particles as efficiently as a smaller diameter separator. The efficiency of the separator increases as the cone diameter decreases. Thus, reducing the cone diameter enables the removal of finer particles. A small diameter cone extracts much finer particles from a gas stream than a larger diameter cone.

iv. **Pressure drop**-All cyclone separators have an associated **pressure drop**. The pressure drop can be thought of as the amount of energy required to move the gas through the separator, or, the amount of resistance the cyclone separator adds to the system flow. The pressure drop is a product of the gas flow rate, gas density and cyclone geometry. Pressure drop can be expressed as:

DP = Pa Inlet - Pa Outlet; When: DP = Cyclone Pressure Drop, Pa = Absolute Pressure.

v. **Accept port diameter**- Another way to increase a separator 's efficiency is to reduce the accept port diameter. This changes the separator body to accept port diameter ratio and has the effect of only allowing finer particles to leave the separator through the accept port.

1.8 Large or Small Separator- a selection criteria

Small cyclone separators have a higher efficiency rating, but the associated pressure drop is high and the volumetric flow rate is low. Gas velocity through small separators is also very high which leads to a high level of erosion if the gas stream contains abrasive particles.

Large cyclone separators have a lower efficiency rating, but the associated pressure drop is low and the volumetric flow rate high. A large diameter separator is not suitable for removing fine particles from a gas stream. Reducing the gas-outlet duct diameter also increases the collection efficiency.

1.9 Advantages and Disadvantages of cyclone separator

1.9.1 Advantages

There are many advantages associated with cyclone separators such as:

a) Low cost
b) Low maintenance
c) Suitable for high temperatures
d) Suitable for liquid mists
e) Less space for installation.

1.9.2 Disadvantages

The disadvantages associated with cyclone separators are:

a) Increased operating costs associated with the pressure drop (assuming large pressure drop).
b) Inefficient when handling small/fine particles.
c) Not suitable for 'sticky' substances.

1.10 Material Selection

Material selection is a very important to prevent erosion. So it is necessary to add a protection layer to the cyclone's internal surfaces.

Suitable materials for protecting the separator within **erosive systems** include materials such as ceramic or some form of enamel.

1.11 Applications

Cyclone separators are utilised in many applications due to their low cost, simple design and high efficiency. Cyclone separators require no bags or filters and require only low maintenance.

Dust is drawn into the main extraction system by a negative pressure created by a fan, usually a centrifugal fan is required for the purpose, fig-11.7. The dust laden air then passes through a cyclone separator where most of the particles is separated from the air stream; the clean air is then discharged directly to ambient air whilst the dust particles is recycled or disposed of.

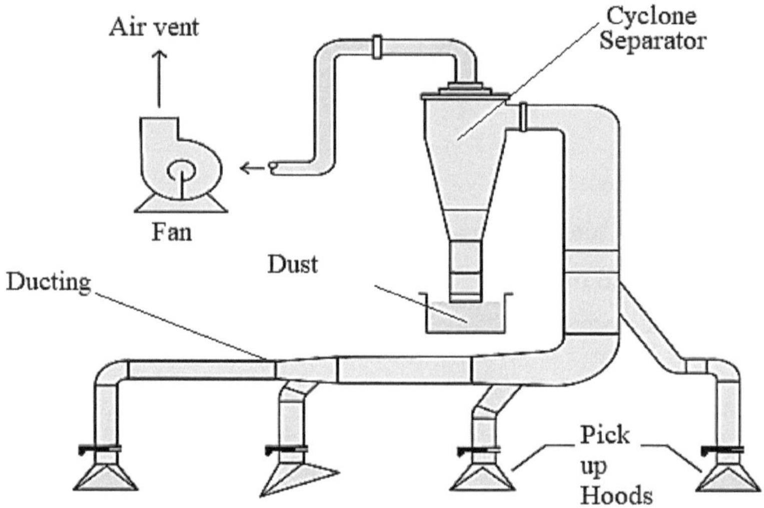

▲ **Fig- 11.7: Set up of cyclone separator**

2.0 Bag Filter- Working and Performance Assessment

2.1 Introduction

To have a better efficiency than the cyclone separator, bag filter has been introduced in industry for dust separation. Its capturing efficiency is ≈99%.

2.2 Particle Collection mechanism in Fabric Filters

Due to the fact that dust particles are always larger than gas molecules, fabric filters are used to capture dust in fabric filters. Hence, when dust laden gas is passed through a filter, the particles are captured on the filter while the clean gas escapes.

Fabric filters captures dust particles through a combination of mechanical processes as discussed below.

2.2.1 Straining

Straining is the process of removing larger particles, which are larger than filter pore/opening. Particles just retained on the surface due to its size.

2.2.2 Impaction

If any restriction is placed on the gas flow path, the stream would bump (knock) into it, here it is fiber. In case of large particles, because of their too much inertia, they can't make turn around the fiber with other particles and keep going straight ahead until they impact on the fiber's surface and stay there. As a filter is having tiny pores in it, it allows the gas molecules to flow through it. This behavior is called **impaction**.

2.2.3 Interception

Medium sized particles having less inertia tend to go around the fiber with the gas stream. So, instead of hitting the fiber head on, they end up scratching it on the side or being "intercepted". This behavior is called direct **interception**.

2.2.4 Diffusion

Particles <1 μm move along with the gas in random motion, called Brownian motion. The particles have a different velocity than the gas stream and at some point could come in contact with the fiber and be collected. This behavior is called **diffusion**.

Impaction and direct interception account for almost 99% collection of the particles > 1 (μm) in aerodynamic diameter in fabric filter system.

2.3 Fabric Filtration Mechanisms

A fiber is any long, thin hair-like material. Woven fabrics are the traditional textile fabric. Fibers are first formed into yarns (threads) and the yarns are then woven together to make a woven fabric.

Woven fabrics have a definite repeated weave pattern. Various weaving patterns increase or decrease the amount of space between yarns and affects the permeability of the fabric. A tighter weave has lower permeability and, therefore, captures fine particles more efficiently.

A nonwoven (felted) fabric is made from long fibers, bonded together by chemical, mechanical, heat or solvent treatment. Most bags are either completely or partially made by weaving since nonwoven fabrics are generally attached to a woven base called a scrim.

Fabric filtration can be subdivided into two mechanisms/modes as describe below.

2.3.1 Cake Filtration

In a woven fabric filter, particles get collected by a 'sieving' mechanism in which particles too large to pass through the mesh of the fabric are caught and retained on the surface of the filter. The caught particles gradually build up a cake on the fabric surface so that the gas flow path becomes complicated while the effective mesh size decreases. Initially some particles escape through the filter until the cake is formed. The collection efficiency of the filter gradually improves with its use but, the pressure drop across the filter increases. Hence, optimum operating cake layer is an important aspect between collection efficiency and pressure drop. A regular cleaning is essential to maintain the pressure drop at an operational level. As filtering takes place primarily due to presence of the dust cake, this type of filtering is called cake filtration.

Practice should be to have the gas flow rate to be half of the standard for about two hours and then to increase stage by stage to prevent the finer particles from clogging the filter. As the dust cake is formed on surface of the woven fabric, ***the low energy cleaning methods such as shaking or reverse air flow through the filter bags is sufficient.***

2.3.2 Non-cake/Depth Filtration

In a nonwoven fabric, particles are mostly caught by their impingement on the fibers within the fabric. The actual flow paths followed by the gas passing through a depth filter are extremely twisting and a particle unable to follow these paths which sooner or later brings it into contact with a fiber where it adheres, largely as a result of Van der Waal's forces. Since particles have to be removed from depth, a ***high energy cleaning technique such as pulse jet is required for their cleaning***.

2.4 Performance difference between woven & non-woven fabrics

Table-1: Performance difference between woven & non-woven fabrics

SL. no	Woven fabric	Non-Woven fabric
1	Dust cake formation is faster due to the flow is relatively uniform across the fabric surface.	Dust cake forms slower in a nonwoven fabric. Dust cake never becomes fully established; hence, this type of filtering is called non-cake filtration.
2	Lower cleaning frequency	The relatively high cleaning frequency,
3	General use for filtration	A nonwoven fabric is generally used for filtering very fine particles.
4	Such fabric can be used in any humidity condition	Felted bags should not be used in high humidity situations to avoid clogging or blinding.

2.5 Pre-coating / Seeding

For a baghouse to operate efficiently and effectively development of dust cake is essential. Unprotected new fabric openings work like miniature venturis to accelerate airflow through the fabric, causing particulate impingement resulting in blinding. To make bag filters effective immediately after use and not to wait for cake building, pre-coating is applied on the filter media.

Pre-coating is the application of a relatively coarse, dry dust to a filter element before startup to provide an initial filter cake for immediate high efficiency and to protect filter elements from blinding. Pre-coating of the filter media with a layer of an inert dust of known particle size distribution, such as lime, calcium carbonate ($CaCO_3$) or fly ash, can minimize problems associated with these type of dusts.

2.6 Assessment of bag filter performance

The best indicator of fabric filter performance is the particulate matter outlet concentration, which can be measured:

a) With a particulate matter Continuous Emissions Monitoring System (CEMS) or a bag leak detection system used to monitor bag breakage and leakage.

b) Opacity monitoring is also an indicator of fabric filter performance.

c) Other indicators of performance include pressure differential, inlet temperature, temperature differential, exhaust gas flow rate, cleaning mechanism operation and fan current.

2.7 Cleaning Mechanism

Baghouses are generally classified by the cleaning method used. The three most common types of baghouses are a) mechanical shakers, b) reverse gas and c) pulse jet.

Mechanical Shaker and Reverse Air Baghouses

2.7.1 Mechanical Shaker Baghouses

i. **Working:** In mechanical shaker baghouses dirty gas enters the bottom of the baghouse and particles are caught on the insides of the fabric bags as the gas flows upwards through the unit, fig-11.8(a).

   Cleaning of the mechanical shaker baghouse is carried out by shaking the top horizontal bar from which the bags are suspended. Shakings are imparted to the bags to have a rippling movement (simple harmonic or sinusoidal) along the fabric. As the fabric moves outward from the bag centerline during portions of the rippling movement, accumulated dust on the surface moves with the fabric. When the fabric reaches the limit of its extension, the patches of dust have enough inertia to dislodge from the fabric and fall into the collecting hopper.

Parameters that affect cleaning include the amplitude and frequency of the shaking motion and the tension in the mounted bag. The first two parameters are part of the baghouse design and generally are not changed easily. Typical values are about 4 Hz for frequency and 50 to 75 mm for amplitude (half-stroke). The tension is set when bags are installed.

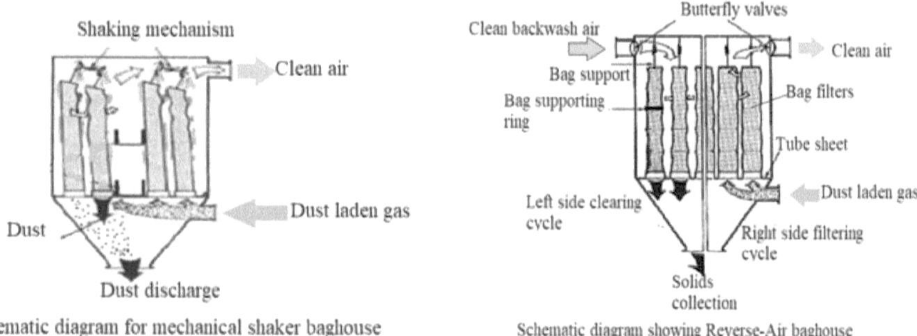

▲ **Fig-11.8 (a) & (b): Mechanical shaker baghouse & Reverse-Air baghouse**

ii. Advantages & disadvantages of shaker bag-house

a) The cleaning does not function properly when on load, and so the filter can only be shaken effectively at the end of a conveying cycle in the absence of gas/air and material flow.

b) There must be no positive pressure inside the bags during the shake cycle because pressures as low as 0.5 mm WC can interfere with cleaning. Hence, though the cleaning system is cheaper, its application is restricted to batch operations.

c) It is also restricted to installations handling materials which readily form a caked layer on the surface of the filter fabric.

d) As the air to cloth ratio for shaker baghouses is relatively low (due to use of fabric type filter), the space requirements are quite high.

2.7.2 Reverse Air Baghouses

Reverse air cleaning was developed as a less intensive way to impart energy to the bags. As shown in Fig-11.9(b), in a reverse air baghouse, the bags are fastened to the tube sheet at the bottom of the baghouse and suspended from adjustable hangers (for adjusting bag tension) at the top.

Dirty gas flow normally enters the baghouse and passes through the bags from the inside, and the dust collects on the inside of the bags. For continuous operation, reverse air baghouses are compartmentalized. Before a cleaning cycle begins, filtration is stopped in the compartment to be cleaned. Bags are cleaned by injecting clean air into the dust collector in a reverse direction, which pressurizes the compartment. The pressure gently collapses the bags partially toward their centerlines, which causes the dust cake to crack and detach from the fabric surface causing the dust cake to fall into the hopper below.

The flow of the dirty gas helps maintain the shape of the bags. However, to prevent total collapse and fabric wear due to rubbing during the cleaning cycle, rigid rings are sewn into the bags at intervals.

2.7.3 Pulse Jet Baghouses

Pulse jet baghouses are also commonly called pulse jet bag filters or pulse jet filters. In a pulse jet filter, the bags are cleaned by introducing a high-pressure pulse of compressed air at the top of each bag.

i. **Working of Pulse Jet Baghouses/Filters**

   The term "pulse jet" probably originated with the use of a venturi (jet pump) to create a high pressure bag cleaning pulse, so is the term "pulse jet cleaning" which was initially fitted with a venture. This usage is no longer universal, and the term pulse jet cleaning as used here refers to cleaning action brought about by short (50 to 150 ms.), high pressure pulse created by any technique, with or without a venturi.

   As shown in, Fig-11.9, if a venturi is installed on top of the tube sheet, the entire bag length can be utilized for effective filtering as compared to alternative design using a venturi within the bag. A major difference between reverse air baghouse cleaning and that of pulse jet is that the reverse air pulse used in pulse jet baghouse cleaning is a fraction of a second in duration while the reverse air cleaned baghouses use reverse air flow of much longer durations (tens of seconds).

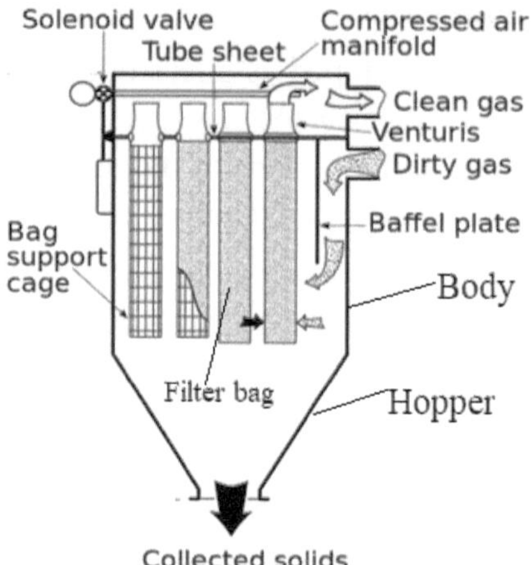

▲ **Fig-11.9: Pulse jet baghouse**

A pulse jet bag filter, individual bags are supported by a metal/wire/filter cage, which are fastened to the tube sheet at the top of the casing. The metal cage prevents collapse of the bag. Dust laden air/gas enters from the bottom of the casing and flows from outside to inside the bags. The air passes through the bags (filter media) while solids are retained on the outside surface of the bags. A signal from a pre-set timer actuates the opening of solenoid valve. Opening of the solenoid allows air to flow from the compressed air header into a compressed air manifold. A row of bags is cleaned by a short burst of compressed air injected through the compressed air manifold (also called blow pipe or purge pipe).

A short burst of compressed air creates a rapidly moving air bubble (shock wave) which results in flexing of the bags. This flexing of the bags breaks the dust cake, and the dislodged dust falls into a storage hopper below. The separated dust from the hopper is removed through an airlock (airlock prevents air from entering the hopper through its outlet) or other airtight device like rotary valve. As air flows in the reverse direction of gas flow during cleaning operation, these types of filters are also called reverse pulse jet filters.

ii. **Advantages & disadvantages of pulse jet bag filter**

a) Pulse jet bag filters can be operated continuously and cleaned without interruption of flow because the burst of compressed air last for only a very short period of time, typically less than a second. Because of this continuous cleaning feature, maximum utilization of the fabric area can be achieved.

b) Pulse jet cleaning is more intense and occurs with greater frequency than the other fabric filter cleaning methods (mechanical shaker and reverse air) to keep the unit from having a high pressure drop across the filter.

c) The intense cleaning dislodges nearly all of the dust cake each time the bag is pulsed. As a result, pulse jet filters do not rely on a dust cake to provide filtration. Usually felted (nonwoven) fabrics are used in pulse jet filters because they do not require a dust cake to achieve high collection efficiencies.

d) If bag rows are cleaned in a sequential order, material from one row of filter bags simply gets moved to the row previously cleaned. This not only results in poor cleaning but tends to retain very fine particles on the filter bags and can lead to premature blinding. Cleaning in a random or non-sequential order distances rows from each other and greatly reduces both re-entrainment and potential premature blinding of bags.

e) Typically bag life in a pulse jet filter is 3 to 6 years. In the case of a traditional pulse jet system with ventures, there is clearly a limit to the length of filter bag that can be effectively cleaned, and a reduction in cleaning efficiency must be expected if very long bags are used. The phenomenal growth of the Cement Industry has seen the advent of plants of much bigger size.

2.8 Air-to-cloth ratios

Typical Air to Cloth Ratio (Filtration Velocity) Comparisons for three cleaning mechanisms as per table-2.

▼ **Table 2: Comparison for three cleaning mechanisms**

Cleaning mechanism	Airflow/ cloth ratio (cc per s/cm^2)	Filtration velocity (cm/s)
Mechanical shaking	1-3	1-3
Reverse air	0.5-2	0.5-2
Pulse-jet	1-7.5	1-7.5

2.9 Bag Failure – the reasons

Reasons for bag failure are:

i. **High temperature in flue gas**- The most important selection criteria is the upper temperature limit of the fabric, or thermal durability. Hence, the process exhaust temperature determines which fabric material should be used for the dust collection.

ii. **Abrasion**- Another problem frequently encountered in baghouse operation is abrasion resulting from bags rubbing against each other and from the type of bag cleaning method as explained below:

   a) In a shaker baghouse, vigorous shaking may cause premature bag deterioration, particularly at the points where the bags are attached.

   b) In pulse jet units, the continual, slight motion of the bags against the supporting cages can seriously affect bag life.

iii. **Chemical attack**- Bag failure can also occur from chemical attack to the fabric. Hence, proper fabric selection and good process operating practices can help eliminate bag deterioration caused by chemical attack.

   It may be noted that changes in dust composition and exhaust gas temperatures (lowered to its dew point) from industrial processes can greatly affect the bag material.

3.0 Flue gas cleaning: Acoustic/Sonic Cleaning

For any type of cleaning, enough energy must be imparted to the fabric to overcome the adhesive forces that holds dust to the bags. In sonic cleaning system vibrational energy is required for cleaning the bags.

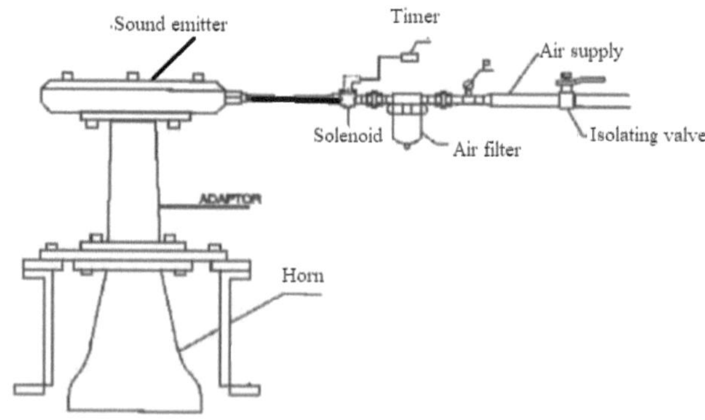

▲ **Fig-11.10: Sonic cleaning system**

3.1 Description of sonic cleaning system

Acoustic / Sonic horns powered by compressed air are a typical means of applying this energy. A horn-shaped outlet attached to an inlet chamber containing a diaphragm. Compressed air at 3 to 5 kg/cm^2 enters the chamber, vibrates the diaphragm, and escapes through the horn. Sound waves leaving the horn contact and vibrate dust containing fabric with sufficient energy to loosen or detach patches of dust that fall through the bag to the hopper below. The horns also can be suspended inside the baghouse structure. Though sonic horns are used as supplemental equipment for applications that require added energy for adequate cleaning, these are also used as the only source of cleaning energy.

3.2 Working of sonic cleaning system

As shown in fig- 11.10, as compressed air is introduced to the driver assembly of an acoustic horn, the pressure builds up rapidly, causing the diaphragm plate to flex. As the air escapes past the flexed titanium diaphragm plate into the horn, reducing the air pressure in the driver. The pressure reduction allows the diaphragm to snap back quickly, creating a strong pulse in the horn body. As the pulses travel through the horn, they are amplified by the bell shape and become powerful bursts of acoustic energy capable of blasting particles away from fabric surfaces.

3.3 Design Considerations

Following are the considerations for designing Acoustic horns.

a) Typically dislodging dust and particulate is more effective at low frequencies, but acoustic cleaning horns operate at 75 Hz or higher i.e. above the structure's natural frequency to ensure that structural components are not damaged.

b) The length of the horn's "bell" determines frequency of the horn. The longer the bell, the lower the frequency.

c) The horns typically operate in the range of 75 to 550 Hz (commonly in the 75 to 160 Hz range) and produce sound pressures of 145 to 150 db.

d) Sonic horns are activated for approximately 10 to 30 seconds during each cleaning cycle.

3.4 Application

In addition to fabric filters, acoustic horns can be used to produce powerful pulses of sound energy to dislodge dust from boiler tubes, heat exchangers, collecting plates of electrostatic precipitators (ESPs) and hoppers.

4.0 Electro Static Precipitator

While flue gas exits from APH and enters into ESP, it contains ash in suspension. Function of ESP is to remove the entrained ash from flue gas so that, the flue gas becomes clean before escapes into the atmosphere.

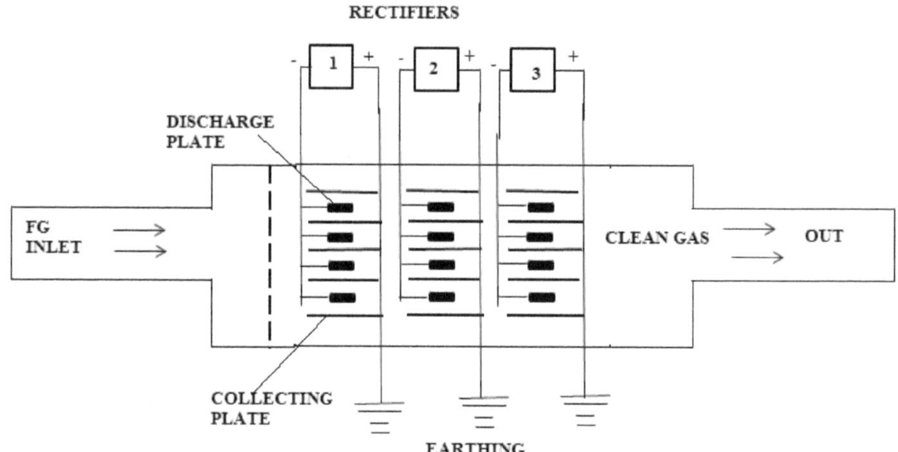

▲ **Fig-11.11: Diagram for 3-field ESP**

4.1 Working Process

The process of separation of ash from flue gas happens in following steps:

a) Suspended particles get charged by negative field of ESP having voltage at the range of 45-120KV

b) Charged particle gets attracted by positive field of ESP, which is grounded

c) Attracted particles gets dislodged from positive field by rapping or hammering

d) Dislodged ash is collected in hopper and disposed to ash silo or nearby ash collection pond.

4.2 Efficiency Assessment for ESP

4.2.1 Important Definition related to ESP efficiency

i. **Particle-Migration Velocity** - This is the speed at which a particle, once charged, migrates toward the grounded collection electrode. Variables affecting particle velocity are particle size, the strength of the electric field, and the viscosity of the gas. How readily the charged particles move to the collection electrode is denoted by the symbol, V, called the particle-migration velocity, or drift velocity. The migration-velocity parameter represents the collectability of the particle within the confines of a specific ESP. The equation for migration or drift velocity (V) is:

$$\mathbf{V = d_p E_o E_p / 4\pi\mu}$$

When,

a) d_p-diameter of particle in micron

b) E_o - Field strength in which particles are charged (V/m)

c) E_p-Field strength in which particles are collected(V/m)

d) μ-gas viscosity (Pa-s).

ii. **Corona Power Ratio**- It is the ratio of power consumed in Watts to flue gas flow in CFM.

This tells about energy consumed in filtering on Ft^3 of flue gas/min. Higher the CPR, higher is the efficiency of ESP.

iii. **Specific collecting area**-It is the ratio of total collecting area of ESP to flue gas flow rate in $m^2/(m^3/s)$.

iv. **Aspect Ratio**- It is the ratio of length to height of ESP, normally it is 0.5-2.0.

4.2.2 Equation for ESP efficiency

$$\boldsymbol{\eta = [1-e^{(-AV/Q)}] x100.}$$

When,

a) A- total collecting surface area

b) V- migration velocity of dust particle(m/s)

c) Q-flue gas flow(m^3/s).

4.2.3 Examples showing numerical problems

i. **Example-1**

Calculate the Aspect Ratio (AR) of an ESP having total length 12m and collecting plate height is 11 m.

Solution: AR= Length of ESP/Height of ESP= 12/11=1.09.

ii. **Example-2**

What is the treatment time in ESP having gas flow rate 90 m^3/s at a velocity 0.7 M/s; ESP is having 4 numbers of field each of length 3.5 m.

Solution:

Total length= 3.5x4=14 m; Velocity= 0.7 m/s.

Hence, treatment time=14/0.7=20 seconds.

iii. **Example-3**

ESP consumes 7 kW for treating gas of flow 17 m^3/s, calculate corona power ratio.

Solution:

Corona power ratio (CPR) =power consumed in Watt/flue gas flow CFM.

Flow=17m^3/s=17x35.3 ft^3/s=600.1x60 CFM=36006 CFM

CPR=7x1000/36006=0.2.

iv. **Example-4**

An ESP handles flue gas flow 61920 $m^3/hr.$, specific collecting surface area is 132 $m^2/(m^3/s)$. Calculate efficiency of ESP, if migration velocity is 5.46 cm/s.

Solution:

Gas flow (Q) = 61920 $m^3/hr.$ =17.2 m^3/s

Then collecting area (A)= Specific collecting area x gas flow =132x17.2=2270 m^2

Migration velocity (V) =5.46 cm/s=0.0546 m/s.

AV/Q=2270x0.0546/17.2=7.2

Efficiency = $1\text{-}e^{(-AV/Q)}$ x100=$1\text{-}e^{(-7.2)}$ x100=99.9%

4.2.4 Power consumption

Standard power consumption is 10-30 W for the air flow rate of 1 000 $m^3/hr.$; at gas velocity in the filter section 2 m/s, resistance to air-flow 10-50 Pa.

4.2.5 Factors affecting ESP Performance

Following factors largely affects the performance of ESP:

i. **Flow rate of gas i.e. gas velocity**: Higher is the gas velocity, lower is the performance of ESP

ii. **Temperature of flue gas**: Higher the gas temperature, higher is the viscosity of gas. Higher gas viscosity causes negative impact of ESP performance. Due to high temperature resistivity of ash particle also increases which affects the ESP performance

iii. **Dust concentration in flue gas:** If ash concentration is increased beyond design value, collection efficiency in ESP also reduces.

iv. **Size of ash particle:** Higher size of ash particle has a negative impact on ESP efficiency

v. **Composition of ash particle**: if, C content is ash is more due to bad combustion or high fixed carbon (FC) fuel, ESP performance is badly affected.

vi. **Operating Voltage**: ESP works at a higher efficiency when higher voltage is maintained.

4.3 Advantages and Disadvantages of ESP

4.3.1 Advantages of electrostatic precipitator

a) The durability of the ESP is high.

b) It can be used for the collection of both dry and wet impurities.

c) It has low operating costs.

d) The collection efficiency of the device is high even for small particles.

e) It can handle large gas volumes and heavy dust loads at low pressures.

4.3.2 Disadvantages of electrostatic precipitator

a) Can't be used for mitigation of gaseous emissions.

b) Space requirement is more.

c) Capital investment is high.

d) Not adaptable to changes in operating conditions.

5.0 FGD- Working and Performance Assessment

5.1 Introduction

Discharge of SO_2 from coal-fired boilers is an environmental threat resulting pollution in the atmosphere and acid rain. Flue Gas De-sulfurization (FGD) system is adopted in fossil fuel-based boilers and many industrial processes to remove SO_2 and to mitigate environmental pollution. This system is of two kinds- dry system and wet system. Subsequent discussions on wet FGD system.

5.2 FGD System Description

Flue gas desulfurization (FGD) consists of the following systems as shown in Fig-11.12.

1. Limestone Preparation System
2. Flue Gas & Absorber System
3. Gypsum Dewatering System

5.2.1 Limestone Preparation System

The limestone preparation system consists of the following sub-systems:

a) **Limestone Handling System-** This system is for bulk handling of lime and storage.

b) **Reagent Preparation and Storage System-** Here, lime is finely ground and the solution is prepared by adding water. The acts as a reagent for absorbing SO2 and is stored in a tank fitted with an agitator to prevent settling.

5.2.2 Flue Gas System

Flue gas from the boiler ID fans enters the flue gas desulfurization (FGD) absorber where SO_2 is removed. From the absorber, treated flue gas flows back to the main FG duct and finally vents to the atmosphere through a stack.

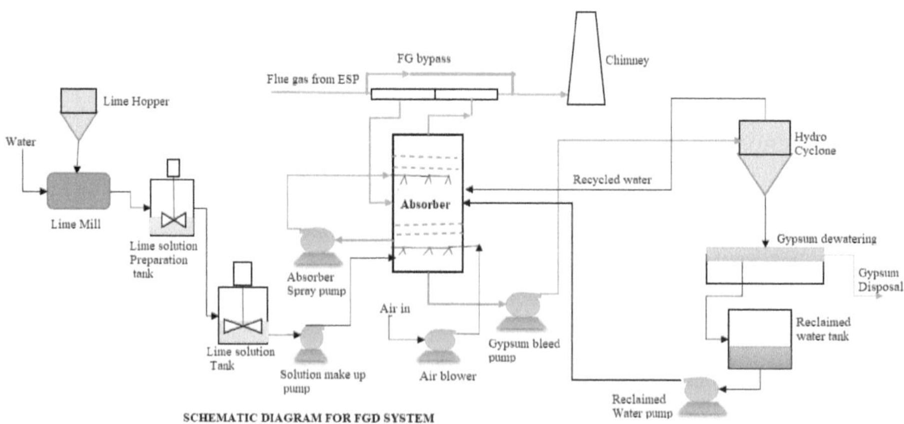

▲ **Fig:11.12: FGD system**

5.2.3 Absorber System

Flue gas enters the absorber near the bottom of the spray tower and flows upward to react with a descending spray of finely divided droplets of recycle slurry. The slurry contains the alkali needed to react with SO_2.

Intimate contact of the flue gas with the alkaline limestone slurry is achieved in successive spray zones. Each spray bank is provided with a series of spray nozzles designed to achieve proper atomization of slurry.

The absorption system in flue gas desulfurization (FGD) consists of the following equipment:

a) **Absorber-**The absorber reaction tank volume is adequately sized to provide sufficient retention time for limestone dissolution, oxidation of calcium sulfite to calcium sulfate (gypsum), and promote gypsum crystal growth.

b) **Absorber Recirculation Pump-**Absorber recirculation pump is for spray solution at the top of the absorber and to impart adequate agitation to ensure optimum utilization of limestone and crystallization as well as precipitation of gypsum.

c) **Oxidation Air Blower-**Oxidation of calcium sulfite to calcium sulfate or gypsum is also accomplished in the reaction tank. Oxidation fan supplies air to lances submerged in the recycle tank for the oxidation reaction.

d) **Mist Eliminator-** A two-stage mist eliminator is provided at the top of the absorber removes entrained liquid droplets from the gas stream. The mist eliminators are washed intermittently to maintain clean surfaces and low gas-side pressure losses.

5.2.4 Gypsum Bleed Pump

The main purpose of the gypsum bleed pump is to bleed off gypsum slurry from the absorber reaction tank to the gypsum dewatering system.

5.2.5 Absorber Area Sump

Drainage and flush water from pumps is collected in the absorber area sump through trenches. One absorber slurry sump and one absorber area sump pump is provided for one absorber. The agitator is provided to keep the slurry solids in suspension during tank usage. The collected slurry in the absorber area sump is returned to the absorber reaction tank or to the emergency storage tank.

5.2.6 Gypsum Dewatering System

A stream of slurry is taken from the absorber by gypsum bleed pump to remove the gypsum produced as a result of SO_2 absorption and oxidation. The gypsum bleed from the absorber is fed to the primary hydro-cyclone and then primary hydro-cyclone underflow is delivered to the vacuum belt filter.

The majority of the filtrate from the gypsum dewatering system is returned to the FGD system. With this flow, the unreacted limestone is thus sent back to the absorber to increase limestone utilization. Chloride concentration in the FGD system is controlled less than 6,000 ppm.

Gypsum dewatering system in flue gas desulfurization (FGD) consists of following equipment:

a) **Hydro-cyclones-**The hydro-cyclone is operated at a constant flowrate. Underflow from hydro-cyclone to be 40~50% solids. The overflow of primary hydro-cyclone is return to absorber reaction tank.

b) **Vacuum Filter-**The vacuum belt filter produces the dewatered gypsum cake, which contains less than 10% moisture. The vacuum filter system consists of a vacuum receiver, vacuum pump, cake wash tank, cake wash pump, cloth wash tank, and cloth wash pump.

c) A belt and belt washing system provides cleaning of the belt after the dried cake has been discharged.

d) **Reclaim Water Tank-**Reclaim water tank is provided for absorber. The reclaim water tank pump is to reclaim filtrate water to the limestone slurry preparation system and absorber reaction tank.

5.2.7 Gypsum Handling System

The gypsum containing less than 10% moisture discharged from the vacuum belt filters to a conveyor belt when it is transported to the gypsum storage shed.

5.3 Process Chemistry of Flue Gas Desulfurization (FGD)

In a wet limestone forced oxidized scrubbing system, a series of reactions occur in the integral absorber and reaction tank. These chemical reactions occur in the gas, liquid, and solid phases. The reactions may be stated in an overall expression as:

$$\mathbf{CaCO_3 + SO_2 + 2\ H_2O + 1/2\ O_2 \rightarrow CaSO_4{,}2H_2O + CO_2.}$$

5.4 Achievable Emission Limits/Reductions

FGD system is capable of reduction efficiencies in the range of 50% to 98%. The highest removal efficiencies are achieved by wet scrubbers, greater than 90% and the lowest by dry scrubbers, typically less than 80%.

6.0 Selective Catalytic Reduction (SCR) System

6.1 Introduction

The selective catalytic reduction (SCR) removes nitrogen oxides (NOx) from flue gas emitted by power plant boilers and other combustion sources. The location of the system within the flue gas cleaning process depends on the type of fuel involved. In a **thermal power plant,** SCR is located between economizer and air preheater.

6.2 Basic Working Principle of Selective Catalytic Reduction

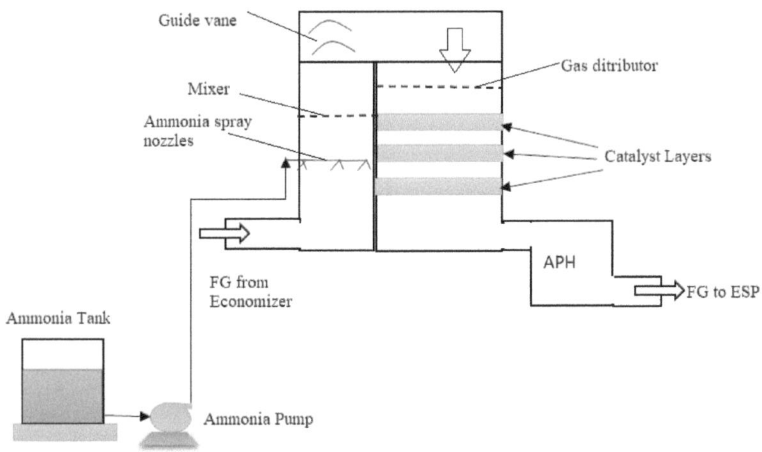

▲ **Fig-11.13: SCR system**

The flue gas entering the reactor passes through the ammonia injection grid When ammonia is injected into the flue gas, fig-11.13. The mixture passes through the **catalyst** layer where conversion of the NOx into N_2 and H_2O take place. SCR catalysts are made from various porous ceramic materials used as a support, such as titanium oxide, and active catalytic components are usually either oxides of base metals (such as vanadium, molybdenum and tungsten), zeolites.

6.3 Chemical Reactions

a) $4NO + 4NH_3 + O_2 + (catalyst) \rightarrow 4N_2 + 6H_2O$

b) $6NO_2 + 8NH_3 + (catalyst) \rightarrow 7N_2 + 12H_2O$

6.4 Application of SCR

a) Power plants

b) Energy from waste plants

c) Industrial Plants (Steel industry, Oil Industry, Pulp and paper industry)

6.5 SCR efficiency

Though theoretically, SCR systems can be designed for NOx removal efficiencies up close to 100 percent, in practice, commercial coal, oil, and natural gas fired SCR systems targets achieving over 90%.

Questions:

1. Describe working of reverse flow cyclone separator.
2. How to assess the efficiency of a cyclone separator?
3. What are the factors that affect efficiency of cyclone separator?
4. What the mechanical processes that cause the filtration through bag filter?
5. How to assess the performance of bag filter?
6. Explain pulse jet bag filter.
7. What are reasons for failure of bag filter?
8. Discuss in brief about sonic cleaning system.
9. Define related to ESP: particle migration velocity, Corona Power ratio, specific collecting area, Aspect ratio.
10. What are the factors that affect ESP efficiency?
11. What are the advantages & disadvantages of ESP?
12. Explain process chemistry for FGD & SCR for flue gas purification.

SECTION 12

OPERATION, PERFORMANCE AND EFFICIENCY ASSESSMENT FOR PSA, HVAC & LIGHTING SYSTEM

1.0 Introduction

In industries separation of O_2 & N_2 from air is conventionally adopted in air liquefaction process by using cryogenic technology. This is a complex system and is meant for mass production or consumption. The PSA (Pressure Swing Adsorption) process for generating nitrogen & oxygen is very much suitable for small & medium industries that require less quantity of nitrogen and also for hospitals that require medical grade oxygen for patients.

An HVAC system is an important system in industry & office that maintains comfort for the occupants and also it maintains a low temperature for critical instruments.

The lighting system in any industry and office installation consumes a considerable amount of power. Proper selection of lamps and luminosity is vital for people working in an office.

This section is included to have a discussion on PSA, HVAC & lighting systems.

1.1 Pressure Swing Adsorption (PSA) -Gas generator

This is the process by which a gas mixture is separated into its components. This method is used mostly for the separation of oxygen & nitrogen from air, called Pressure Swing Adsorption process.

Carbon Molecular Sieves (CMS) is used to produce nitrogen and Zeolite is used for oxygen production. An external buffer tank stores the pure gas.

1.2 Components of a PSA Plant

Fig-12.1 shows different components of a PSA plant as described below.

i. **Air Compressor** -The lubricated screw type air compressor is typically used in PSA plants to compress the air at approximately 5-6 bar.

ii. **Air Dryer**- It is used to remove moisture from compressed air. The dried air eliminates operational issues in the molecular sieves and solenoid valve. Typically, refrigeration dryers are used in PSA plants.

iii. **Air Filters**- In a PSA plant, various filters such as pre-filter, carbon filter, fine filter, bacterial filter, etc. are used to remove solid particles such as dust, dirt, and other foreign material; fine particles, oil particles of varying sizes (5 microns to 0.01 microns), vapour, moisture, and bacteria across various stages of oxygen generation. Bacterial filters are used at the final stage of oxygen production just before the oxygen is allowed for medical purpose.

iv. **Air Storage Tank/Receiver**- It is a pressure vessel where the air from the dryer and filters is collected (pressure:4.5-5 bar) before it is sent to the adsorption towers.

v. **Adsorption Towers**- There are two adsorption towers installed in the PSA plants.

vi. **Adsorbent-** The adsorption towers contain adsorbent are:

   a) **For Oxygen production**: Zeolite molecular sieves and activated alumina/activated carbon. Zeolites are crystalline alumina-silicates with a unique porous structure. They have a high affinity for cations, including ammonium ion (NH_4^+). In the adsorption towers, nitrogen is removed from the air through adsorption and 93% (+/-3%) pure oxygen is supplied to the oxygen storage tank.

   b) **For Nitrogen production**: Carbon molecular sieve (CMS), taps oxygen from the compressed air stream using adsorption. Adsorption takes place when oxygen molecules bind themselves to the adsorbent.

vii. **Valves** -Different types of valves are used across a PSA plant. These valves are solenoid operated and gets operated through sequential controller. Fig: 10. Shows the valves in different positions.

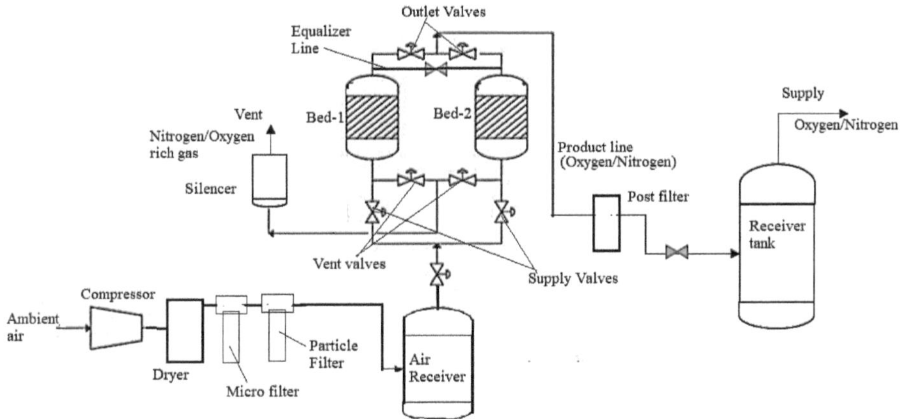

▲ **Fig- 12.1: Pressure Swing Adsorption Process**

1.3 Working of PSA process

i. The working process for either oxygen generation or nitrogen generation is the same. The process is adsorption of unwanted gas (UG) (nitrogen & argon are unwanted gases for oxygen generation and Oxygen & argon are unwanted gases for nitrogen generation) by adsorbent and production of required gas.

ii. The PSA process starts as clean and dry compressed air enters the first cylinder (Bed-1).

iii. The unwanted gas is adsorbed by the pellets at high pressure, but the required gas passes through the bed. The generating high-purity gas is stored in the product tank.

iv. During this step of the cycle, the second cylinder (Bed-2) is under regeneration cycle by:

   a) De-pressurizing

   b) Scavenging through product gas

   c) Ready to take air for generation of gas.

v. Next, the pressure between the two cylinders is equalised.

vi. Bed changeover happens and production is started through bed-2, when the pressure drop increase in the left cylinder (bed-1) and the pellets are saturated.

vii. Exhaust valve for bed-1 opens and waste products are blown out through the exhaust system by desorption process.

viii. The process is repeated and a constant production of product gas is established and cycle time is 2-10 minutes.

ix. An automatic system called sequential controller, takes care of bed change over, venting, scavenging automatically so that, constant production is achieved without less human interference.

1.4 Adsorbent characteristics- affecting efficiency of PSA

Following are the characteristics of adsorbents which affects the efficiency & performance of PSA system.

1.4.1 Nature of Adsorbent

A gas can be absorbed in varying amounts on various absorbent surfaces. Carbon Molecular Sieve (CMS) has played an important role in the nitrogen separation from air pressure swing adsorption(PSA) technique. The key differences between them and activated carbon (AC) are pore size distribution and surface area. Comparatively, CMS has a significant impact on the effectiveness of PSA for nitrogen separation from air.

1.4.2 Surface Area

Very high pore size is necessary for liquid-phase adsorption and such materials can be produced from a wide range of carbonaceous materials using both thermal and chemical activation processes. The pores of activated carbon employed in gas adsorption are typically smaller, with a significant portion of the overall porosity in the micro-pore range. High area small-pore carbons is often too weak for PSA applications. As the surface area of the adsorbent rises, so does the adsorption of gases. This is because the number of adsorbing sites grows as surface area increases. As a result, finely divided materials and some porous compounds work well as adsorbents.

1.4.3 Pressure

At a constant temperature, the amount of gas absorbed by metal adsorbent should increase with the increase in pressure, according to Le Chatleier's principle. Over a narrow range of pressures, a gas's adsorption is directly proportional to its pressure. When the pressure is increased, the rate of adsorption increases at first because the number of gas molecules striking on the surface increases. As a result, increasing the pressure linearly increases the rate of adsorption. However, after a while, the pressure will have no influence on the rate of adsorption since the number of adsorption sites is set and no more adsorption may occur in those locations. Hence, the extent of adsorption is independent of the pressure at that site.

1.4.4 Nature of the gas

A gas will be more easily adsorbed if it is more liquefiable. For example, liquid gases such as NH_3, HCI, Cl_2, CO_2 are more readily adsorbed on the solids surface than permanent gases such as O_2, H_2, and so on.

1.4.5 Exothermic nature

Exothermic adsorption occurs when energy is freed during a process of gas adsorption on a solid surface. Furthermore, the residual pressures on the adsorbent's surface decrease during this process, resulting in a reduction in surface energy. As the temperature rises, the kinetic energy of the gas molecules increases, resulting in more collisions between the molecules and the surface. As a result, the exothermic nature of the adsorbent influences PSA efficiency.

1.5 Advantage of PSA system

Following are the advantages of PSA system.

a) Round-the-clock availability

b) The purity and flow of generated gas are controlled and monitored through the adjustment of the PSA system's cycle time.

c) The plant makes use of air from the atmosphere, so there is no need for any raw material. This ensures, that the plant is cost-effective.

d) Reduced risk from potential mishandling of high-pressure cylinders or containers and ensures the reduction in explosions.

e) Energy efficient, low energy consumption through the use of the PSA technique.

f) The PSA gas generation plant is cost-effective. It saves up to 40-70% over liquid supply systems and 80% compared to cryogenic plants.

2.0 HVAC- Operation, Performance & Efficiency Assessment

Full form of HVAC is Heating, Ventilation, and Air Conditioning. The term HVAC is used to describe a complete home comfort system that can be used to heat and cool an office & home, as well as provide improved indoor air quality.

To be noted that AC simply refers to air conditioning on its own, while HVAC refers to the broader system and does more functions than AC.

2.1 Working of HVAC

Fig-12.2 shows the schematic flow diagram of HVAC system with vapour compression refrigeration as cooling source. Air is the working fluid that is supplied to the ventilation areas after conditioning at a required thermal level maintaining certain level of quality. The working process is briefed as below:

i. Air is sucked by fan from mixing plenum through filter

ii. It passes through the coil chambers. Coil chamber contains pre-heating, cooling and heating coils.

iii. Then it passes through humidifier to maintain humidity in air.

iv. The conditioned air is suppled to ventilation chambers through duct.

v. Recirculation air from ventilation chambers are taken through grills.

vi. A part of recirculated air is vented through vent and part of fresh air is added in plenum from atmosphere to maintain CO_2 level in circulated air system.

2.2 Purpose for HVAC system

HVAC systems are more used in different types of buildings such as industrial, commercial, residential and institutional buildings. The main purpose of HVAC system is to satisfy the thermal comfort of occupants by adjusting and changing the outdoor air conditions to the desired conditions of occupied buildings.

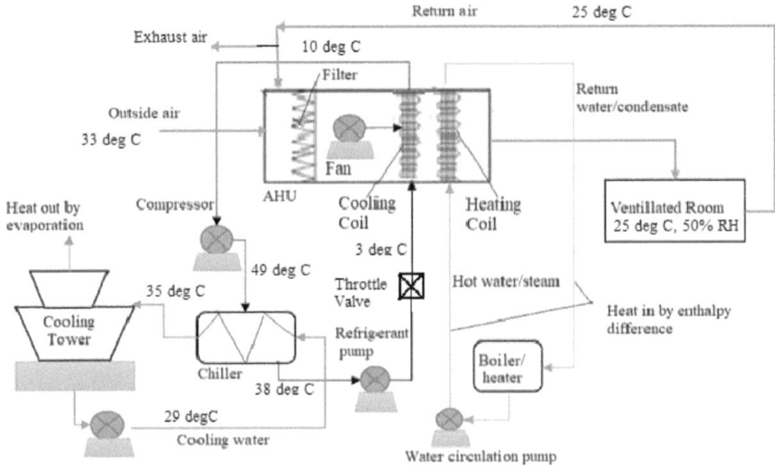

▲ **Fig-12.2: HVAC system**

Depending on outdoor conditions, the outdoor air is drawn into the buildings and cooled or heated before it is distributed into the occupied spaces, then it is exhausted to the ambient air or reused in the system.

2.3 Basic components of an HVAC system

The basic components or equipment of an HVAC system, fig-12.3, that delivers conditioned air to satisfy thermal comfort of space and occupants and to achieve the indoor air quality are:

i. Mixed-air plenum and outdoor air control-In this zone air from ventilated area is recirculated and a small quantity of atmospheric is mixed.
ii. Air filter- Entire air is passed through the filter to get rid of any dust.
iii. Supply Fan-It supplies the entire quantity of air required for the system.
iv. Exhaust or relief fans and an air Outlet-Part of recirculated air is vented through this system.
v. Ducts- The conditioned air is supplied to the ventilation area using ducts.
vi. Terminal devices
vii. Return air system- Return air is brought to the mixed air plenum using return air duct.

viii. Heating and cooling coils- Coils are placed in the duct, after fan discharge, which can cool or heat the air being supplied to places.

ix. Cooling tower- It is provided to supply cooling water in chilling unit.

x. Compressor-It is required for vapour compression refrigeration system.

xi. Boiler- Boiler supplies hot water/ steam to impart heating effect in the supplied air.

xii. Water chiller- Water is used as intermediated fluid that is cooled by the refrigerant and it is done in water chiller.

xiii. Humidification and dehumidification equipment- This system is meant for maintaining a comfort humidity in the air.

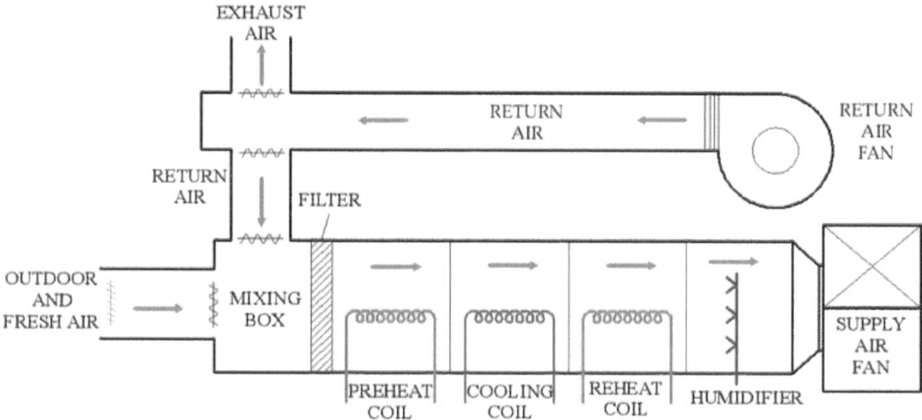

▲ **Fig-12.3: Basic components of HVAC system**

2.4 Classification of HVAC systems

Air conditioning systems are generally characterized by their thermo-fluid distribution medium, air or water, and by their means of controlling heating and cooling. They may also be classified as central or local depending on the plant employed.

Four major categories of air conditioning systems, may therefore, be identified as below:

i. Centralized all air

ii. Centralized air and water

iii. Centralized all water

iv. Localized

a) Unitary system

b) Wind System

c) Split system

d) Packaged roof top

2.4.1 Centralized all air systems

In centralized all air system, conditioned air flows in the space or out from space; no liquid is allowed to flow out of central plant room. These systems consist of a central plant room where incoming air is conditioned to a controlled supply state and then distributed throughout the building. They are classified [ASHRAE (1992)] into single duct, dual duct and multi-zone systems which are then subdivided into constant and variable air volume systems, fig-12.4.

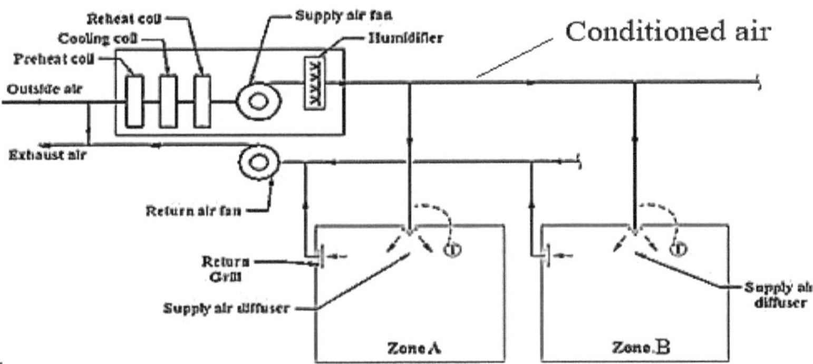

▲ **Fig- 12.4: Centralized air system**

Constant volume systems meet changing air conditioning loads by varying the supply state of the air. Thus, the heat gain to the space rises so the supply temperature is reduced. On the other hand, variable air volume systems meet changing loads by keeping the same supply state but varying the volume supplied.

2.4.2 Centralized air and water

In centralized air-water system, chilled or hot water and conditioned air are generated in a central plant room. These are distributed throughout the building, Fig-12.5. However, the volumes of air distributed are usually much less than the all air systems because, only the intake of fresh air for ventilation is conditioned centrally. Additional room air is treated within the conditioned room by passing it through a terminal plant served by the chilled or heated water distributed from the central plant room.

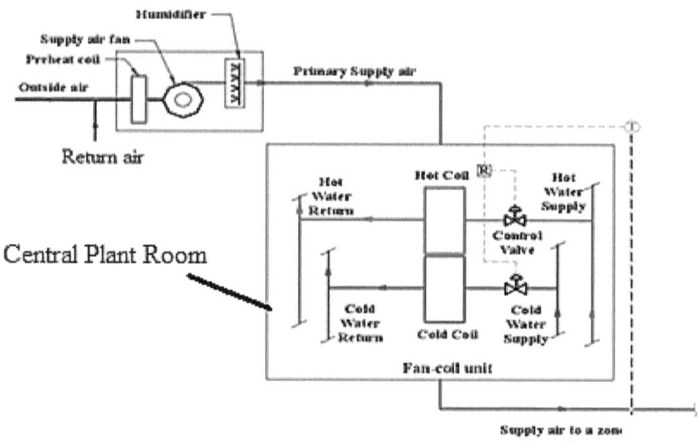

▲ **Fig-12.5: Centralized air-water system**

2.4.3 Centralized all water systems

These systems operate on the distribution of chilled and heated water from a centralized plant room to terminal plant such as Fan Coil Units (FCU) when heating and cooling of the air takes place locally, Fig-12.6. The requirement for fresh air ventilation must be provided locally either by natural infiltration or by an opening through the wall.

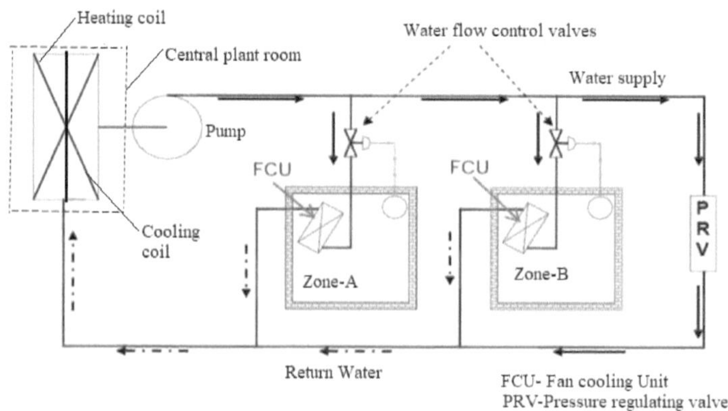

▲ **Fig-12.6: Centralized all water system**

It is not possible with these systems to control the humidity within the space as the terminal plant includes only sensible heat exchangers.

2.4.4 Localized air conditioning systems

As the name suggests, these systems are stand-alone units that are located in or near the space to be conditioned. They are based upon the vapor compression cycle. In this case the evaporator is housed within the conditioned space, cooling the room air that is drawn across it. The condenser on the other hand is positioned outside the conditioned space dissipating the heat to atmosphere. These are sub-classified as:

i. **Unitary system:** If the roles of the two heat exchangers, evaporator and condenser, are reversed the unit can be used as a heat pump heating the inside air from the external source.

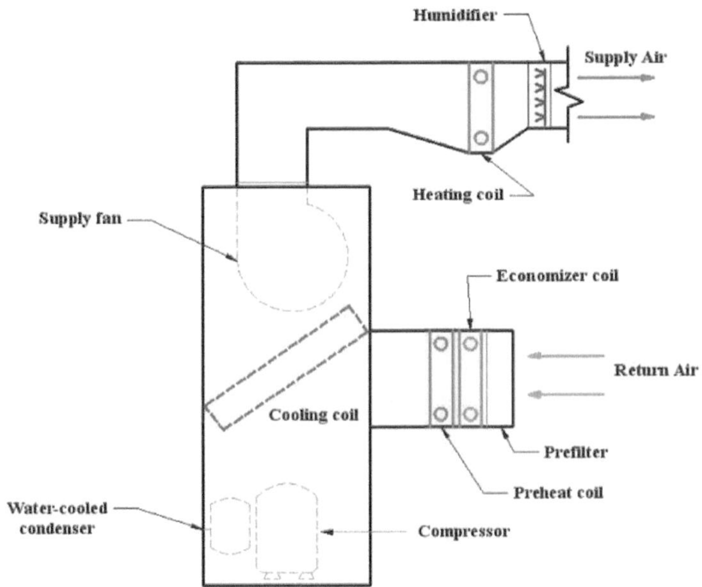

▲ **Fig: 12.7: Unitary system**

ii. **Window air-conditioner:** This system is a packaged device consisting of a vapor compression refrigeration cycle, a fan, a filter, control system and housing. Window air-conditioners can be installed in a framed or unframed opening in building walls and in window openings without any ductwork and distribution the cooling or heating air effectively inside the conditioned space. The air conditioning contains the condenser is located outside the space while the evaporate is inside the space. The heating process can be achieved by adding electric resistance coil in the air conditioning or reversing the refrigeration cycle to act as a heat pump.

iii. **Split systems:** The split systems contain two central devices- the condenser is located outdoor and the evaporator located indoors. The two devices are connected by a conduit for refrigerant lines and wiring. This system solves some issues of small-scale single-zone systems since the location and installation of window, unitary or rooftop air conditioners may affect the aesthetic value and architectural design of the building. The split systems can contain one condenser unit and connected to multiple evaporator units to serve multiple zones as possible under same conditions or different environmental conditions.

iv. **Packaged rooftop air-conditioner:** It consists of a vapor compression refrigeration cycle; heat source such as heat pump and electric resistance; an air handler such as dampers, filter, and fan; and control devices, as shown in fig-12.8. This system may be connected to ductwork and serve a large-size single zone that cannot be served by unitary or window air conditioners.

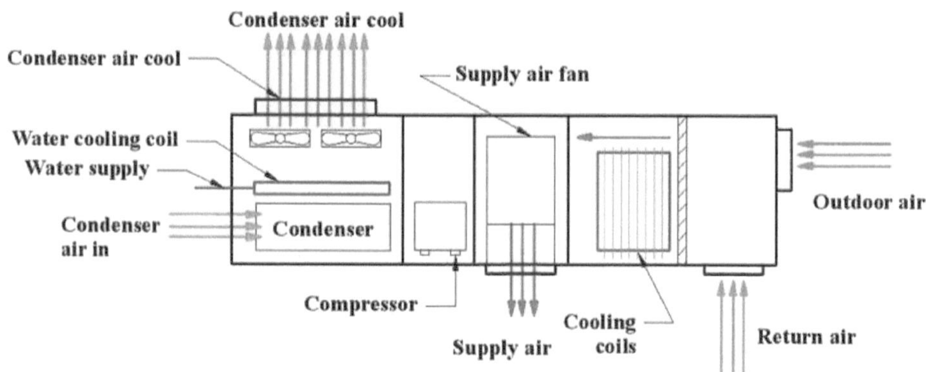

▲ **Fig-12.8: Packaged roof top air conditioner**

2.5 Working of Vapour Compression Refrigeration System & chilling water system

Refrigeration section in HVAC system is very vital section and it cools the air required for HVAC system. In this system the refrigerant undergoes phase changes thereby heat transfer takes place from one medium to the other for cooling. It is widely used method for air-conditioning of buildings, automobiles and several industrial processes.

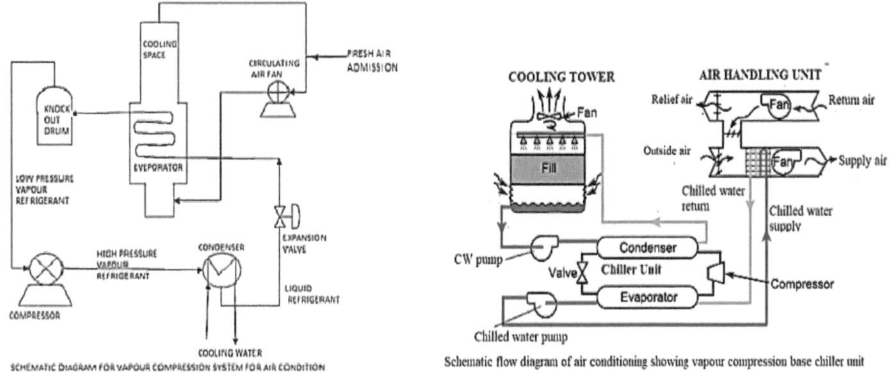

▲ **Fig-12.9(a)&(b): Vapour compression systems for air condition**

Figure-12.9(a) shows different equipment involved in vapour compression refrigeration system for cooling air, Fig-12.9 (b) shows the chilling unit for cooling water.

Dry saturated refrigerant vapour is compressed in compressor at an elevated pressure. The compressed gas is condensed in a condenser at a constant pressure by using cooling water or by air in radiator.

Liquid from condenser is passed through the expansion valve when the pressure is reduced to evaporator pressure. The liquid gets evaporated in evaporator by taking away heat from hot medium (here it is air); thereby the medium gets cooled. The cooled air is circulated to the space to be cooled down for air conditioning. The vapour generated in the evaporator again gets sucked in the compressor and thus cycles go on.

Chiller Unit- If water gets cooled in place of air, this unit is called chiller unit. The chilled water is used to cool air which is used for chamber cooling. Fig-12.9(b).

2.6 Important definition & terminology related to HVAC and system load assessment

i. **CLTD- Cooling Load Temperature Difference:** This is defined as temperature difference between external temperature and internal temperature being maintained by air conditioning system.

ii. **CLF- Cooling Load Factor:** It is radiant solar energy that affects the cooling space and is a function on solar time and orientation.

iii. **SCL- Solar Cooling Load factor:** The factor considered for heat gain from glass.

iv. **U-factor:** Over all heat transfer coefficient for wall, floor & roof.

v. **Room Sensible Heat Ratio** (**RSHR**) = Sensible heat /(Sensible heat + Latent heat).

vi. **ASHRAE standards:** While the ISO 7730 focuses on thermal comfort and expands upon the variables mentioned in it, the ASHRAE (American Society of Heating, Refrigerating and Air Conditioning Engineers) series has more documents pertaining to other HVAC applications.

Sample data from ASHRAE table: CLTD at 17 hrs: for flat roof =44°C, for wall 12°C, for glass =7°C.

2.7 Load calculation using CLTD, CLF& SCL method

HVAC system load calculation is vital before designing a system. Commonly ASHRAE hand book is followed to evaluate the loading and designing of HVAC system. Some basic equations are shown here to have a glimpse about the design calculation.

2.7.1 External heat gain

Following equation can be used to calculate external heat gain.

1. **Conduction heat gain through wall**

   Equation:

   a. $Q = U \times A \times \Delta T$ [wall is completely shaded, ΔT- temperature difference between outside DBT & inside room]

   b. $Q = U \times A \times CLTD$ [shining on wall]

2. **Conduction heat gain through roof**

   Equation: $Q = U \times A \times CLTD$ [shining on roof]

3. **Conduction through window**

   Equation: $Q = U \times A \times CLTD$ [shining on window]

4. **Conduction through glass**

   Equation: $Q = A \times SC \times SCL$ [SC-shading coefficient, SCL-Solar cooling load factor].

2.7.2 Internal heat gain

1. **Heat gain from people**

   a) Sensible heat gain (Qs) = No. of people x sensible heat gain per person x CLF

   b) Latent heat gains from people (Q_L)=No. of people x Latent heat gain per person

   [Fresh air requirement =0.01 m^3/s per person]

2. **Heat gain from lighting**

   Equation: (Q) = Watts x ballast factor x CLF

   [ballast factor: 1.2 for fluorescent lamp,1.0 for incandescent lamp; If light is on for 24 hrs, AC off in night, CLF-1]

3. **Heat generated by computer & office equipment**

   Equation: 5.4 Watt/ m^2 x office area (m^2)

4. **Heat gain through infiltration**

   a) Calculation of infiltration air quantity = volume of area x air change/hr x density of air [air change factor =0.3]

   b) Sensible heat gains: (Q_s) = infiltration air quantity(Kg/s) x Cp x ΔT

   c) Latent heat gain (Q_L) = Air flow x Δh x L [Δh –difference of humidity ratio (kg of water/kg dry air), L- latent heat of water at partial pressure of moisture in air,600 kcal/kg].

5. **Volume of flow rate**

   The volume flow rate can be determined from the following equations using Q_s & Q_L.

   $V= Q_S / (t_r - t_s) \times T/ 358$; Or, $V= Q_L / (g_r - g_s) \times T/856$.

   When,

   a. V= volume flow rate at temperature t (m^3/s)

   b. t_r= room temperature (°C)

   c. t_s= supply temperature (°C)

   d. T= temperature at which the flow rate volume flow rate is calculated (k)

   e. g_r= room moisture content (g/kg)

   f. g_s= supply moisture content (g/kg).

2.8 Other Features in HVAC

2.8.1 Heat recovery

Whenver air conditioning systems employ return air ductwork there is a potential for heat recovery from the exhaust air before disposal to atmosphere. It is accomplished through the use of air-to-air heat exchangers. Under certain climate conditions, when the return air is nearer the supply conditions than the external air, maximum recirculation will reduce energy consumption. However, there is always be a need to exhaust some air as there is always a need for fresh air intake for ventilation and the wellbeing of the occupants.

2.8.2 Control and instrumentation

Since the loads in various zones of a building vary with time, control systems are used to match the output of the HVAC system to the loads. A HVAC system is designed to meet:

a) The extremes in the demand

b) Control at part load conditions.

c) Good indoor air quality and comfort under all anticipated conditions

d) Lowest possible cost of operation.

Controls may be pneumatic, electric, and electronic or they may even be self-contained, analogue and digital.

2.9 Overall Performance assessment of HVAC

To assess the overall performance of HVAC system, performance of following systems are assessed to get over all assessment of HVAC system.

a) **Refrigeration system**

b) **Air handling system**

c) **Pumping system**

2.9.1 Performance assessment for refrigeration system

I. **Terms and Definitions**

Before progressing further following terms & definitions need to be discussed.

a) **Ton of refrigeration (TR)**-The amount of heat absorbed by the system is the refrigerating effect. A machine which can produce a refrigerating effect that can make 1 short ton (2000 pound) of water at 0 °C to ice at 0 °C in 24 hrs., is called 1-ton Machine or capacity is 1 TR.

1TR = 2000 x 0.454 kg x 80 /24 kcal/hr. (latent heat of melting of ice =80 kcal/kg)

=3024 kcal/hr. = 50.44 kcal/ min= 210.85 KJ/ min.

b) **Coefficient of Performance (COP)-**It is the ratio of heat absorbed or abstracted by the system and work done on the system. Value remains between 2.2 to 2.6.Chiller efficiency measured in Kcal output (cooling) divided by Kcal input (electric power).

c) **Specific humidity or humidity ratio -**Specific humidity or humidity ratio of an air sample is the ratio of the weight of water vapor contained in the sample compared to the weight of the dry air in the same sample.

d) **Net Refrigerating Capacity**- A quantity defined as the mass flow rate of the evaporator water multiplied by the difference in enthalpy of water entering and leaving the cooler, expressed in kcal /h, Tons of Refrigeration. M x ΔH [M= quantity of water flow kg/hr, ΔH enthalpy difference].

e) **kW/ton rating**- Commonly referred to as efficiency, but actually power input to compressor motor divided by tons of cooling produced, or kiloWatts per ton (kW/ton). Lower kW/ton indicates higher efficiency.

f) **Energy Efficiency Ratio (EER):** Performance of smaller chillers and rooftop units is frequently measured in EER rather than kW/ton. EER is calculated by dividing a chiller's cooling capacity (in Btu/h) by its power input (in Watt) at full-load conditions. The higher the EER, the more efficient is the unit.

Air conditioning and refrigeration consume significant amount of energy in buildings and in process industries. The energy consumed in HVAC is sensitive to load changes, seasonal variations, O&M, ambient conditions etc. Hence, the performance evaluation is essential for this system.

II. **Performance assessment of refrigeration parameter**

The purpose of performance assessment of refrigeration system is to verify the performance of the followings.

a) Net cooling capacity (tons of refrigeration)

b) Energy requirements, at the actual operating conditions.

c) To estimate the energy consumption at actual load vis-à-vis design conditions.

1. **Determination of net refrigeration capacity (TR)**

   To determine the net refrigeration capacity, that is the net heat removed from the water as it passes through the evaporator by determination of the following:

   a) Chilled Water flow rate (m), kg/s

   b) Temperature difference between entering and leaving water (t_{in}-t_{out}),^{0}C.

   Equation for net refrigeration capacity in tons:

   TR= m C_p (t_{in}-t_{out})/3024. When, m= Flow of chilled water, t_{in}- Inlet temperature, t_{out}- outlet temp.

2. **Measurement of compressor power (kW)**

   The compressor power can be measured by a portable power analyzer directly in kW. If not, the ampere has to be measured by the available on-line ammeter or by using a tong tester. The power can then be calculated by assuming a power factor of 0.9 Power (kW) = $\sqrt{3}$ x V x I x cos φ.

3. **Performance calculations of refrigeration system**

   The energy efficiency of a chiller is commonly expressed in one of the three following ratios:

   a) Coefficient of Performance (COP)= kW refrigerant effect (TR)/kW input

   b) Energy Efficient Ratio (EER)=Kcal/ hr. refrigerant effect /Watt input

   c) Power / ton, (kW/ ton) =kW input/Ton of refrigeration effect.

First to calculate the kW/ton rating from the measured parameters.
Equation: kW/ton rating = Measured compressor power, kW/ Net refrigeration Capacity (TR)

Use this data, other energy efficiency parameters are calculated with the following relations:

COP = 0.293 EER

EER = 3.413 COP

kW/Ton = 12 / EER

EER = 12 / (kW/Ton)

kW/Ton = 3.516 / COP

COP = 3.516 / (kW/Ton).

Example: In a brewery chilling system, ethylene glycol is used a secondary refrigerant. The designed capacity is 40 TR. A test was conducted to find out the operating capacity and energy performance ratios. The flow was measured by switching off the secondary pump and measuring the tank level difference in hot well.

Measurements data: Temperature of ethylene glycol entering evaporator = (-) 1°C

Temperature of ethylene glycol leaving evaporator = (-) 4°C

Ethylene glycol flow rates = 13200 kg/hr.

Evaporator (ethylene glycol) pressure drop (inlet to outlet) = 0.7 kg/cm^2

Power input to compressor electrical power, kW = 39.5 kW

Specific heat capacity of ethylene glycol = 2.34 k Cal/kg-°C

a) Net refrigeration capacity (TR)= 13200x2.34x [-1- (-4)]/3024 = 30.65 TR
b) kW/ ton of Refrigeration, (kW/ TR) = 39.5/30.65=1.29
c) COP=3.516/1.29=2.73. Or, 30.65x3024/(860x39.5)=2.73
d) EER=12/1.29=9.3.

2.9.2 Performance evaluation of air handling systems (AHU).

For centralized air conditioning systems, the air flow at the air handling unit (AHU) can be measured with an anemometer. The dry bulb and wet bulb temperatures can be measured at the AHU inlet and outlet. The data can be used along with a psychrometric chart (Figure 9.1) to determine the enthalpy (heat content of air at the AHU inlet and outlet)

Heat load (TR) = m x (H_{in} – H_{out}) 4.18 / 3024

a) m – mass flow rate of air, kg/hr.
b) H_{in} – enthalpy of inlet air at AHU, kJ/kg or Kcal/kg
c) H_{out} – enthalpy of outlet air at AHU, kJ/kg or Kca/kg.

[If enthalpy is in Kcal/kg, factor 4.18 is to be ignored]

2.9.3 Performance Assessment of Pumping Systems

i. **Performance assessment for chilled water pumping system**
   a) Power Measurements for Chilled Water Pumping System.
   b) Chilled Water Main Supply Flow Measurements
   c) Calculation & Assessments of Specific Power Consumption

ii. **Performance Assessment of Cooling Water Pumping System**

a) CW inlet & outlet temperature in Cooling tower

b) Cooling Water Pumping System & fan Power Supply Measurements

c) CW main supply flow

d) Calculation & Assessments of Cooling Tower Efficiency and Specific Power Consumption.

2.9.4 Assessment of results

a) Assessments of HAVC System for Losses and Specific Power Consumption

b) Identification of Energy Conservation opportunities

c) Running –load, flow in primary circuit and temperature difference between inlet and outlet of evaporator are measured to calculate present load on Chiller and its specific power consumption.

d) Flow and load of each chilled-water supply pump are measured to assess performance of primary and secondary chilled water pumps, power consumption and efficiency.

e) Performance of condenser water pumps are assessed using data for flow, power and efficiency.

f) Cooling tower performance can be assessed with respect to air flow, inlet and sump water temperature, loading on cooling towers.

g) Performance of AHUs are studied for load and effectiveness.

h) Opportunities for Energy Conservation can be explored.

2.10 Routine HVAC Inspection

a) Clean air intake when necessary

b) Change air filters monthly or bi-annually

c) Check for excessive noise or vibration when blower motors are running

d) Check if condensate drain pans are draining properly

e) Ensure motors and ductwork are clean

f) Inspect flexible duct connectors for cracks and leaks

g) Check screws, latches, and gaskets in need of repair or replacement

h) Check pumps; bearings must be lubricated once a year

i) Inspect condition of all electrical hardware as well as connections

j) Make sure safety controls and equipment are working properly
k) Make sure all guards and access panels remain secure
l) Check operation of interior and exterior units
m) Clean damper operators
n) Ensure any mineral buildup inside water heater is minimized
o) Drain boiler to remove any accumulated sediment
p) Clean or replace boiler's oil filter once a month
q) Ensure thermostat is calibrated correctly.

2.11 HVAC Inspection

During a routine HVAC tune-up, checks are:

a) Check HVAC controls, refrigerant levels, electrical connections, and thermostats.
b) Clean condensation drains and drip pans.
c) Clean, adjust, and replace HVAC coils.
d) Check surge protectors and fail-safes.
e) Lubricate and adjust moving parts.
f) Evaluate HVAC system airflow.
g) Adjust belts and blower motors.
h) Replace air filters.
i) Straighten fins.

3.0 Performance assessment for lighting system

3.1 Introduction

Lighting is provided in industries, commercial buildings, indoor and outdoor for providing comfortable working environment. The primary objective is to provide the required lighting effect for the lowest installed load i.e. highest lighting at lowest power consumption.

3.2 Efficacy of the system

The purpose of performance test is to calculate the installed efficacy in terms of lux/Watt/m^2 (existing or design) for general lighting installation. The calculated value can be compared with the norms for specific types of interior installations for assessing improvement options.

3.3 Performance Terms and Definitions

i. **Lumen**- It is a unit of light flow or luminous flux. The lumen rating of a lamp is a measure of the total light output of the lamp. The most common measurement of light output (or luminous flux) is the lumen. Light sources are labeled with an output rating in lumens.

ii. **Lux** – It is the metric unit of measure for illuminance of a surface. 1 lux = 1 lumen/m^2. [*Lumens are how much light is given off. Lux is how bright the surface will be. Candela measures the visible intensity from the light source*].

iii. **Circuit Watts**- It is the total power drawn by lamps and ballasts in a lighting circuit.

iv. **Installed Load Efficacy**- It is the average maintained illuminance provided on a horizontal working plane per circuit Watt with general lighting of an interior. Unit: lux per Watt per square meter [(lux/(W/m^2)]

v. **Lamp Circuit Efficacy**- It is the amount of light (lumens) emitted by a lamp for each Watt of power consumed by the lamp circuit, i.e. including control gear losses. This is a more meaningful measure for those lamps that require control gear. Unit: lumens per circuit Watt (lm/W)

vi. **LPD (Lighting Power Density)-**It is defined as the ratio of total lighting load (Watt) at the specific zone to built-up area (m^2) of the specific zone. Unit: Watt/m^2.

vii. **Load efficacy**= Lux/LPD

viii. **Installed Power Density-** The installed power density per 100 lux is the power needed/m^2 of floor area to achieve 100 lux of average maintained illuminance on a horizontal working plane with general lighting of an interior. Unit: Watts per square meter per 100 lux = (W/m^2) /(100 lux) = (LPD/100lux).

ix. **Installed load Efficacy Ratio (ILER)** =Actual load efficacy/Target load efficacy = [actual lux/LPD] / [Target lux/LPD]

x. **Average maintained illuminance** is the average of lux levels measured at various points in a defined area.

xi. **Color Rendering Index (CRI)**- is a measure of the effect of light on the perceived color of objects. To determine the CRI of a lamp, the color appearances of a set of standard color chips are measured with special equipment under a reference light source with the same correlated color temperature as the lamp being evaluated. If the lamp renders the color of the chips identical to the reference light source, its CRI is 100. If the color rendering differs from the reference light source, the CRI is less than 100. A low CRI indicates that some colors may appear unnatural when illuminated by the lamp.

xii. **Room Index (RI)** - is the ratio of horizontal area to vertical area. $=L \times W/[(L+W) \times H_{m.}]$ [L- length, W-width, H_m- mounting height]

3.4 Procedure for Assessment of Lighting Systems

Following procedures can be followed to assess lighting system.

3.4.1 Calculation of the Room Index (RI)

Equation: $RI = L \times W /(L+W) \times H_m$.

When, L = length of interior; W = width of interior; Hm = the mounting height, which is the height of the lighting fittings above the horizontal working plane.

[*The working plane is usually assumed to be 0.75m above the floor in offices and at 0.85m above floor level in manufacturing areas.*]

For example, the dimensions of an interior are: Length = 9m, Width = 5m, Height of luminaires above working plane (Hm) = 2m, calculate RI.

$RI = (9x5)/[(9+5) x2] =1.6$.

3.4.2 Calculation of installed load efficacy and Installed load efficacy ratio

Following steps can be followed to evaluate *installed load efficacy and Installed load efficacy ratio* of a general lighting installation in an interior room.

a) Measure the floor area: Lx W (m^2)

b) Calculate Room Index (RI): $(L \times W) /[(L+W) \times H_m]$

c) Determine total load: Watt

d) Calculate: LPD= Watt/ m^2

e) Calculate average illumination: E_{av}= Summation of illumination of areas measured part by part/Total area

f) Calculate Lux/ LPD= Lux/(w/m^2)

g) Obtain Target Lux /LPD from the table-1 against RI= Target Lux/LPD

h) Installed load efficacy= [Calculated Lux/LPD]/[Target Lux/LPD]

i) Compare ILER as per table-2.

▼ Table-1: Target lux/W/m² values for maintained illuminance on horizontal plane

Target lux/W/m² values for maintained illuminance on horizontal plane for all room indices and applications

Room Index	Commercial lighting (Offices, Retail stores, etc.) & very clean industrial applications, Standard or good colour rendering. Ra: 40-85	Industrial lighting (Manufacturing areas, Workshops, Warehousing etc.) Standard or good colour rendering. Ra: 40-85	Industrial lighting installations where standard or good colour rendering is not essential but some colour discrimination is required. Ra: 20-40
5	53	49	67
4	52	48	66
3	50	46	65
2.5	48	44	64
2	46	42	61
1.5	43	39	58
1.25	40	36	55
1	36	33	52

Ra : Colour Rendering Index

▼ Table-2: Assessment (Indicators of performance)

Assessment (Indicators of Performance)

ILER	Assessment
0.75 or over	Satisfactory to Good
0.51 – 0.74	Review suggested
0.5 or less	Urgent action required

Questions:

1. Describe PSA plant working.
2. What are the adsorbents used for O2 & N2 production?
3. What are the basic components of a HVAC system?
4. Explain centralised all air system HVAC system.
5. Define the terms related to HVAC: CLTD, CLF, U-factor& RSHR.
6. What are the external heat gain and relevant equations to calculate?
7. What are the internal heat gain and write relevant equations to calculate?
8. Define terms related to performance of HVAC: TR, COP& EER.
9. What are the routine checks for HVAC system?
10. How to proceed for assessing lighting system?

SECTION

13 MOTOR, GENERATOR & TRANSFORMER

1.0 Introduction

In industry and in our daily life use of electrical is very much imminent. It has wide range of application in modern days. The industrial equipment do all process activities with the help of motor which is electrical power driven. Other than drive of equipment, electricity is also used for heating. Transmission of power from generating station to consumer is done using transformer. Other than drive & heating, everywhere we require light when sun sets. This section is a study on performance and efficiency of all the stated electrical equipment and systems that are involved in all electrical activities.

1.1 Motor – Its working and performance assessment

An electric motor is an electric machine that converts electrical energy into mechanical energy. Most electric motors operate through the interaction between the motor's magnetic field and electric current in a wire winding. This interaction generates a force in the form of torque which is applied to the motor's shaft.

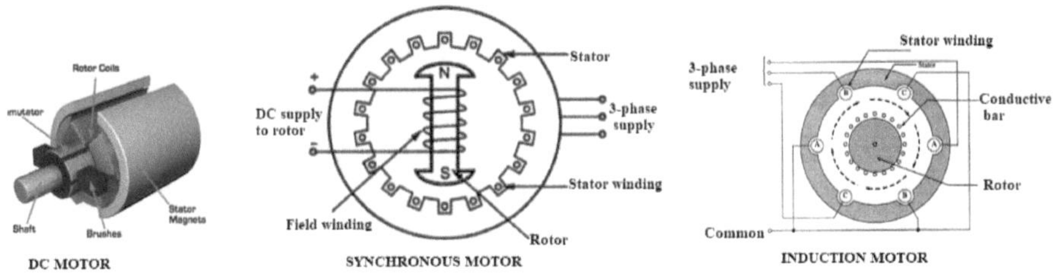

Schematic diagram showing DC motor, Synchronous motor & Induction motor

▲ **Fig: 13.1: DC motor, Synchronous motor & Induction**

1.2 Types of Electric Motors

The various types of motors, fig-13.1, can be briefed as:

a) **DC Motors**- It is driven by direct current. It's the most primitive version of the electric motor when rotating torque is produced due to the flow of current through the conductor inside a magnetic field. (Different types of DC motors are not further discussed as not much relevant).

b) **Synchronous Motors**-It is driven by AC supply. In synchronous motor, the rotor is an electromagnet (produced by externally supplied DC power) that is magnetically locked with a stator rotating magnetic field and rotates with it. The speed (N_s) of such motor depends on the frequency (f) and the number of poles (P), as $N_s = 120\ f/P$.

c) **Asynchronous or Induction motor**- It is one of the types of AC motor where Rotating Magnetic Field (RMF), generated by stator, cuts the rotor conductors, hence, circulating current induced in these short-circuited rotor conductors and in turn rotor produces a rotating magnetic field. Due to the interaction of magnetic fields the rotor starts to rotate and continues its rotation. This is an induction motor, which is also known as the asynchronous motor, runs at a speed which is less than its synchronous speed, and the rotating torque and speed is governed by varying the slip, which gives the difference between synchronous speed N_s and rotor speed N. Slip (s) = $(N_s\text{-}N)/N_s$.

d) Discussions of Other Types of motor are excluded due to less relevancy of the subject matter.

1.3 Energy performance assessment of motors

The two parameters of importance in a motor are a) Efficiency and b) Power factor. The efficiencies of induction motors remain almost constant between 50% to 100% loading.

i. Efficiency: The efficiency of the motor is given by: Load output/ load input

ii. Motor Loading% = Actual load/ Load rating.

1.4 Field Tests for assessment of losses in motor

To derive the efficiency of a motor, it is evident that losses can be used to derive efficiency as: Efficiency (%) =100-% losses. The losses are discussed below.

1.4.1 Stator I^2 R Loss

While current flows through the coil of stator, energy is lost in terms of heat which is generated due to the resistance of the coil. The loss is also called as copper loss and expressed as I^2R. Steps for determination of stator I^2R loss as described below.

1. The motor copper loss is derived from measured current & resistance of stator winding. The stator winding resistance is directly measured by a bridge or volt amp method, fig-13.2 (a).
2. The resistance is corrected to the operating temperature as per equation:

   $\mathbf{R_2/R_1 = (235 + t_2)/(235+t_1)}$
3. Corrected stator resistance (R_2) = Measured Resistance R_1 (at ambient temperature) x $(235+t_2)/(235+t_1)$.

   [When, R_1 is the resistance at t_1 at ambient temperature, °C &R_2 is the resistance at t_2 in operating temperature, °C.]
4. Actual Loss (Watt) = $I^2 x R_2$

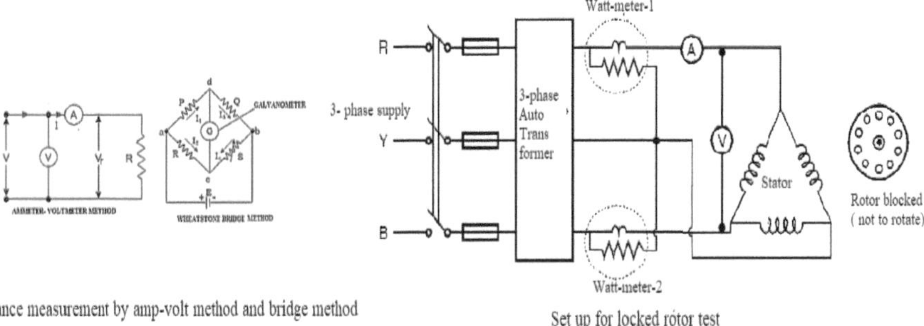

▲ **Fig-13.2 (a) & (b): Resistance measurement & set up for locked rotor test**

1.4.2 Rotor I^2R loss

Rotor copper loss can be derived by the methods as described below.

1. **Rotor I^2R loss by block rotor test**

   The figure 13.2 (b) shows the circuit diagram for the blocked rotor test of an induction motor. The blocked rotor test enables determining the efficiency and the circuit parameters of the equivalent circuit of a 3-phase induction motor.

i. **Test procedure**

a) In the blocked rotor test, the shaft of the motor is locked so that it cannot rotate and the rotor winding is short circuited through slip ring for rotor excitation motor or by shorting the end rings for squirrel cage induction motor.

b) In the blocked rotor test, a reduced voltage at reduced frequency is applied to the stator of the induction motor through a 3-phase autotransformer ensuring rated current flows in the stator winding.

c) Input power (two- Watt meter method) and phase current are measured.

ii. **Readings obtained**

a) Total input power on short circuit is measured by the two Wattmeter method and is given by the algebraic sum of the two Watt-meter readings. Due to application of reduced voltage to the stator and rotation of the rotor is not allowed, the core and mechanical losses are negligible.

b) Therefore, the total input power in the blocked rotor test is equal to the sum of stator copper losses and rotor copper losses for all the 3-phases.

c) The ammeter reads the value of line current (I_{sc}) with blocked rotor which is corresponds to the short circuit condition.

d) The voltmeter reads the value of reduced line voltage with blocked rotor (V_{sc}).

iii. **Derivation of rotor I^2R loss**

a) The input power under blocked rotor condition is given by:

$P_{sc} = \sqrt{3}V_{sc}I_{sc}\cos\varphi_{sc}$ [cos φ_{sc} = Power factor in short circuit condition, normally-0.9]

b) Stator copper loss =3x Isc^2xR

c) Hence, Rotor I^2R loss= P_{sc}-Stator copper loss.

iv. **Cares during test**

a) In order to obtain the accurate results, the blocked rotor test is performed at a frequency 25% less of the rated frequency.

b) Induction motors of less than 20kW rating, the effect of the frequency is negligible and hence, the blocked rotor test can be performed directly at the rated frequency.

2. **Derivation of rotor I^2R loss by slip method**

   i. **Equation derivation for slip**

   Rotor gross output (Pm) = Rotor input - Rotor copper loss.

   Now, rotor gross output= $2\pi N T_g$ (T_g shaft gross torque)

   If there were no copper losses in the rotor, then rotor output = rotor input and the rotor would run at synchronous speed:

   Rotor input = $2\pi N_s T_g$ [N_s-synchronous Speed]

   As rotor copper losses = Rotor input - Rotor output, Rotor Copper losses= $T_g 2\pi (N_s - N)$

   Rotor Cu. Loss / Rotor Input = $(N_s - N)/ N_s$

   Rotor Cu. Loss = s x Rotor input [s= slip=(Ns-N)/Ns].

   Rotor gross output = Rotor input - Rotor Cu. Loss = (1-S) Rotor input

   Rotor efficiency = Rotor gross output / Rotor Input = (1 - S) Rotor Input / Rotor Input

   = 1-S = 1 - (Ns-N)/Ns = N / Ns

   Therefore, rotor efficiency = actual speed of rotor / Synchronous speed.

   Example: A 4 pole induction motor is operating at 1470 rpm, find efficiency by slip method neglecting all other losses. Supply frequency is 50 Hz.

   Synchronized Speed (N_s) =120xf/p =120x50/4=1500.

   Hence, rotor efficiency N/ N_s x100= 1470/1500=98%.

ii. **Procedure**

   a) Operate the motor over 80% load.

   b) Measure supply power (Pi) to the motor.

   c) Measure rotor speed by a non-contact type tachometer (stroboscope)

   d) Evaluate slip using equation: $(N_s$-N)/N_s and correct the same at operating temperature.

   e) Rotor I^2 R losses = Slip (s) x [(Stator Input power, (Pi) - Stator I^2 R Losses (Copper loss) – Core Loss (Iron loss)]

   [When, s=slip, Pi- input power, I= stator current].

   f) When the motor slip increases, the motor's temperature also increase. Hence, Rotor copper loss is to corrected to temperature.

1.4.3 F&W (Friction & Windage) and core losses

F&W loss is derived by no load test condition.

i. **Procedure**

a) The motor is run at rated voltage and frequency without any shaft load (decoupled condition).

b) Input power, current, frequency and voltage are noted.

c) From the input power, stator I^2 R losses under no load are subtracted to get the sum of Friction and Windage (F&W) and core losses.

d) F&W and core losses = No load power (Watts) – (No load current)^2 x Stator resistance (R).

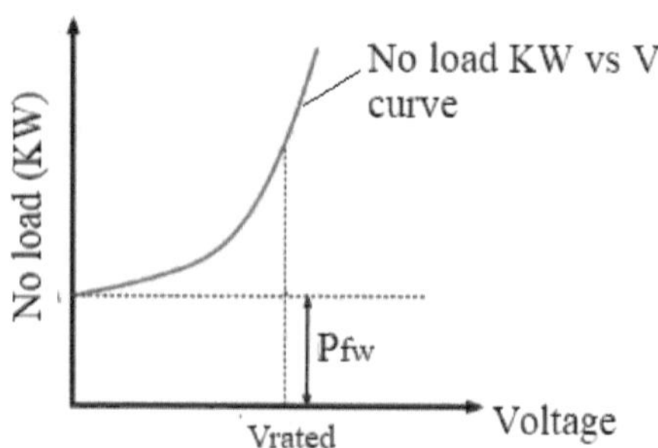

▲ **Fig-13.3: Derivation of friction and windage loss**

To separate core and F & W losses, test is repeated at variable voltages. If no-load input kW versus Voltage is plotted, the intercept is (P_{fw}) F & W kW loss component, Fig-13.3.

1.4.4 Stray Load Losses:

IEEE - 112 specifies values from 0.9% to 1.8%.

Motor Rating & Stray Losses: 1 – 125 HP: 1.8%, 125 – 500 HP: 1.5%, 501 – 2499 HP: 1.2%, 2500HP and above 0.9%.

1.4.5 Estimation of motor efficiency

Estimation of efficiency in the field can be summarized as follows:

a) Measure stator resistance and correct to operating temperature. From rated current value, I^2 R losses are calculated.

b) From rated speed and output, rotor I^2 R losses are calculated by using slip method.

c) From no load test, core and F & W losses are determined for stray loss.

Efficiency = [Input power – (stator copper loss+ Rotor copper loss+ Core and F&W loss)]/Input power x100%.

The method is illustrated by the following example:

1.4.6 Example showing derivation of losses, pf

Given: Motor Specifications:

Rated power = 34 kW/45 HP, Voltage = 415 Volt, Current = 57 Amps, Speed = 1475 rpm Insulation class = F, Connection = Delta.

No load test Data:

Voltage, V = 415 Volts, Current, I = 16.1 Amps, Frequency F = 50 Hz, Stator phase resistance at 30°C = 0.264 Ohms, No load power, P_{nl} = 1063.74 Watts.

Calculate:

a) Iron loss & F&W loss

b) Calculate stator Resistance and copper loss at 120°C

c) Rotor input

d) Full load

e) Motor input, assuming 0.5% of motor power

f) Full load efficiency and full load power

Solution

i. **Iron loss and F&W loss are derived during no load test condition.**

Let Iron Loss =Pi, Friction & Windage loss=P_{fw}

No load power=1063.74 Watt

At no load, Line current=16.1 amp; phase current= 16.1/√3=9.29 amp.

Full load line current = 57 amp; phase current=57/1.732=33 amp

Stator copper loss at 30°C at (no load condition) = 3x I^2R=3x 9.29^2x0.26=67.3 Watt

Answer to (a): (Pi+ Pfw) = P_{noload}- Copper loss at 30 0 C.=1063.74-67.3=996.44 Watt.

ii. **Full load test**

Answer to (b): Stator resistance at 120 ^{0} C= 0.26 x(120+235)/(30+235)=0.35 ohm/phase

Stator copper loss at full load= 3x I^2R_{120}=3x33^2x0.35=1143 Watt.

Answer to (c): Full load slip=(1500-1475)/1500=0.0167=1.67%.

Now, rotor output= Rotor input x(1-s)

Rotor input=34/(1-0.0167)=34.577 kW

Answer to (d): Motor Full load input =Rotor input +Stator copper loss+(Pi+ P_{fw}) +stray loss (0.5% of rated power) =34577+1143+996.44+0.005x34000=36886.44 Watt.

Answer to (e): Motor efficiency at full load= Rated kW(output)/full load input =34000/36886.44=92.2%.

Answer to (f): Now, Full load input power =√3 VI x Pf

Hence, Full load Pf= 36886.44/(√3 x415x57)=0.90.

1.5 Determining Motor Loading

1.5.1 By Input Power Measurements

a) First measure input power (Pi) with a hand held or in-line power meter Pi = Three-phase power in kW

b) Note the rated kW and efficiency from the motor name plate. The figures of kW mentioned in the name plate is for output conditions.

c) *Rated Input power at full-rated load*=Name plate full load power (Output) / efficiency at full load= $P_{name\ plate}/\eta$ at full load.

d) The percentage loading = measured Input power (Pi) /Rated Input power at full rated load.

1.5.2 By Line Current Measurement

The line current load estimation method is used when input power cannot be measured and only amperage measurements are possible. The amperage draw of a motor varies approximately linearly with respect to load, from 75%- full load.

Loading% = Input load current /Rated load current x100

*[**Note: As below the 75% load, power factor degrades and the amperage curve becomes increasingly non-linear, current measurements are not a useful indicator of load.**]*

1.5.3 Slip Method

Slip method can be used to find percentage loading.

a) Slip Load = Slip speed (Sync speed-actual measured speed)/ (Sync speed-name plate rated speed) x100%. Slip remains in 3-5%.

b) Slip also varies inversely with respect to the motor terminal voltage squared. A voltage correction factor can, also, be inserted into the slip load equation.

c) The voltage compensated load can be calculated as:

Slip speed / [(Sync speed-name plate rated speed) x (Name plate voltage/ actual measured voltage)^2].

1.6 Losses in motor Operation & Remedial measures

i. **Voltage & frequency**- Motor performance is affected considerably by the quality of input power, that is the actual volts and frequency available at motor terminals.

   As per Indian Standards (BIS) a motor should be capable of delivering its rated output with:

   a) voltage variation of +/- 6%

   b) frequency variation of +/- 3%.

ii. **Voltage unbalance**- The condition when the voltages in the three phases are not equal, is detrimental to motor performance and motor life. It happens due to supply of disproportion single-phase loads from one of the phases. It can also result from the use of different sizes of cables in the distribution system.

   Percent unbalance in voltage is defined as = $(V_{max} - V_{avg}) / V_{avg}$ x100%.

   When, V_{max} and V_{avg} are the maximum and the average of the three phase voltages, respectively.

iii. **Motor Loading**- Motor running at lower load results in lower efficiency.

iv. **Over capacity Design**- Original equipment manufacturers tend to use a large safety factor in motors they select. Under-loading of the motor may also occur from under-utilisation of the equipment.

1.7 Power Factor Correction

Induction motors are characterized by power factors less than unity, leading to lower overall efficiency. Capacitors connected in parallel (shunted) with the motor are typically used to improve the power factor. The impacts of PF correction include:

a) Reduces kVA demand

b) Reduces I^2 R losses and voltage drop in cables upstream of the capacitor

c) Increases overall efficiency of the plant electrical system.

2.0 Performance and efficiency assessment for generator

2.1 Introduction

An **alternator** is an electrical machine that converts mechanical energy into electrical energy in the form is alternating current, AC. It is also known as a **synchronous generator** or **AC generator**. It generates a specific voltage at a specific frequency.

2.2 Working of Synchronous Generator

DC generator comprises of a rotating armature winding and a stationary magnetic field where as, an alternator is made of a stationary armature winding and a rotating magnetic field. The field windings are placed in the rotor while the armature windings are placed in the stator.

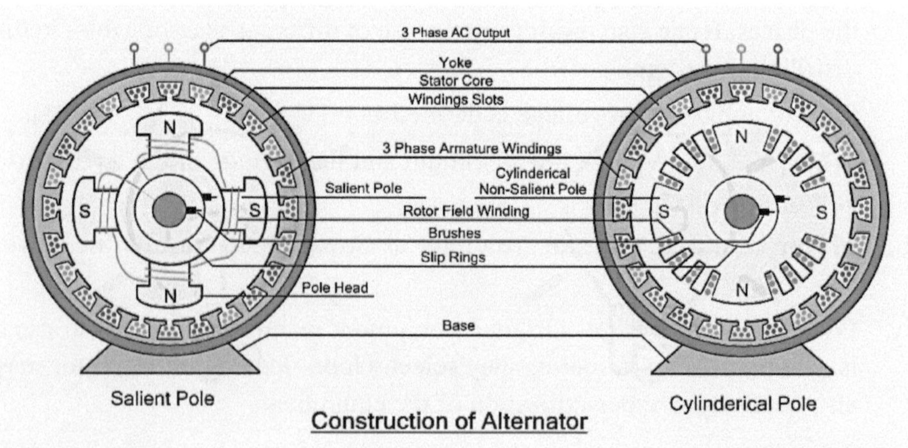

▲ **Fig-13.4: Construction of Alternator**

The rotor field windings are connected to an external DC supply with the help of slip rings and brushes, called static excitation or through rotating diode, called brushless excitation. A prime mover, either an engine or turbine, rotates the rotor using a pulley and belt or coupled directly with the rotor shaft. The rotating rotor generates rotating magnetic field. This varying field generates voltage in the armature windings and supplies it to the load or circuit through its terminals.

2.3 Components of Alternator or AC Generator

The alternator is made of different stationary and moving components as discussed below, Fig-13.4.

2.3.1 Rotor

The rotor is the rotating part of the alternator. It is made in a cylindrical shape that has copper windings also known as field winding. The field windings are electromagnets that generate the rotating magnetic field when rotated. Rotor shaft that is rotated is connected with the prime mover. It can be anything such as an engine, water turbine, wind turbine, etc.

There are two types of rotors used in alternators or synchronous generators as described below.

i. **Salient Pole Type**

It is a type of rotor that has a large number of protruding or projecting poles mounted on a core made of magnetic laminated steel or cast iron. The term salient refers to protruding or projecting as shown in the figure.

The salient poles are made of laminated steel or iron cast of good magnetic properties to reduce the eddy current losses. The pole shoes have multiple slots for damper winding that helps in preventing hunting. The field coils are wounded across the poles and then connected in series. The field coil is energized by connecting its ends to a separate DC source from outside.

The salient pole rotor has a large diameter and small axial length. They are used in low and medium-speed alternators such as in hydropower stations. They are not suitable for high speed due to the increased Windage loss at high speed due to their design (salient poles). Its design does not have enough mechanical strength to handle high speed.

iv. **Cylindrical Type**

Such type of rotor has very few 2 or 4 poles. It is made up of a laminated steel cylinder. The cylindrical rotor has slots for field winding that is connected in series. Since the poles are not protruding out of the core, it is also known as a non-salient pole or round rotor. Its rotor diameter size is small while its axial length is longer than the salient pole rotor.

2.3.2 Stator

A stator is the stationary part of an electrical machine. In an alternator, it is used for holding the armature winding that generates the induced EMF. The core itself is made of laminated steel or cast iron of good magnetic quality to reduce eddy current losses. The rotor that carries the field windings rotates inside the stator without physically touching it.On the contrary, the DC generator's stator holds the magnets to generate the necessary magnetic field.

2.3.3 Yoke

The yoke is the outermost part of the alternator that is used to provide mechanical support and protect the inner parts from environmental conditions that can damage it.

2.3.4 Slip Ring and Brushes

A slip ring is a component that transfers electrical power between stationary and rotating parts of a machine. In an alternator, it is used to supply DC power to the rotor field windings from a DC source using brushes that slide over the slip ring. It is made of concentric discs placed on the shaft of the rotor. As it supplies DC, the alternator only requires two slip rings.

The DC current flow through the field winding generating the magnetic field that varies with the rotation of the rotor.

2.3.5 Diode Rectifier

Diode rectifier is two terminal semiconductor component used for the conversion of alternating current AC into DC. There are 6 diodes used two per phase to convert into smooth DC.

2.3.6 Voltage Regulator

A voltage regulator (AVR- Automatic Voltage Regulator) is used to monitor the output of the alternator and adjust its voltage by adjusting the excitation current to the rotor. The alternator's output is feedback into the rotor through the voltage regulator. It maintains constant output voltage regardless of the rotor speed of the alternator.

2.4 Advantages & Disadvantages of Alternator

2.4.1 Advantages

Some advantages of an alternator

a) An alternator has a stationary armature thus the output is taken directly from its terminals without brushes and slip rings.

b) There are no electric sparks, wear & tear due to friction between the slip ring and brushes, thus it requires less maintenance.

c) It has higher efficiency and higher voltage than a DC generator.

d) The rotor field winding is powered by low DC voltage; thus they last longer.

e) It has a lower weight, is more compact, and is smaller in size.

f) Its design allows it to be used for high-speed and smooth operation.

g) It uses a diode rectifier that has a lower voltage drop than a commutator with noiseless and smooth DC output.

h) It has a simple and robust design and cheaper than a DC generator.

2.4.2 Disadvantages

Some disadvantages of an alternator are:

a) It requires an efficient cooling system as the large current can overheat it which reduces its performance

b) It requires a diode rectifier to convert AC into DC whereas, the generator can generate both AC as well as DC.

2.5 Applications of Alternator

An alternator is mainly used for converting mechanical energy into electrical energy in various applications such as:

a) In automobiles

b) In locomotives

c) Power generation plants

d) In marine and navy boats

e) Radio frequency transmission

2.6 Performance and efficiency assessment

Generators/ alternators are highly efficient electrical equipment. Efficiency of generator = electrical power out of the alternator/ mechanical power put into x100%.

Two standards guide the method of calculating efficiencies: NEMA (typically used in the US) and IEC (typically used everywhere else).

However, both the NEMA and IEC standards allow efficiency to be measured by an indirect method. This is also known as summation of losses measurement. Instead of measuring power in and out of the machine, separate tests are used to quantify the five types of losses in the generator. These losses can be categorised into:

a) Fixed losses

b) Variable losses.

Fixed losses are independent of load.

2.6.1 Fixed loss

i. **Friction and Windage losses** – These losses are due to the friction in bearings, fans, windage and the rotor itself. Optimising the fan design is the best way to reduce these losses.

ii. **Core losses** – These are losses caused by hysteresis and eddy currents. Core losses can be reduced by using a better grade of lamination steel.

   Core or iron Losses, Pi = Hysteresis Loss(P_h)+Eddy Current Loss (P_e).

   a) Hysteresis Loss, $P_h = K_h (B_{max})^{1.6}$ f xV (Watt)

   b) Eddy Current Loss, Pe = $K_e (B_{max})^2 f^2 t^2 V$ (Watt).

[K_h- hysteresis constant, K_e- Eddy constant, B- flux density, t-thickness, V-voltage, f-frequency]

2.6.2 Variable losses

Variable losses are dependent on load. The losses can be divided in three categories, such as:

i. **Stray load losses** – These are losses caused by the load current changes in the magnetic flux distribution, eddy currents and harmonics. They can be reduced by using better materials and thinner laminations.

ii. **I^2R losses in the armature and field** – I^2R loss results heating and the heat losses are due to resistance in the windings and can be reduced by using more copper or by running at cooler temperatures. Armature winding cu loss=I^2aRa, Rotor winding cu loss=I^2rRr.

iii. **Brushless exciter loss** – This small loss is due to electrical energy going into the exciter and rectifier.

$\eta\% = P_{out}/(P_{out} + \text{Losses})x100$

3.0 Performance and efficiency assessment for Transformer

3.1 Introduction

Transformer is a static electrical device that is used to transfer electrical energy from one alternating-current circuit to another circuit or multiple circuits *at a constant frequency*, through the process of electromagnetic induction.

It is used for either increasing or decreasing the voltage level of the AC supply with a corresponding decrease or increase in the current at constant frequency.

3.2 Parts of a Transformer (fig-13.5)

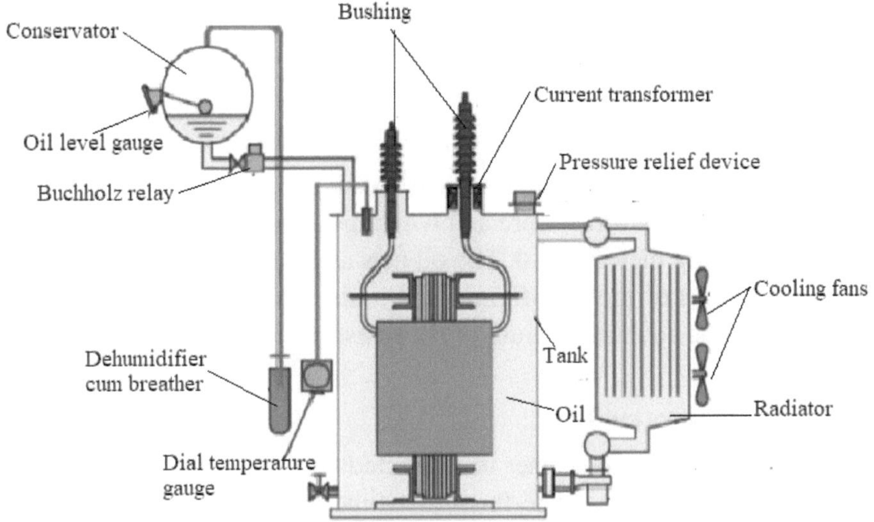

▲ **Fig-13.5: A typical transfer with different parts**

3.2.1 Magnetic Core

The core of the transformer is made up of magnetic materials having high permeability. As the transformer is subjected to the AC supply, thus its core is built up of thin lamination stacked together to reduce the eddy current losses in the core. The windings of the transformer are wound on the core of the transformer. The core of the transformer mainly serves two purposes as,

a) It provides support to the windings.

b) It provides a low reluctance path to the magnetic flux.

3.2.2 Transformer Windings

The transformer consists of two windings viz. the *primary winding* and the *secondary winding*. The winding connected to the source of AC supply is called as the primary winding while the winding to which the load is connected is known as the secondary winding. The AC voltage V_1 whose magnitude is to be changed is applied across the primary winding.

Depending upon the number of turns in the primary and secondary windings, an alternating EMF (E_2) being induced in the secondary winding of the transformer. This induced EMF (E_2) results a load current I_2, hence, a terminal V_2 would appear across the load.

If $V_2 > V_1$, the transformer is said to be *step-up transformer*. On the other hand, if $V_1 > V_2$, the transformer is said to be *step-down transformer*.

3.2.3 Transformer Body / Tank and Dielectric Oil

The transformer tank provides protection to the core and the windings of the transformer. The transformer core and windings arrangement are immersed in the tank containing dielectric oil. The oil acts as an insulating medium for the core and windings of the transformer and it also absorbs the heat generated, hence, it works as cooling medium of the transformer also.

3.2.4 Oil Conservator Tank

The main tank of the transformer is connected through a pipe to a small tank, called as *conservator tank*. The function of the conservator tank is to ensure the transformer tank completely filled with the dielectric oil at all operating conditions. The conservator tank is designed to act as a reservoir for the transformer oil. When the temperature of the transformer is increased, the oil inside the transformer being expanded. The conservator tank provides space for this expansion of the dielectric oil.

3.2.5 Dehumidifier cum Breather

The breather of the transformer consists of silica gel, which prevents any atmospheric moisture from entering the tank of the transformer with air. When the temperature of transformer oil decreases, oil volume reduces and generates a negative pressure. Hence, the atmospheric air gets inhaled by the transformer. While entering the moisture of the air gets absorbed by silica gel of the breather.

3.3 Types of transformers

There are primarily two types of Transformer based on the operating voltage. The following are some of them:

i. **Step-down Transformer:** In such type of transformer, primary voltage is converted to a lower voltage across the secondary output. The number of windings on the primary side of a step-down transformer is more than that of secondary side. As a result, the overall secondary to primary winding ratio is always be less than one. Step-down transformers are used in electrical systems that distribute electricity near the consumer.

ii. **Step-up Transformer:** In step-up transformer, secondary voltage is raised from the low primary voltage. In such transformer, the primary winding has fewer turns than the secondary winding, the ratio of the secondary to primary winding is be greater than one. A step-up transformer is used in long distance transmission and distribution of electrical power. In the grid, a step-up transformer is used to raise the voltage level prior to distribution.

3.4 Working Principle of a Transformer

Mutual induction between the two coils is the fundamental principle that how a transformer functions. The laminated silicon steel core of the transformer is covered by two distinct windings. Magnetic flux generated by primary coil induces the core and in turn core flux cuts the secondary coil to produce secondary voltage. Only alternating current can be used because mutual induction between the two windings requires an alternating flux.

Transformer equation is: $E1/E2 = N1/N2$

[E1 & N1 are primary voltage & number of winding respectively and E2 & N2 are for secondary coil].

From the expression above, it is clear that the size of EMFs E1 and E2 is dependent on the number of turns in the transformer primary and secondary windings, respectively. If $N2 > N1$, then $E2 > E1$, and the transformer is a step-up transformer; if $N2 < N1$, then $E2 < E1$, and the transformer is a step-down transformer.

3.5 Efficiency assessment

The **Efficiency** of the transformer is defined as the ratio of useful power output to the input power, the two being measured in the same unit. Its unit is either in Watts (W) or kW.

3.5.1 Maximum efficiency equation

It can be proved that, Efficiency is maximum in a transformer when Copper losses = Iron losses.

Let, for secondary coil; V= terminal voltage, I= Full load current & Cos ϕ= Power factor.

Pi= Iron loss= (Eddy current loss + Hysteresis loss) is a constant loss.

Pc= I^2R=Full load Copper losses.

The efficiency is a function of load i.e. load current I. Assuming power factor, cos Φ & terminal voltage, V are constant.

Efficiency (η) =VI cos Φ /(VI cos Φ +Pi+I^2R)

For maximum efficiency, $d\eta / dI=0$

$d\eta / dI$ [VI cos Φ /(VI cos Φ +Pi+I^2R)]=0 **[Reference: d/dx(u/v)= [v(du/dx)-u(dv/dx)]/v²]**

Hence, (VI cos Φ+Pi+I^2R)] d/dI [VI cos Φ]- [VI cos Φ] d/dI (VI cos Φ+Pi+I^2R) =0

Or, [(VI cos Φ +Pi+I^2R) (V cos Φ)]- [(VI cos Φ) (V cos Φ +2IR)] =0

Or, (VI cos Φ +Pi+I^2R) (V cos Φ) = (VI cos Φ) (V cos Φ +2IR) [cancelling V cos Φ from both terms]

(VI cos Φ +Pi+I^2R) = (VI cos Φ +2 I^2R); Pi= I^2R (Copper Loss).

Hence, at maximum efficiency, Iron loss=Copper loss.

3.6 Transformer loss

Transformer losses consist of two parts: No-load loss and Load loss

i. **No-load loss**- It is also called core loss or iron loss and this power is consumed to sustain the magnetic field in the transformer's steel core. Core loss occurs whenever the transformer is energized; core loss does not vary with load. Core losses are due to hysteresis and eddy current losses. Hysteresis loss is that energy lost by reversing the magnetic field in the core as the magnetizing AC rises and falls and reverses direction. Eddy current loss is a result of induced currents circulating in the core.

ii. **Load loss** – This loss is also called copper loss and it is due to current flow in the transformer windings. Copper loss is power lost in the primary and secondary windings of a transformer due to the ohmic resistance of the windings. Copper loss varies with the square of the load current. ($P = I^2 R$).

Total transformer loss,

a) At any load level, transformer load can then be calculated from: $P_{total} = P_{no\text{-}Load} + (\%\ Load/100)^2 \times P_{Load}$

b) If transformer loading is known, the **actual transformers loss** at given load can be computed as: = No load loss+ (Actual KVA/Rated KVA)^{2}x Full load loss.

4.0 Performance and efficiency assessment for PCC, MCC & transmission

Power generated at power generating station reaches to consumer through transmission system, it may be over head transmission line or underground cable. While supplying, power has to pass through switchgears, transformers and conveying conductor, Fig-13.6, incurs loss. This section is included to discuss different losses and recommendations to reduce losses.

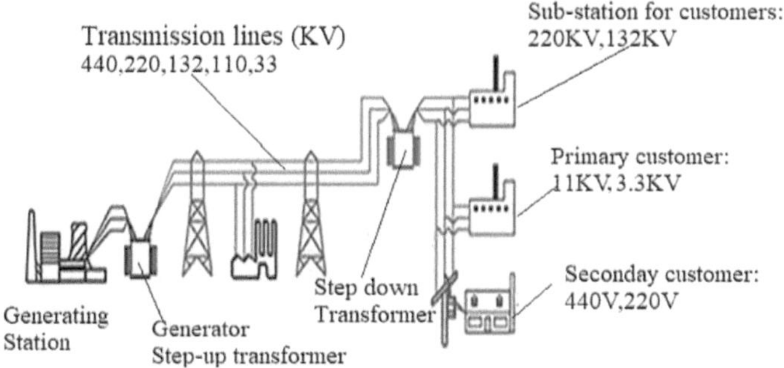

▲ **Fig-13.6: Typical power distribution system**

4.1 Definition and nomenclature related to power transmission & distribution System

4.1.1 Switchyard

A switchyard is the set of facilities outside a power plant in which voltage is transformed (commonly increased) and power flows onto transmission lines. The switchyard comprises transformers and a series of switches, breakers and other protective devices that can be manually or remotely opened and controlled to energize or de-energize specific transmission circuits leaving the power plant. To disconnect a transmission line leaving the plant, the appropriate switches or breakers are opened in the switchyard at the plant and to connect a line the same switches or breakers are closed.

4.1.2 Substation

The electrical substation of a power system is the system in which the voltage is transformed from high to low or vice-versa for transmission, distribution, transformation and switching. The power transformer, circuit breaker, bus-bar, insulator, lightning arrester are the main components of an electrical substation.

4.1.3 PCC (Power Control Centre)

PCC Panels are used to supervise and control the voltage and power of the power system. It is used to power feeder. These are the most essential part of electrical system of an industry from where the power of the industry is controlled.

PCC Panel is a power distribution board to control the electrical power supplied to HT Panels, MCC panels and transformers that play vital role in all electrical control system. Normally Power Control Centres are installed near power source hence, fault level is high.

Various protections such as short circuit, overload, earth fault, under voltage etc. are provided to protect source and equipment. PCC Panels widely used in refineries, chemical plants, pharmaceuticals and Solar Industries.

4.1.4 MCC

Motor Control Centre (MCC) Panel is used to distribute the power through the panels to the different motor loads located at various parts of the factory system. MCCs are typically found in large commercial or industrial buildings where there are many electric motors that need to be controlled from a central location, such as a mechanical room or electrical room.

A motor control centre (MCC) consists of multiple enclosed sections having a common power bus with in-comer and with each section containing a motor starter, fuses or circuit breaker.

4.1.5 Switch gear

Electrical switchgear in electrical system is a device that regulates, protects, and isolates a power system with a variety of controls housed in a metal enclosure.

Switchgear contains fuses, switches, and other power conductors. However, circuit breakers are the most common component found in switchgear. During an electrical fault, a circuit breaker senses abnormality and interrupt the power flow, effectively limiting damage to the system. Because it is designed to control the flow of power, switchgear plays a role in enhancing energy efficiency and safety of a facility.

4.1.6 Motor Starter Panel

A starter panel is the basic type of motor control panel. It houses the starter and associated controls for an electric motor. The motor starter panel may include a circuit breaker to protect the motor from overload.

A typical motor starter panel consists of incoming power either (Single or three phase) connected to a protection system such as fuse block or a circuit breaker with or without locking option, contactor switchgear.

Motor starter panel includes control elements of a motor such as selector switches as Auto/ Manual, Local / Remote and push buttons as Motor Start/ Stop and indicators for the status.

4.1.7 Circuit breaker

A circuit breaker is an electrical switch designed to protect an electrical circuit from damage caused by overcurrent/overload or short circuit. Its basic function is to interrupt power supply getting signal from protective relays which detect a fault.

4.1.8 Relay

A protective relay is a device that detects the fault and initiates the operation of the circuit breaker to isolate the defective element from the rest of the system.

Protective relays detect the abnormal conditions in the electrical circuits by constantly measuring the electrical quantities which are different under normal and fault conditions. Relay gets the input from CT, PT & other protection signal from system to measure the electrical parameters.

4.2 PCC& MCC Loss

Losses in HT & LT breaker and starter panel is 0.025-0.05%

4.3 T&D Losses

Distribution sector is considered as the weakest link in the entire power sector. Transmission losses are approximately 17% while distribution losses are approximately 50%.

Equation: T&D Losses (%) = [Energy Input to feeder (kWh)-Billed Energy to Consumer (kWh)] / Energy Input kWh x100.

There are two types of transmission and distribution losses:

a) Technical Losses

b) Non-Technical Losses (Commercial Losses).

[*Discussion on non-technical losses is excluded from subject matter.*]

4.3.1 Technical Losses in Industries

The technical losses are due to:

a) Energy dissipated in the conductors

b) Equipment used for transmission line

c) Transformer loss

d) Sub- transmission line and distribution line.

Technical losses are normally 22.5%, and directly depend on the network characteristics and the mode of operation. The major losses in a power system is in primary and secondary distribution lines. There are two types of technical losses.

4.3.2 Permanent / Fixed Technical Losses

i. **Fixed technical loss**

20-30% of technical losses on distribution networks are fixed losses. Fixed losses on a network are as below:

a) Corona Losses

b) Leakage Current Losses

c) Dielectric Losses

d) Open-circuit Losses

e) Losses caused by continuous load for measuring & control elements.

ii. **Variable technical losses**

Variable losses vary with the amount of electricity distributed and proportional to the square of the current. Variable loss is 60%-75% of total losses on distribution networks.

By increasing the cross-sectional area of lines and cables for a given load, losses can be reduced.

4.3.3 Reasons for technical Losses

i. **Lengthy distribution lines-** 11 KV and 415-volt lines extended over long distances to feed loads scattered over large areas, results in high line resistance and high I^2R losses in the line.

ii. **Inadequate size of conductors of distribution lines-**The size of the conductors should be selected on the basis of KVA x KM capacity of standard conductor for a required voltage regulation.

iii. **Installation of distribution transformers away from load centres-**If distribution transformers are not located at load centre on the secondary distribution system, rather located centrally with respect to consumers, results in low supply voltage at the farthest consumers even though a good voltage level maintained at the transformers secondary. This leads to higher line losses due to decreased voltage at the consumer's end.

The distribution transformer should be located at the load centre to keep voltage drop within permissible limits.

iv. **Low power factor of primary and secondary distribution system-** In most LT distribution circuits normally the power factor ranges from 0.65 to 0.75. A low power factor (PF) contributes towards high distribution losses. For a given load, if the power factor is low, the current drawn is high and losses (proportional to square of the current) is more.

v. **Installation of capacitor to improve power factor-** Line losses owing to the poor PF can be reduced by improving the power factor. This can be done by providing Shunt capacitors in secondary side (11 KV side) of the 33/11 KV power transformers or at various point of distribution line. Fig-13.7 shows connection of capacitor banks with load.

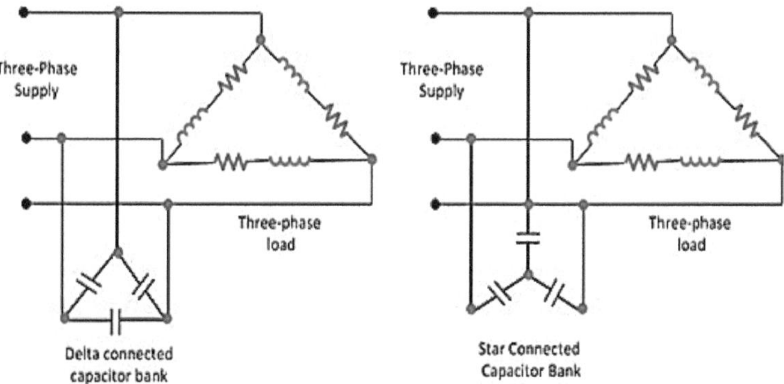

▲ **Fig-13.7:Delta & Star connections of capacitor banks**

a) **Rating of capacitor**- The optimum rating of capacitor banks for a distribution system is 2/3rd of the average KVAR requirement of that distribution system. The view point is at 2/3rd the length of the main distributor from the transformer. A more appropriate manner of improving this PF of the distribution is to connect capacitors across the terminals of the consumers having inductive loads. By connecting the capacitors across individual loads, the line loss is reduced from 4 to 9% depending upon the extent of PF improvement.

b) **Selection of capacitor- calculation**

As per Power Triangle; KVAR=Tan ϕ, When, ϕ= Cos^{-1} pf. If running pf =pf1 and pf to increase to pf2.

Then required KVAR required = KVR_1-KVR_2= kW x (Tan ϕ1- Tan ϕ1).

c) **Example**

Let in a 500 kW line pf is 0.8(lagging) and its pf to be increased to 0.9, what is the rated capacitor to be installed to do so?

Cos ϕ= 0.8, ϕ= $cos^{-1}0.8=36.9^{0}$, tan ϕ1=0.75;

New power factor = 0.9, ϕ=25.84 ^{0}, tan ϕ2= 0.48

Requiredcapacitor(KVR)=500x(0.75-0.48)=135.Note:[0.48/0.75=0.64=2/3].

vi. **Bad workmanship and maintenance-** Joints are a source of power loss. Therefore, the number of joints should be kept to a minimum &proper joining techniques should be used to ensure firm connections. Connections to the transformer bushing-stem, drop out fuse, isolator, and LT switch etc. should be periodically inspected and proper pressure maintained to avoid sparking and heating of contacts. Replacement of deteriorated wires and services should also be made timely to avoid any cause of leaking and loss of power.

vii. **Feeder Phase Current and Load Balancing-** One of the easiest loss savings of the distribution system is balancing current along three-phase circuits. Feeder phase balancing balances voltage drop among distribution feeders results lower losses assuming similar conductor resistance.

viii. **Transformer Sizing and Selection-**Distribution transformers use copper conductor windings to induce a magnetic field into a grain-oriented silicon steel core. Therefore, transformers have both load losses and no-load core losses.

Transformer copper losses vary with load based on the resistive power loss equation

(P_{loss} = I^2R). For some utilities, economic transformer loading means loading distribution transformers to capacity or slightly above capacity for a short time in an effort to minimize capital costs and still maintain long transformer life.

ix. **Switching off Transformers** - One method of reducing fixed losses is to switch off transformers in periods of low demand. If two transformers of a certain size are required at a substation during peak periods, only one might be required during times of low demand so that the other transformer might be switched off in order to reduce fixed losses.

4.3.4 Heat loss from electrical equipment

Heat loss to the ambient air from some typical electrical equipment are indicated below:

i. **Transformers**

Transformers are in general highly efficient and large power transformers - 100 MVA and larger, can be more than 99% efficient. Smaller transformers - like used in consumer electronics - may be less than 85% efficient.

Typical Heat loss for transformer:

a) 150 kVA and smaller: 50 Watt/kVA (approx. 5%)
b) 150 - 500 kVA: 30 Watts/kVA (approx. 3%)
c) 500 - 1000 kVA: 25 Watts/kVA (approx. 2.5%)
d) 1000 - 2500 kVA: 20 Watts/kVA (approx. 2%)
e) larger than 2500 kVA: 15 Watts/kVA (approx. 1.5%).

ii. **Losses in Switch Gear - Circuit Switch Breakers**

Typical losses in switch gear

1. LV/ MV breaker
   a) 0-40 Amps: 10 Watts
   b) 50-100 Amps: 20 Watts
   c) 225 Amps: 60 Watts
   d) 400 Amps: 100 Watts
   e) 600 Amps: 130 Watts
   f) 800 Amps: 170 Watts
   g) 1600 Amps: 460 Watts
   h) 2000 Amps: 600 Watts

i) 3000 Amps: 1100 Watts

j) 4000 Amps: 1500 Watts

2. Medium / high voltage breaker/switch

   a) 600 Amps: 1000 Watts

   b) 1200 Amps: 1500 Watts

   c) 2000 Amps: 2000 Watts

   d) 2500 Amps: 2500 Watts

iii. **Motor Control Centers**

   a) Medium voltage starters size 200 Amp: 400 Watts

   b) Medium voltage starters size 400 Amp: 1300 Watts

   c) Medium voltage starters size 700 Amp: 1700 Watts

iv. **Variable Frequency Drive**

   a) Variable frequency drive: 2 - 6% of kVA rating

   b) Bus duct: 0.015 Watts /ft- Amp

   c) Capacitors: 2 Watts / KVAR

Question review

1. What are main three types of motor based on supply and field excitation.
2. How to measure stator winding loss?
3. What is slip and how it is used to find rotor copper loss.
4. What are the common types of alternator rotor enumerating the advantages and disadvantages?
5. Describe brush-ring excitation and brushless excitation system.
6. What are the losses in an alternator?
7. Briefly describe different parts of a transformer.
8. When is the maximum efficiency of a transformer, prove it?
9. What are causes for technical fixed loss in power transmission?
10. How capacitor banks can reduce power loss, discuss how to select the same.

SECTION 14

PERFORMANCE & EFFICIENCY ASSESSMENT PROCEDURE

1.0 Introduction

Saving is Gain is a well-known proverb and applicable for all type of activities in industries & power generation units. Performance and efficiency assessment is undertaken through energy audit which highlights the losses & wastage of energy, equipment by equipment & system by system, also reveal about ways and means to plug and reduce the losses. It also suggests about alternative pathway to save energy in plant process.

The assessment is undertaken for the purpose of energy conservation and safety concerns. The audit may be undertaken in a process plant, system or residential place to reduce the energy input into the system and to control energy losses without affecting the output.

1.1 Objective of assessment

The objective of performance & efficiency assessment of a plant operation is to:

a) Achieve energy management
b) Maintain optimum energy procurement and utilisation, throughout the plant operation
c) Ascertain safety of man & machineries
d) Minimise energy costs / wastes without affecting production & quality
e) Reduce environmental effects.

1.2 Types of performance assessment

Performance assessment of a production plant to be undertaken depends on- type of industry, extent to which final audit is needed and potential and magnitude of cost reduction desired. Hence, performance assessment can broadly be classified as:

a) Preliminary assessment
b) Detail assessment

1.2.1 Preliminary assessment/audit

i. **Procedure:** Such audit process is a walk-through inspection and checks are undertaken such as:

a) SOP & SMP being followed
b) Deficiency in equipment, if any
c) To identify actual energy consumption and losses
d) To perform economic calculations
e) Spares issue
f) Safety aspects.

ii. **Outcome from audit**

a) To rectify deficiencies in SOP & SMP, if any
b) An energy balance; an input output balance
c) To make a list of measures, which are derived from performance analysis, to increase energy efficiency
d) To identify energy saving opportunities
e) An analysis of the implementation of energy saving measures
f) An estimation of its potential energy saving
g) Any unsafe conditions prevailing in plant
h) To categorize and prioritize the implementation of the stated measures.

1.2.2 Detail assessment and investment grade audit

i. **Procedure:** Such type of audit is undertaken in addition to preliminary audit:

a) A detailed account of energy use and it is undertaken equipment-wise and section-wise.

b) A quantitative study of the implementation of measures with detailed investments.

c) Operational and maintenance costs on new investment.

ii. **Expectations/ outcomes from audit**

Outcomes from detail assessment are:

a) The real energy demand and an energy balance
b) It suggests a number of energy saving measures
c) It proposes correction measures with a financing plan and implementation
d) It also shows the savings verification plans.

2.0 Definitions and equations related to performance assessment & energy audit

Following definitions are commonly used while assessing energy audit of any equipment or any plant.

2.1 Active Power

i. **Direct Current** (DC)

   Power = V x I =I^2R=V^2/R; when, V= Potential Difference (volt), I= Current (ampere),

   R= Resistance (Ohm).

ii. **Alternating Current (AC)**

   Single Phase AC power = VI Cos φ.

   Three phase AC power = √3 VI Cos φ; when, Cos φ is power factor.

2.2 Reactive Power

Single Phase AC power = VI Sin φ.= VI √ (1- Cos^2 φ).

Three phase AC power = √3 VI sin φ= √3 VI √ (1- Cos^2 φ). When, Cos φ is power factor.

2.3 Power Factor

In AC circuit there is a phase difference between voltage & current. The Cosine angle between them i.e. Cos φ is called power factor. Lesser the angle, better is power factor or pf approaches towards 1.

2.4 Efficiency

Efficiency of a system = Output/Input= (Input-loss)/Input= (1-loss/Input).

2.5 Specific Fuel Consumption

Fuel required to produce unit product is called specific fuel consumption.

Unit: Kg of Fuel/ kg of Product. For Power generation it is kg fuel/ kWhr.

2.6 Specific heat energy consumption

It is the ratio of total heat including electrical energy to produce unit product. Unit: Kcal/ Kg.

2.7 Specific electrical power consumption

Electrical power required to produce unit mass of product is called Specific Electrical Power Consumption. Unit: kW hr / kg.

2.8 Heat Rate

This unit is applicable for power generating industries. Heat required to generate unit power is called heat rate. Unit: Kcal/ kWhr.

2.9 Auxiliary Power

In power generation unit, several auxiliaries are required to operate the power plant. Power consumed by the stated auxiliaries is called auxiliary power. It remains between 7-10% of total power generation.

2.10 Gross Power

Total power generation that is metered at generator terminal is called Gross power.

2.11 Net power

Power that is supplied to consumer is called net power. Normally it is metered at export feeder.

Net Power = Gross Power – Auxiliary power.

2.12 Gross Calorific Value (GCV)

Total heat value of any fuel is called Gross Calorific Value. It is also called GCV-ARB (As received basis).

Unit: Kcal / Kg or KJ/ Kg; It is obtained by testing in bomb calorimeter.

2.13 Net Calorific Value (NCV)

Calorific value excluded for moisture of hydrogen is called NCV. NCV= GCV-53 x H%.

2.14 Instrument error/ accuracy

The measurement error is defined as the difference between the true or actual value and the measured value.

2.15 Calibration

Calibration of an instrument refers to the activities on an instrument such that instrument's output value is accurate corresponding to its input value throughout a specified range. This can be checked only if output signal value is measured against a known input signal value. This means a standard known value is to be used as a known input condition and to measure output signals.

2.16 Calibrator

A calibrator is an instrument used for calibration of field instruments. A typical calibrator contains following features:

i. Pneumatic calibrators can generate regulated air pressure for pressure calibration.

ii. Block calibrators are used for temperature probes. Signal reference is used to calibrate panel meters & controllers.

iii. It can generate a known signal such as voltage, current, frequency as reference. Once known signal is fed as reference, the output value of the instrument under test is matched and corrected, if required. The simulator can generate and read the signal.

2.17 Performance Test

To ascertain the performance of any equipment, system and the plant as a whole a test is conducted as per industry standard. Such test is termed as performance test. Following highlighted criteria are maintained during performance test.

a) Operating Load = > 80%.

b) All process Input quality are near about the design specification.

c) All measuring instruments are calibrated.

d) Constant load operation is of minimum of 4-6 hrs.

e) Additional activities, which are undertaken during normal Operation, are to be suspended during the period of test.

f) All calculations are to be done as per standards.

3.0 Study of plant process and associated sub-processes

Understanding of plant process is an important aspect in performance assessment and energy audit. It is required to study and review the main process flow diagram along with sub-processes showing the operating parameters like pressure, temperature, flow, concentration, Enthalpy at each input-output stage.

The entire plant process can be divided into sub- process or sub-systems which facilitates to study the process more closely.

Table-1 shows a few industrial sectors and sub-process of the sectors.

▼ Table-1: Sub-systems under different industrial sectors

Sector	Sub-Processes
Thermal Power Plant (coal/ liquid fuel/ gaseous fuel)	1. Steam Generation System 2. Turbo-Generator system 3. WTP & Water make up system 4. Coal handling Plant 5. Ash Handling System 6. CW system 7. ESP 8. Compressed Air System 9. Power distribution system 10. Control System
Combines Cycle Power Plant (Gas)	1. Gas Turbine system 2. Heat Recovery Steam Generator (HRSG) 3. STG system 4. CW system 5. Compressed Air System
Cement	1. Raw Material Preparation System 2. Cement Kiln 3. Grinding Section 4. Captive Power Plant(CPP) 5. WHRS-power plant
Iron & Steel (Integrated)	1. Coke oven 2. Sintering Plant 3. Blast Furnace 4. BOF 5. Casting Section 6. Rolling Mills 7. CPP
Iron & Steel (Sponge Iron)	1. Sintering Plant 2. DRI section 3. EAF (electric arc Furnace) section / IF (Induction furnace) Section 4. Casting section 5. Rolling Mill 6. CPP

Sector	Sub-Processes
Fertilizer	1. Hydrogen / synthesis gas generation plant 2. Oxygen/Nitrogen Plant 3. Ammonia Plant 4. Urea Plant 5. Steam Generation System 6. CPP
Pulp & Paper	1. Raw Material Preparation 2. Pulping Section 3. Paper Machine Section 4. CPP 5. Co-Generation Plant
Textile	1. Spinning section 2. Drying Section 3. Steam Generation System 4. CPP

[The above sub-processes are indicative only. The sub-process based on the material & energy flow and quantum of energy consumed can be suitably selected]

3.1 Study of Sub-Processes

By studying sub-Processes, following data are collected:

a) Detail about the energy & material flow into and out of each sub-process

b) Process parameters for system in the sub-process under study

c) Metering system for all energy sources in the sub-process under study

d) This should be in the form of block diagrams, Fig-14.1

Table-2 shows a sample of chart for parameters of sub-process.

▼ **Table-2:Parameters of sub-systems**

Brief Description of the Sub-Process	**Parameters**	**Unit (UOM)**	**Design**	**operating**	**Remarks**
Material Flow	Name				
	Flow Rate	Kg/hr.			
Energy Input	Electricity	kWh			
	Solid Fuel (Coal/Lignite)	Kg/hr.			
	Liquid Fuel (FO/HSD/ LHHS…)	L/hr.			
	Gaseous Fuel (NG/LPG…)	Sm^3/hr.			
	Others	Kg or L/ hr.			
Major Equipment	Tag Number				
	Technology				
	Numbers	Numbers			
	Technical Specifications				
	Specific Energy Consumption	kWhr /Kg			
Factors Affecting Energy Efficiency	Technical Parameters				
	Operating Parameters				
Major Auxiliaries Associated (Pumps, Fans, Compressors, Furnaces, HVAC system etc.)	Numbers Operating	Numbers			
	Technical Specifications				
	Per-cent loading or Capacity Utilization	%			
	Efficiency	%			
	Specific Energy Consumption	kWhr /Kg			

[*In order to get the operating values suitable measurements to be done by energy auditing instruments or observation from installed instruments.*]

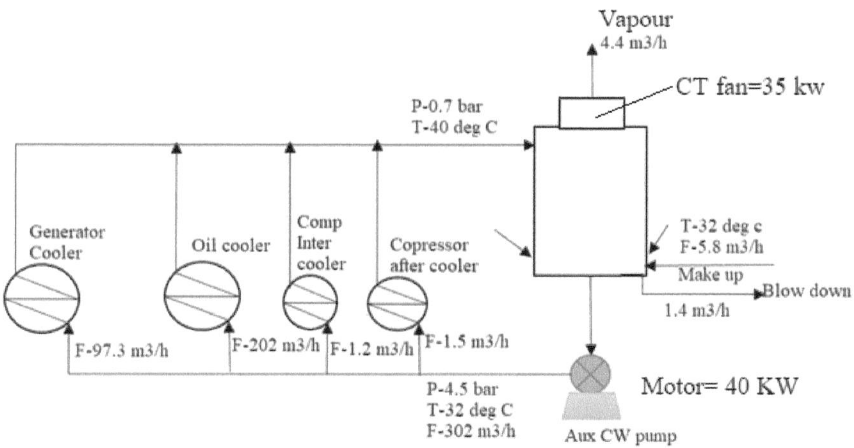

▲ **Fig-14.1: Auxiliary CW system**

3.2 Study of technology up gradation in last 3 years

i. Technology Change

ii. Process Modification, if any

iii. O & M Practices

iv. Resulting Energy Saving

v. Any investment done.

3.3 Study of energy flow in plant processes

Heavy industries like steel, cement, fertilizer & chemical etc. need uninterrupted power supply otherwise industries face huge loss due to interruption of power. From the point of uninterrupted power availability at a lower price, it is decided whether to go for power import from grid or plant should have its power generation unit or combination of two.

Energy flow process in some common plant configurations are described below.

3.3.1 All kind of energy including electricity are purchased from outside

(Electricity is purchased from the Grid)

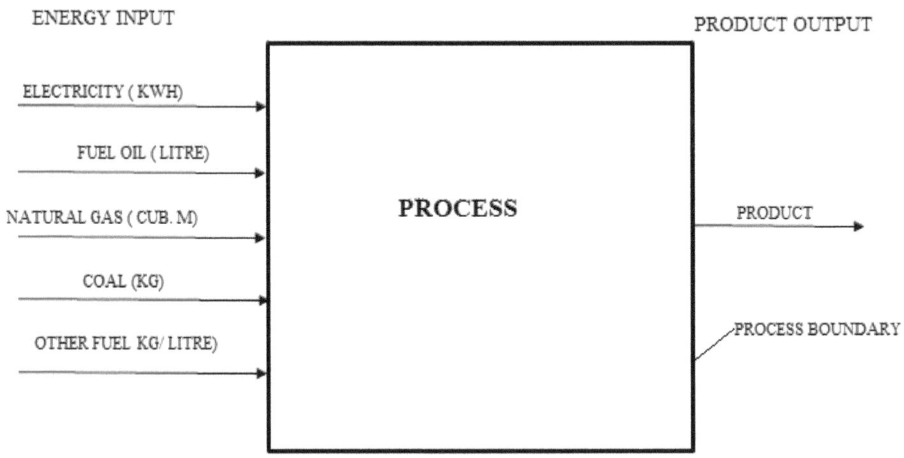

▲ **Fig-14.2: Flow of electricity in energy system**

i. **Energy input**

Fig-14.2 shows that the input energy for process are:

a) Electrical energy- purchased from grid

b) Thermal energy- FO, NG, coal & other fuels

ii. **Related Equations for energy consumption**

a) Total electrical energy (E_{ELEC}) consumption = kWh = kWh x860 Kcal.

b) Total thermal energy ($E_{THERMAL}$) consumption= [FO x GCV_{FO}+ Natural Gas x GCV_{NG}+ Coal x GCV_{COAL} + Other Fuel x GCV_{OTHER}] Kcal.

c) Total energy (E_{TOTAL}) consumption = [Electrical Energy (E_{ELEC}) (kWh) x 860 + ($E_{THERMAL}$)] Kcal.

iii. **Derivations**

a) Specific electrical power consumption = Total Electrical Energy (E_{ELEC}) Consumption (kWh) ÷ Product (Kg) = kWh/kg.

b) Per cent of electrical energy (%) = [Total Electrical Energy (E_{ELEC}) Consumption (kWh) x 860 ÷ Total Energy (E_{TOTAL}) Consumption] x100.

c) Gate to Gate Energy Consumption = Total Energy (E_{TOTAL}) Consumption ÷ Product (Kg)

= Kcal/kg.

3.3.2 Electrical energy partial generated by DG/ TPP, balance energy purchased and consumed

(Electricity is purchased from the Grid & generated by DG set)

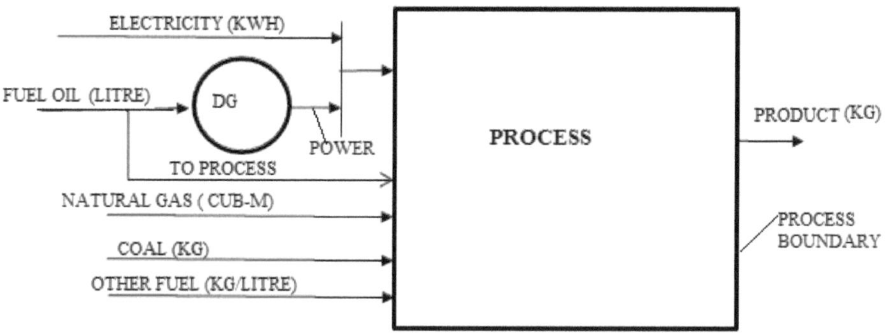

▲ **Fig-14.3: Flow of electricity system including DG/TPP**

I. **Energy Input**

Fig-14.3 shows different types of energy input in the process as below

a) Electrical energy: Purchased (P)+ own generation by DG (X)

b) Thermal energy: FO for power generation (A)+ process requirement, coal, NG & other fuel

II. **Related Equations for energy consumption**

Electricity consumption

i. Power purchase =P kWh

ii. Power generated =X kWh

a) Specific Fuel Consumption (SFC) for DG Power (Litre/ kWh) (Commonly 0.25-0.35)

b) FO Consumption for power generation (A) (Litre) = SFC x Power Generation

iii. Total Electrical Energy Consumption (EL_{total}) = [Purchased Electricity (P) + X] kWh

Thermal energy consumption

iv. Total Thermal Energy Consumption (TH_{total}) = [(FO-A) x GCV_{FO} + Coal x GCV_{coal} + NG x GCV_{NG} + Others x GCV_{each}] Kcal

III. **Derivations**

a) Total Energy Consumption (E_{TOTAL}) = [(EL_{total}) Kcal +(TH_{total})] Kcal.

b) Gate –to-Gate Specific Energy Consumption (SEC)= [E_{total} /kg of product] Kcal/kg

3.3.3 Electricity generated and other energy purchased & consumed; electricity partially sold to grid

(Electricity is generated by Coal Based CPP, partially sold to grid)

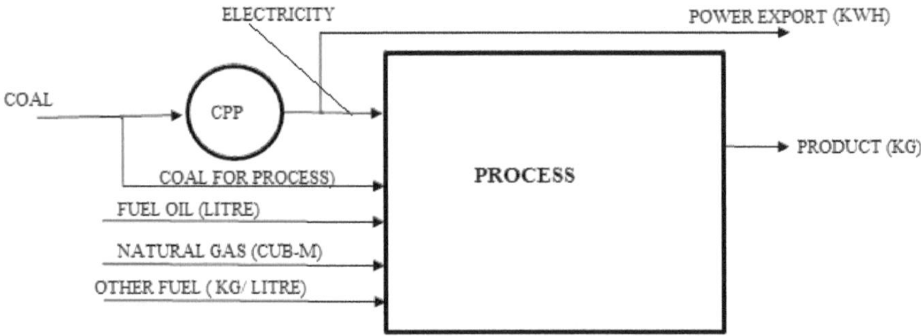

▲ **Fig-14.4:Flow of electricity and energy**

i. **Fuel input**

From fig-14.4

a) Coal for Power generation & power export

b) FO, NG, Other fuel

ii. **Power generation**

Let, X=Total power generation (kWh), Y= Power export (kWh), A= Quantity of coal used for power generation. Total coal input = Q

PHR= (A x GCV_{coal})/X Kcal/kWh; X= (GCV_{coal} /PHR) x A kWhx860 kcal [PHR-plant heat rate].

iii. **Energy consumption**

a) Electrical energy (EL_{total}) = (X-Y) kWhx860 kcal

b) Thermal energy (TH_{total}) = [FO x GCV_{FO} + (Q-A) x GCV_{coal} + NG x GCV_{NG} + Others x GCV_{each}] Kcal.

c) Total energy input (including power export) (E_{total}) = [FO x GCV_{FO} + Q x GCV_{coal} + NG x GCV_{NG} + Others x GCV_{each}] Kcal.

iv. **Derivation**

Gate –to-Gate SEC = [E_{total}/kg of product] Kcal/kg.

3.3.4 Electricity generated by CPP, other energy purchased & consumed; electricity partially sold to grid, CPP is separate entity.

Electricity is generated by Coal based CPP, partially sold to grid. CPP is in separate boundary.

Here, in case of two separate boundaries, two different Norms / Targets are to be established.

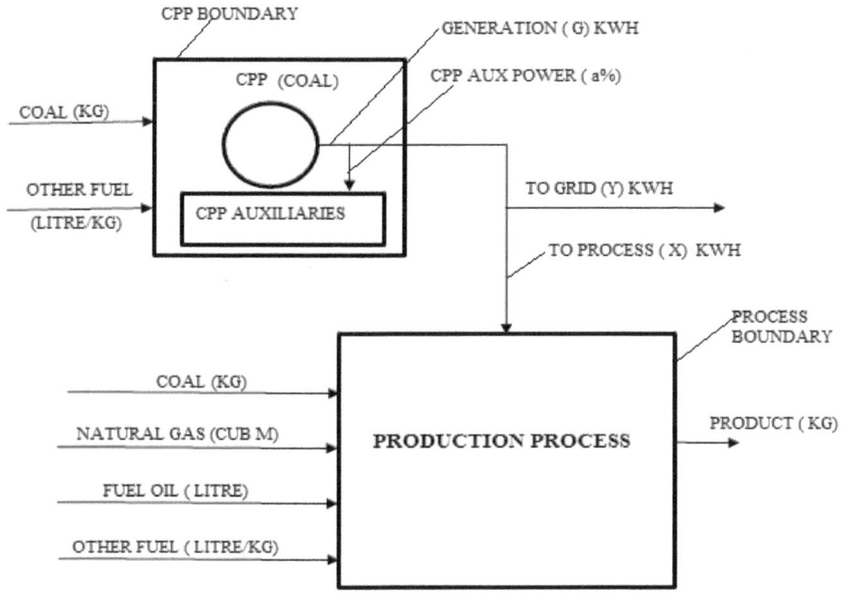

▲ **Fig-14.5: Flow of electricity and energy in auxiliaries and production processes**

i. **For the 'Production Plant' Boundary (A):**

a) Total Electrical Energy Consumption (EL_{total}) = X kWh

b) Total Thermal Energy Consumption (TH_{total}) = [FO x GCV_{FO} + Coal x GCV_{coal} + NG x GCV_{NG} + Others x GCV_{each}] Kcal.

c) Total Energy Consumption (E_{total}) = [X *860 + FO x GCV_{FO} + Coal x GCV_{coal} + NG x GCV_{NG} + Others x GCV_{each}] Kcal.

d) Gate –to-Gate SEC = [E_{total} /kg of product] Kcal/kg

ii. **For the 'Captive Power Plant' Boundary (B):**

a) Total Energy Consumption (E_{total}) =[FO x GCV_{FO} + Coal x GCV_{coal} + Others x GCV_{each}] Kcal

b) Gross Heat rate = E_{total} / G Kcal/ kWh [G- total power generation]

c) Total Electrical Energy Consumption by auxiliary (EL_{total}) = (G) x (a/100) kWh [a- auxiliary%]

d) Gate –to-Gate SEC (i.e. Net Heat Rate) = E_{total} / (G – EL_{total}) Kcal / kWh

3.4 Instruments for performance assessment

Several readings on process parameter and electrical parameters are required to be logged during auditing as normal activities. The parameters that are logged must be error free to derive a real result. The parameters logged are available from:

i. Existing field Instruments

ii. Instrument are brought from outside.

Whether to use the existing field instrument or brought out, it is advisable to do the calibration of existing field instruments or last calibration date of brought out instruments is valid as on audit date.

Typical list of instruments that are required for energy audit of a process plant is as per table-3.

▼ **Table-3: List of instruments for energy audit of a process plant**

SL No	Measurement	Instrument	Purpose
1	Temperature	Temp Gauge TC/ RTD Portable Non-Contact Temperature sensor	To measure the Operating temperature
2	Pressure	Pressure Gauge Pressure Transmitter	To measure the Operating Pressure.
3	Flow	Flow transmitter	To measure the Operating Flow

SL No	Measurement	Instrument	Purpose
4	Differential Pressure	DP gauge DP Transmitter	Differential pressure across filter, pipe flow etc.
5	Acid/ Alkali	pH Meter	
6	Oxygen	O_2 analyser	O_2 in flue gas to derive combustion efficiency.
7	CO, CO_2	CO, CO_2 Analyser	To derive combustion efficiency
8	SOx, NOx	SOx, NOx Analyser	Pollution
9	TDS	Conductivity meter	In liquid
10	Speed	Speedometer Tachometer	
11		Thermal insulation scanner	To measure loss of heat from insulation surface.
12	Ambient condition	Thermo-Hygrometer	To measure ambient humidity
13	SPM	Opacity Meter	Pollution/ efficiency
14	Active power	Energy Meter	Power consumption/ Generation
15	Reactive Power	VAR Meter	PF
16		Harmonic Analyser	For analysis of harmonics in power System
17	Illumination	Lux-meter	For measurement of illumination level.
18		Clip on i. Watt Meter ii. PF meter iii. Amp-Meter	
19	Electrical parameter	Multi-Meter	For measurement of Voltage, Current & Resistance.
20		Frequency Meter	Quality

3.5 Approach for plant performance and efficiency assessment

For assessing the performance of a plant, physical audit is vital. So, during this period plant should run at least 80% load and the assessment team must have a thorough plant visit to have a realistic information about running parameter. The team should log few sets of readings related to plant process, power consumption, analytical parameter etc. The team should collect information about past interruption/ break down and maintenance along with spares inventory. Safety systems and its implementation in the plant is also an important aspect of assessment of a plant performance. The step-by step approach for performance assessment is described below.

3.5.1 Brief description about the plant

a) History of establishment, ownership (public / private / co-operative sector)

b) Present installed capacity and stages of capacity addition (if any)

c) Trend of capacity utilization in last 5 years

d) Trend of monthly production and capacity utilization for last one year.

3.5.2 Review of process flow diagram and Present Metering System to assess energy scenario

i. Study and review the process flow diagram (PFD). Develop the schematic of the PFD depicting the operating parameters like flow, temperature, pressure etc. at each input-output stage.

ii. Divide the entire process into key sub-processes or sub-systems, table-1.

iii. Define the energy & material flow into and out of each sub-process, process parameters and metering system for all energy sources in the sub-process under study. (in the form of block diagrams, fig-14.2).

iv. Investigate the following in each sub-process.

   a) Different types of energy sources being used and their % share of contribution

   b) Primary Energy: Solid fuel, Liquid fuel, gaseous fuel etc.

   c) Secondary Energy (Electricity): Purchased from grid or, Self-generated (CPP/DG sets/Co-Generation)

   d) Energy from internally generated by-products or waste heat

   e) Renewable energy.

3.5.3 Collect relevant data

Data collection is an important activity in performance assessment. The data should be related to operation, maintenance, interruptions/break down, safety checks, testing etc. The source of data may be:

a) Documents, records, log books, installed meters and gadgets etc.

b) Instantaneous/hourly/daily/monthly/yearly

c) Manual reading or recording /SCADA/automatic MIS

3.5.4 Review of collected data

Collected data needs to review as follows:

a) Eliminate inconsistent data.

b) Reconcile the data with respect to production, typical parameters like heat rate, specific fuel consumption or any other observations.

c) Convert different units of same input (such as, all energy in Kcal or KJ) into a common unit.

d) Clearly define the inclusions / exclusions of the energy consumption of colony power, temporary constructions, internal transportations etc. in the reported data.

3.5.5 Establish the trend for data of last 2/3 years as below

1. Share of **purchased electricity** from grid to that of total electricity consumed (%)
2. Share of **self-generated electricity** to that of total electricity consumed (%)
3. Share of **internally generated thermal energy** from waste-heat, co-generation or by-product of total consumed energy (%)
4. Share of **other fuel purchased** for production (%)
5. Share of **renewable energy** to that of total energy consumed (%)
6. Specific fuel consumption (Kcal/kg)
7. Cost of energy to that of total production cost
8. Generate trend curve of monthly production and energy consumption (Electrical & thermal) of last one year
9. Share of energy in auxiliaries (%)
10. Share of energy in various Purposes (Lighting, HVAC etc.) (%).

3.5.6 Establish the energy scenario in gate-to-gate

1. Develop the gate-to-gate schematic of the plant depicting the energy input to the plant boundary, internally generated energy and the product output
2. Trend the energy consumption in last 5 years
3. Establish the Specific Energy Consumption (SEC) in gate-to-gate concept.

3.5.7 Review the future energy efficiency improvement strategy

1. Specific Projects to be undertaken
2. Investment Plan
3. Human Resource Planning in Energy Management
4. Any target fixed by management as per their policy.

3.5.8 Investigation of the all possible energy efficiency improvement option

Following energy efficiency equipment & systems may be reviewed.

1. **Waste Heat Recovery Options**
    a. Source
    b. Comment on Feasibility
    c. Quantum of Possible Extracted Energy.
2. **Co-Generation Option**
    a. Comment on Feasibility
    b. Approximate Thermal & Electrical Energy output.
3. **Use of Energy Efficient Equipment**
    a. Energy efficient motors, Pumps, Fans, Compressors, Lighting system etc.
    b. Variable Frequency Drives / Variable Speed Drives
    c. Any other.
4. **O & M Practices**
    a. Waste minimization (Reuse, Recycle, Resizing)
    b. Control Mechanism
    c. Data Monitoring & Accounting.
5. **Fuel switch options**
6. **Retrofit options.**

3.6 Preparation of plant assessment report with following structure/ format

After plant assessment a consolidated report is needed to submit. The contents that should cover the report are as follows.

1. Acknowledgement
2. Team Members associated
3. Executive Summary
4. Introduction
5. Methodology Adopted
    a) Preliminary assessment
    b) Detail assessment
6. Instruments Used for Energy audit
    a) Testing instrument
    b) Calibrated plant instrument
7. Brief Description about the plant (Process Flow Diagram)
8. Description about sub-processes
9. Energy scenario & usage pattern
10. Data collection
11. Calculation of energy consumption
    a) Section-by- section/ sub-processes
    b) Plant as a whole
    c) Specific energy consumption
    d) Efficiency assessment for high- energy consuming equipment.
12. Compare actual vs design
13. Identifying energy losses- causes and their remedial recommendations
14. Exploring energy consumption with other fuel or fuel mix
15. Description about future energy consumption strategy
16. Identify any safety lapses and its remedies
17. Conclusion & critical comment
18. Annexures

3.7 Performance Assessment of 18 MW Power Plant- A Sample

The auxiliaries of the stated power plant include some HT drives (6.6 KV) and balance LT drives (440 V). During the performance assessment, auxiliary power readings for individual HT drives were logged and balance auxiliary power consumption, which are LT, was logged through auxiliary transformers. As separate lighting transformer is provided for lighting purpose, lighting load was logged from lighting transformer.

All the data shown below are not corrected to guaranteed conditions. Calculations and derivations that are shown below only to have a feel to the readers that how to approach and the performance of any equipment or system are calculated methodically.

3.7.1 Performance Assessment of Boiler

1. **Boiler Parameter**

▼ **Table-4: Different boiler parameter**

Parameters during assessment		
Description	**UOM**	**Value**
Humidity factor	gm moisture/kg dry air	110
Ambient Temperature	^{0} C	35
Main steam flow	t/h	73.75
Main steam temperature	^{0} C	515
Bed temperature	^{0} C	840
Flue gas temp at Air Pre-heater inlet	^{0} C	220
Flue gas temp at Air Pre-heater outlet	^{0} C	135
FG temp at ESP outlet	^{0} C	126.22
Oxygen in flue gas	%	3.5
Ultimate analysis of coal		
Carbon (C)	%	53.54
Hydrogen (H)	%	4.47
Sulphur (S)	%	1.00
Oxygen (O)	%	10.02
Nitrogen (N)	%	1.57
moisture	%	10.80
Ash (A)	%	18.60
GCV of coal	Kcal/ kg	5,353.00

2. Ash & Unburnt Report

▼ Table-5:

Description	UOM	Unburnt	Ash distribution
Bed Ash	%	0.50	20%
Economizer	%	26.00	4%
Air Pre-heater	%	16.00	6%
Fly ash (ESP)	%	22.00	70%
Average Unburnt	%	17.50	
Average ash temperature	^{0}C	273	

3. Flue gas analysis

▼ Table-6:

Dry flue gas analysis		
Description	**UOM(v/v)**	**Value**
CO_2	%	18.33
O_2	%	2.53
CO	%	0.04
N_2	%	79.10

4. Loss Calculation

▼ Table-7:

SL no	Description	UOM	Value	Equation used
	Derivation of Loss due to dry flue gas			
a)	Theoretical Oxygen required	kg/kg fuel	1.70	(O) = 1/100 (8/3C +8H +S)-O_2 in coal/100
b)	Theoretical air quantity	kg/kg fuel	7.37	Air = Oxygen/0.23
c)	Excess air (EA)	%	20.00	EA (%) = O_2 in FG/(21-O_2 in FG)x100
d)	Actual air supply (AAS)	kg/kg fuel	8.84	Theoretical air (1+Excess air% /100)
e)	Flue gas quantity (m)	kg/kg fuel	9.84	Actual air+1
f)	Loss due to flue gas	Kcal/kg fuel	255.93	M x C_p x(T_f-T_a)
1	**Loss percent**	%	4.78	Loss/GCVx100
	Unburnt loss (U_b loss)	kcal/kg fuel	317.64	[A% x U_b/(100-U_b)]x8050

SL no	Description	UOM	Value	Equation used
2	**Loss percent**	%	5.93	U_b loss/GCVx100
	Hydrogen loss	kcal/kg fuel	252.61	9 x H% x [584 + C_s (T_f - T_a)]
3	**Loss percent**	%	4.72	H-loss/ GCVx100
	Moisture in coal- loss (M)	kcal/kg fuel	67.82	M% x [584 + C_s (T_f - T_a)]
4	**Loss percent**	%	1.27	Moisture loss/ GCVx100
	Loss due to humidity in air	kcal/kg fuel	42.80	AAS x Humidity Factor x C_s (T_f - T_a)
5	**Loss percent**	%	0.80	Loss due to humidity/ GCVx100
	Sensible heat loss through ash	kcal/kg fuel	8.60	[% A x100/(100-Ub)] x C_a x (Av. T_f - Ta)
6	**Loss percent**	%	0.16	Sensible heat loss/GCV of fuelx100
	Loss due to CO	kcal/kg fuel	978.57	[FG quantity/ kg of fuel x CO% /100 x 2410]
7	**Loss percent**	%	0.18	Loss due to CO/GCVx100
8	Heat loss due to radiation and Unaccounted loss	%	0.50	As per AMBA chart
	Total loss (1+2+3+4+5+6+7+8)	%	18.34	-
	Efficiency	%	100- 18.34 = 81.66	-

3.7.2 Other derived performance of boiler

1. **Coal consumption (t/h)**

   Equation: Steam flow x (Enthalpy of steam- Enthalpy of FW)/ (Efficiency of Boiler x GCV of coal).

   = 73.75x(816-199)/(0.8166x5353) = 10.41

2. **Evaporation ratio (kg/kg of coal)**

   Equation = Steam generation (t/h)/Coal consumption (t/h)

   = 73.75/10.41=7.08.

3. **Fly Ash generation (Kg/h)**

   Equation= Coal consumption/hr. x Ash% x 0.8 [fly ash 80%]

   =10.41x18.6%x0.8=1549

3.7.3 Performance Assessment of Turbine-Generator

1. **Parameter Measured**

▼ **Table-8: Flow parameters of Turbine-Generator**

Description	**Unit**	**Value**
Steam flow to turbine (W_s)	t/h	73.10
FW flow to eco inlet (W_f)	t/h	73.86
Steam pressure at turbine inlet	Kg/cm²	87.00
Steam temperature at turbine inlet	°C	508.85
FW temperature at HPH inlet	°C	118.78
FW temp at HPH out let	°C	201.00
Exhaust pressure	kg/cm²	-0.92
Power generated	MW	18.06
Power factor	pf	0.95

5. **Derivation of Turbine Heat Rate**

   Equation for turbine heat rate (Kcal/ kWh) = (W_s x enthalpy of steam at test pressure & temperature - W_f x enthalpy of FW at Economizer inlet)/ kW.

▼ **Table-9: Turbine heat rate**

Description	**UOM**	**Value**
Enthalpy of steam at test pressure & temp.	Kcal/kg	816.00
Enthalpy of water at test pressure & temp.	Kcal/ kg	199.00
Turbine heat rate (THR)	kcal/ kWh	2489.23

6. **Plant heat rate (PHR) (Kcal/kWh)**

   Equation: (PHR) =THR/Boiler efficiency

   =2489/0.8166=3048 kcal/kWh.

7. **Plant efficiency (%)**

   Equation= 860/PHR x100

   =860/3048x100=28.2%

3.7.4 Assessment of auxiliary power consumption

1. **Collected data**

▼ **Table-10: Auxiliary power consumption**

SL. No	DRIVES	UOM	Average value
1	PA Fan current	Amp	119.68
2	HP dosing pump current	Amp	0.92
3	LP dosing pump current	Amp	0.93
4	Air washer blower	amp	20.00
5	ID fan current	amp	111.45
2	Auxiliary trafo-1 power	kW	246.17
4	Auxiliary trafo-2 power	kW	370.67
6	FD power reading	kW	291.83
8	BFP power reading	kW	402.17
10	CWP power reading	kW	298.67
12	Tie-1 power reading to process	kW	9,018.33
14	Tie-2 power reading to process	kW	7,256.67
16	ESP Power reading	kW	61.03
17	HT voltage	KV	6.62
18	LT voltage	Volt	427.20
19	Average Power factor		0.87
20	DM plant MCC reading	kW	38.24
21	STG-1 MCC	kW	42.04
22	Coal feeder-1	kW	0.37
23	Coal feeder-2	kW	0.32
24	Coal feeder-3	kW	0.32
25	Coal feeder-4	kW	0.26
26	Total power consumption by coal feeders	kW	1.27

2. Auxiliary power calculation

▼ **Table-11: Auxiliary power consumption by boiler & auxiliaries**

i. Auxiliary power consumption by boiler & auxiliaries

Description	**UOM**	**value**
Power consumption by FD fan	kW	**291.83**
Power consumption by PA fan	kW	**70.84**
Power consumption by ID fan	kW	**65.97**
Power consumption by BFP	kW	**402.17**
Power consumption by ESP	kW	**61.03**
Power consumption by all Coal Feeders	kW	**1.27**
Power consumption by dosing pumps	kW	**1.09**
Total power consumption by boiler	kW	**894.21**

ii. Power consumed by other auxiliaries

Other auxiliaries Measured through Auxiliary transformer (including CHP load with diversity factor)	kW	649.33
Lighting load (measured through lighting transformer) with diversity factor	kW	27.08

iii. Total auxiliary load (kW) = (894.21+649.33+27.08) =1570.63

3. Table-12: Derivation of other performance values

S No	Description	UOM	Value	Equation used
1	Auxiliary load percentage	%	8.73	Aux load/Gross Generation x100
2	**Total boiler auxiliary**			
a)	Percent of gross generation	%	5	Boiler aux load/gross generation x100
b)	Percent of total auxiliary	%	57%	Boiler auxiliary load/total auxiliary load x100
3	**Fan load**			
a)	Total fan load (FD + PA + ID)	kW	428.6	
b)	Percent gross generation	%	2.4	Fan load/Gross generationx100
c)	Percentage of total auxiliary	%	27.29	total fan load/Aux. loadx100
4	**Boiler Feed Pump**			
a)	Percent of gross generation	%	2.2	Feed Pump load/ Gross Generation x100
b)	Percentage of total auxiliary	%	25.81	Feed pump load/Aux loadx100
5	Specific power consumption by ESP	Watt/1000m^3	0.54	Watt /1000 m^3 of Flue gas

BIBLIOGRAPHY

1. Application of Diversity Factor, H. B. Gear, Proc. Natl. El. Light Assoc., 1915.
2. The Electric Power Industry, McGrew-Hill, 1949.
3. Westinghouse Engineer, January 1950.
4. Economics of Public Utilities, L. R. Nash, McGrew-Hill, 1932.
5. The Electric Power Industry, McGrew-Hill, 1949.
6. Power Economics for Engineering Students, Pittsburgh Printing Co., 1939.
7. Semi-Outdoor Steam Plants, Gourdon, Friend, and Elliot, Combustion, April 1946.
8. Fuel Oils Commercial Standard, U. S. Dept. Commerce, Govt. Printing Office.
9. Diesel Fuels, Standard Oil Co., N. J., 1939.
10. The Process of Combustion in a Furnace, Henry Kreisinger, Combustion, November 1929.
11. Diesel and Gas Engine Power Plants, G.C. Boyer, McGrew-Hill, 1943.
12. Preliminary Cooling Tower Selection, Foster Wheeler Co., Heat Engineering, March-April 1949.
13. Test code for I.C. Engines, ASME.
14. Theory and Design of Gas Turbines and Jet Engines, E.T. Vincent, McGrew-Hill,1959.
15. An Appraisal of Gas Turbines for Power Plants, A.G. Christie, Combustion, November, 1947.
16. Convenient Gas Properties and Charts, C. J. Walker, ASME paper 50-F28, 1950.
17. A Comparison of Gas and Steam Turbines, O. A. F. Munzinger, Combustion. September 1948.
18. Present and Future Trend of Thermal Prime Movers, C. W. E. Clarke, Combustion, July 1949.

19. The Steam Turbine Regenerative Cycle, J.K. Salisbury, General Electric Co.(Collection of 3 reprints)
20. Reheat Cycle (Symposium), Trans. ASME, August 1949.
21. Topping at Sherman Creek, H. Knecht, Combustion, February, 1947.
22. Modern Mercury-Unit Power-Plant Design, Hackett and Douglass, Trans. ASME, January, 1950.
23. Theory of Incremental Rates, Steinburg and Smith, Electrical Engineering, March 1934.
24. Heat Transmission Between Fluids and Solids, W.H. McAdams, Mechanical Engineering, July, 1930.
25. Radiant Heat Transmission, H.C. Hottel, Mechanical Engineering, July, 1930.
26. Elements of Heat transfer and Insulation, Jacob and Hawkins, John Wiley and Sons, 1942.
27. Nuclear Fuel for Power Production, Flagg and Gross, General Electric Review, March 1952.
28. Estimation of Radiant Heat Exchange in a boiler furnace, Orrok and Ortsy, Combustion, April 1938.
29. Combustion Engineering, O. de Lorenzi, Combustion Engineering-Superheater Co., Inc., 1947
30. Steam Generation, Power, December 1946.
31. Economic Factors Involved in Selection of Industrial Boilers, Patterson and Riker, Combustion, August 1949.
32. Refractories, Power, June 1950.
33. Factors affecting Superheat Control, M. Frisch, Heat Engineering. December 1950.
34. Fuels and Firing, Power, December 1948.
35. Test Code for Stationary Steam Generating Units, ASME.
36. Development of Pulverized Coal Firing, C.G.R. Humphreys, Combustions, September, October, November 1948.
37. The Problem of Generating Pure Steam at High Pressures (abstract), Frisch and Lorenzini, Heat Engineering, May 1950.
38. Stationary Steam Engine Lubrication, Socony-Vacuum Oil Co.
39. Steam Turbines, Church, McGrew-Hill, 1950.

40. Steam Turbines and Their Cycles, Salisbury, John Wiley and Sons, 1950.
41. Steam Turbines, Power, December 1945.
42. Relative "Engine Efficiencies" of Large Steam Generator Units, Warren and Knowlton, Trans. ASME, February 1941.
43. Modern Extraction Turbines, L. E. Newman, Power Plant Engineering, January, February, March, April 1945.
44. Steam Turbines for Re-superheat Cycle, E. E. Parkar, Trans. ASME, 1948.
45. Oil System for Industrial Steam Turbines, F. C. Linn, Power Engineering, April, May, June, August 1949.
46. Test Code for Steam Turbines, ASME.
47. Steam Jet Ejectors—- Their selection and operation, combustion, 1946.
48. Standards of the Heat Exchange Institute, Condenser Section, 1952.
49. Power Plant Engineering Fuel Handling, Storage and Preparation Number. January 1, 1929.
50. Dispersion and Spreading of Gases and Dusts from Chimneys, W.F. Davidson, Ind. Hygiene Foundation Bull. 13, 1949.
51. Dust Emission from Coal-fired Boiler Furnaces, E.R. Kaiser, Combustion, May 1951.
52. Fans, Power, October 1951.
53. Standards of the National Association of Fan manufacturers, NAFM Bull. No. 110, 1950.
54. Centrifugal Fans, J.R. Darnell, Southern Power and Industry, June 1948.
55. Pump, Operation & Maintenance, Tyler G Hicks, Tata Mc Graw Hill,1997.
56. Combustion Control, Power, December 1949.
57. Ash Handling Systems, H. R. Pursel, Power, May 1950.
58. Corrosion Identification Chart, Combustion, March 1950.
59. The Prevention of Embrittlement Cracking, A.A. Berk, Trans. ASME, 1951.
60. Trends in Modern Feedwater Conditioning Equipment, R. S. Applebaum, Power Generation, June, July 1948.
61. Boiler Scale, Its Formation and Prevention, R. H. Hayman, Southern Power and Industry, January, February 1945.
62. Feedwater Treatment, Power, December 1947.

63. Standards of the Heat Exchange Institute, Tubular Exchange Section, Heat Exchange Institute, 1948.
64. Standards of the Hydraulic Institute, Hydraulic Institute, 1951.
65. ASA, ASTM, and ASME Pipe Codes.
66. Boiler Construction Code, ASME.
67. An introduction to Thermal Power Plant engineering and operation, Notion Press, P K Das & A K Das,2018
68. Selection and Setting of Safety Valves, J. R. Kruse, Combustion, July 1949.
69. Transmission of Heat Through Insulation, R. H. Heilman, Mechanical Engineering, July 1930.
70. Fluid Mechanics, Dr. A.K. Jain, Khanna Publishers, New Delhi,1986
71. Perry's Chemical Engineers' Handbook, 9th Edition
72. Industrial Maintenance Process (Mechanical),PKDas & AK Das, Notion Press,2020
73. Plant Installation & maintenance, RG Chakraborty, Basu Publishers,1984
74. Power Plant Performance, A. B. Gill, Butterworth-Heinemann, 2016
75. Total Plant Performance Management by R Keith Mobley, ELSEVIER,1999
76. Energy Efficiency Principles and Practices by Penni Mclean-Conner, Pennwell Books,2009
77. Fuel and combustion by Dr. Samir Sarkar,2nd edition, Orient Longman,1990.
78. Industrial Maintenance process (Electrical and C&I) Vol-II, PKDas & AKDas, Notion Press,2020
79. Web Sources
80. Bureau of Energy and Efficiency-India

Reader's Review and Comments

Valuable comments on this book from eminent readers are welcome; which will be incorporated in this next edition

www.ingramcontent.com/pod-product-compliance
Ingram Content Group UK Ltd.
Pitfield, Milton Keynes, MK11 3LW, UK
UKHW041831200726
13854UKWH00002BA/994